ISSUES AND IMAGES
IN THE PHILOSOPHY OF SCIENCE

BOSTON STUDIES IN THE PHILOSOPHY OF SCIENCE

VOLUME 192

AZARYA POLIKAROV

Issues and Images in the Philosophy of Science

Scientific and Philosophical Essays in Honour of Azarya Polikarov

Edited by

DIMITRI GINEV
University of Sofia, Bulgaria

and

ROBERT S. COHEN
Boston University

Springer Science+Business Media, B.V.

A C.I.P. Catalogue record for this book is available from the Library of Congress

ISBN 978-94-010-6443-9 ISBN 978-94-011-5788-9 (eBook)
DOI 10.1007/978-94-011-5788-9

Printed on acid-free paper

TABLE OF CONTENTS

PREFACE

Azarya Polikarov was born in Sofia on October 9, 1921. Through the many stages of politics, economy, and culture in Bulgaria, he maintained his rational humanity and scientific curiosity. He has been a splendid teacher and an accomplished critical philosopher exploring the conceptual and historical vicissitudes of physics in modern times and also the science policies that favor or threaten human life in these decades. Equally and easily at home both within the Eastern and Central European countries and within the Western world, Polikarov is known as a collaborating genial colleague, a working scholar, not at all a visiting academic tourist. He understands the philosophy of science from within, in all its developments, from the classical beginnings through the great ages of Galilean, Newtonian, Maxwellian science, to the times of the stunning discoveries and imaginative theories of his beloved Einstein and Bohr of the twentieth century. Moreover, his understanding has come along with a deep knowledge of the scientific topics in themselves.

Looking at our Appendix listing his principal publications, we see that Polikarov's public research career, after years of science teaching and popular science writing, began in the fifties in Bulgarian, Russian and German journals, developed a dozen books, and became internationally distinguished with 30 papers and essays in English, 24 in Russian, 28 in German, along with 32 in Bulgarian, among our 'selected' list. We see also that Polikarov examines all the central issues of philosophical understanding of science and of nature causality and determinism, perception and observation, ontological prerequisites and ontological entailments of our changing scientific theories, historical contexts and systematic contents of the sciences, and, above all, the methods followed, explicitly or implicitly. He is the master methodologist.

In his first book in English, *Science and Philosophy* (Sofia, 1973), Polikarov set his path with a summary of his earlier works and program for what was to come. A revised version of the first essay in this book was published in the *Boston Studies in the Philosophy of Science* (Vol. 14, 1974) under a title which may serve as his research motto, 'The Divergent-Convergent Method – A Heuristic Approach to Problem-Solving.'

His second volume, again in English, appeared a decade later under a further clarifying full title: *Methodological Problems of Science – The Iteration Cycle: Science – Methodology of Science* (Sofia, 1983). His scope was now encompassing every variant of philosophy of science, and the book presented his major exposition of what he termed 'the historical perspective', following critical analysis of heuristics, hypothetico-deductive classifications, and then in detail the case of the transition to relativistic physics.

Recently, after another decade, a third volume appeared, as yet only in

D. Ginev and R. S. Cohen (eds.), Issues and Images in the Philosophy of Science, ix–xi.

Bulgarian, under the title *Revolutions in Physics* (Sofia, 1995), with a thorough examination of Kuhn's view of the 'structure' of these revolutions, of the subsequent literature concerning Kuhn's views, and of Polikarov's systematic classification of alternative understandings of historical changes and evolution in science (including those of Hanson, Toulmin, Kedrov, Lakatos, Feyerabend). A succinct and fascinating section appeared in English in a recent volume, *Physics, Philosophy and the Scientific Community* (*Boston Studies*, Vol. 163, 1995) under the title 'Some Questions Concerning Limitations of the Range of Validity of Kuhn's Model of the History of Science'.

Meanwhile, Polikarov contributed to many congresses, symposia and institutions. A valuable essay, written with Dr. Ginev in 1989, was prepared for *The Philosophy of A. J. Ayer* (LaSalle, 1992), entitled 'Remarks on Logical Empiricism and Some of A. J. Ayer's Achievements Some Fifty Years Later'. Ayer's friendly response notably accepted Polikarov's suggestion concerning heuristic elements in what might be seen as 'metaphysical' statements within scientific theory.

Professor Polikarov has been a member of scientific centers in many places. His original and continuing work in Moscow (Ph.D., Institute of Philosophy, 1964); his visiting professorships in German universities, East and West; his three years during the later 60s as philosophy program specialist at Unesco in Paris; his appointments as Senior Research Associate in our Center for Philosophy and History of Science at Boston University and at the Center for Philosophy of Science at the University of Pittsburgh in the early 70s; his collaborative work in the USSR and then in the new Russia, and in many other European academic centers. Indeed, Polikarov was always an independent channel of international philosophical communication, accurate, complete, sympathetic. I recall his helpful survey of 'Philosophie und Physik in den Ost-Europäischen Ländern' written for the 1968 issue of Professor Klibansky' *Contemporary Philosophy* (Firenze, 1958). It was a small essay of 17 pages, concise, thorough, admirable, and typical of his quality of mind and purpose.

Another fine paper by Polikarov was soon delivered at the International Congress for the History of Science of 1971 in Moscow: 'From Successive to Parallel Conceptions: An Attempt to Generalize Kuhn's Paradigms Conception'. There he devoted a complex analysis, among other topics, to the creation of Hertz's theoretical system of a force-free kinematics as alternative to Newtonian dynamics, and to the 19th century development of the two competing systems of phenomenological and statistical thermodynamics.

* * *

In this volume, following Dimitri Ginev's lucid introductory essay on Polikarov's philosophical methodology of science, 24 scholars from a dozen countries honor Professor Polikarov by dedicating to him their research papers on theoretical practices of the science, with particular attention to physics. The 'issues and images' will be familiar to Polikarov himself for they parallel or reflect or refract his own concerns: realism and anti-realism, hermeneutics and the ontology of science, determinism and indeterminism, symmetries and alternatives among explanations, cognitive dynamics, the 'new' philosophy of science (with Nietzsche) and the ancient classical philosophy of science (with Aristotle), constructivism and systems and classifications, even putative communication with extra-terrestrial intelligence. Throughout, these authors attend to his great theme: clarity about the dialectic of method and heuristic.

Two favorite citations from Polikarov's beautifully documented publications seem to resonate throughout his thought:

> With accurate experiment and observation to work upon, imagination becomes the architect of physical theory
>
> (Tyndall)

> The development of scientific method . . . is the skeleton which carries the development of the entire body of science.
>
> (Boltzmann)

Boston University,
November 1997

Robert S. Cohen

DIMITRI GINEV

INTRODUCTION

On the Dialectical Nature of Polikarov's Philosophical Methodology of Science

It is not altogether easy to characterize the work of a scholar whose fruitful ideas extend from the methodology of the cognitive sciences and philosophy of quantum mechanics to the history of pre-modern science. But, despite the diversity of themes and methods in his work, Azarya Polikarov has developed a homogeneous project for a philosophical methodology of science. This project has been influential among many philosophers, sociologists, mathematicians, physicists, linguists, psychologists, and experts in scientometrics and science policy. Polikarov has published over a dozen substantial books, and made important contributions to the fields of general philosophy of science, philosophy of physics, models of scientific development, and history of science. He was one of the pioneers in applying heuristic methods from the cognitive sciences to epistemology. At the same time, he has been a prolific initiator in science policy, organizing new academic institutions of meta-scientific studies in his country.

The background of Polikarov's philosophical methodology is Marxism. However, he has rejected all variants of meta-scientific 'dialectical philosophy,' such as dialectical materialism (diamat), materialist theory of categories, materialist 'mirror theory of knowledge' (*Widerspiegelungstheorie*), etc. He has denounced these ideological surrogates of philosophy as a residuum of metaphysics which supports ideologically the political dogmatization of Marxism. This is why the official philosophical administrators in the countries of 'real socialism' never acknowledged his philosophical methodology as a philosophy. They accepted it only as a "non-philosophical methodology of the concrete scientific disciplines." Strangely enough, Polikarov was often characterized by the communist philosophical officials as a positivist. In fact, he has always been an opponent of logical positivism. As early as the beginning of the 60's he criticized the Carnapian idea of 'rational reconstruction', the attempts to achieve a 'unified science' by constructing a universal language of science, the positivist strategy of overcoming metaphysics, the statement (or syntactical) view concerning the structure of scientific theory, the universality of causal explanation, and many other essential aspects of logical-positivist philosophy of science. Nevertheless, in the eyes of the communist philosophical authorities he was a positivist, simply because he denied the tenet that there is a meta-scientific theory called 'diamat'.

In all historico-political situations in the postwar development of Eastern Europe, Polikarov has made no secret of his philosophical and political commitment to the praxis-oriented variant of Marxism. Such a Marxism is one which seeks to explain human knowledge and the whole way of human being

D. Ginev and R. S. Cohen (eds.), Issues and Images in the Philosophy of Science, xiii–xvii.

by a holistic study of man's practical relationship toward the world of her/his existence and toward herself/himself.

Within its widest parameters, Polikarov's philosophical methodology encompasses the tension of competing tendencies in 'post-metaphysical reflection' upon science. Seen against this dialectical backdrop, its main characteristics are:

1. *Scientification of Philosophy of Science Without Naturalizing Epistemology of Science.* Polikarov has subscribed to the view that the philosophical study of science must itself be a science. However, he has been strongly opposed to any form of 'naturalized epistemology'. He has denied the thesis that epistemic properties are reducible to natural properties. He denies this thesis not because he advocates the view that human knowledge, thought and language constitute an 'autonomous reality', or a *sui generis* 'cognitive nature'. In fact, he is unsatisfied with naturalism as well as with traditional anti-naturalism (associated with the claim that there is a 'substantive epistemic relationality' which is essentially normative and can only be studied by logic and probability theory) because both programs are based upon the metaphysics of Cartesian dualism. According to him, all epistemic properties are embedded in the world, and mind cannot be isolated from the environmental context in which it functions.

Polikarov's discontent with naturalism is not based on the view that there is a firm demarcation line between the natural and the human sciences. He has not accepted either the classical Diltheyan hermeneutic distinction between 'explanation' and 'understanding' or the recent attempts to defend the view that there is a philosophically interesting difference between the natural and the human science by referring to the fact that human sciences, which deal with self-interpreting objects, are doubly interpretative. A reconstruction of his views about naturalism would suggest that naturalism is a system of methodological and meta-methodological standards of rationality that are 'derived' from the metaphysics and representationalist epistemology of Cartesian dualism. These are standards of instrumental rationality because to conceive of rationality in terms of an independent knowing subject opposed to an independent object of knowledge amounts to conceiving rationality in an instrumental paradigm of reasoning. The bottom line of his rejection of all kinds of naturalism is that even natural science cannot be fully naturalized. In Polikarov's view, non-naturalism is a paradigm of rationality that is not based upon the dualist interpretation of the subject-object relation. In other words, non-naturalism is not an alternative to the demand to make philosophy of science (and the whole epistemological enterprise) an empirical science. Quite to the contrary, it is the kernel of non-naturalism to empiricize philosophy of science in a non-Cartesian and non-instrumentalist way.

Thus, Polikarov's non-naturalism is in complete agreement with the view that there is no body of meta-scientific, normative, principles which resist translation into any descriptive-explanatory scientific language. Each epistemological dimension of scientific rationality can be studied empirically. However, the scientification of philosophy of science is to be carried out within the frameworks of non-naturalist (or better, non-Cartesian) research programs. (In Polikarov's terms, these are 'heuristic programs.') A typical example are the programs of the so-called 'externalist approach' in AI and cognitive sciences. In his search for both non-substantive and non-instrumental models of rationality he has tended to adopt Herbert Simon's heuristic idea of bounded rationality.

The embeddedness of all forms of cognition in the real world is logically prior to the logical structure of cognition. To put it in other words, Polikarov's philosophical methodology is an attempt to put the logical structure of cognition back into the real life of cognition in the world.

2. *Anti-foundationalism Without Deconstructing Epistemology.* Having rejected the Cartesian conception of mind and the dualist metaphysics of an independent world and an independent mind, Polikarov has always stressed the need for a non-foundational epistemology rather than the need to overcome epistemology. In his view, the theory of knowledge must be non-normative and non-representationalist. The search for invariant norms, criteria, and algorithmic schemes of applying epistemic procedures is to be replaced by a heuristic study of modes of cognitive being in the world. This study defines the scope of a 'heuristic epistemology'. From this perspective, knowledge is not a mirroring of an independent reality. Instead, it is a heuristic process of developing 'fields of possible solutions' of the basic problems arising in our cognitive adjustment in the world, and finding ways of reducing these fields to the optimal solutions. Heuristic epistemology is not a kind of pragmatism, or of what Rorty calls 'epistemological behaviorism'. It is not to be detached from Polikarov's ontological monism. The different modes of cognitive embeddedness in the real world, however incommensurable, reveal the ontological structure of the world. The ontology of the embedded cognition is an alternative both to epistemological foundationalism and to epistemological behaviorism.

3. *Pluralism Without Anarchism and Hypertrophy of Incommensurability.* The rejection of all kinds of epistemic monism is a running theme of Polikarov's work. However, his pluralism is defined entirely in epistemic terms. According to Polikarov, the refutation of any form of ontological pluralism follows from the refutation of the semantico-ontological version of the incommensurability thesis. The dictum of his epistemic pluralism, or, in his terms 'multi-methodologism', is that in facing a problem-situation within a certain cognitive mode of being in the world, one is to explore the possible alternatives of figuring out a solution (like developing a set of competing theories without prescribing in aprioristic manner any epistemic restrictions).

Yet, the very idea of 'multi-methodologism' requires a strategy, which surmounts the anarchistic pluralism. Roughly, this is a strategy of using divergent-convergent heuristics in figuring out the relevance of different systems of methodological rules, norms, standards, and criteria to the main epistemic problem-situations of science. From the perspective of 'multi-methodologism', the field of possible solutions in each epistemic problem-situation can be reduced essentially to the subfield of the best solutions through the heuristic study of the relevance mentioned. Hence, the maxim 'anything goes' proves to be misleading.

4. *Historical Approach Without Historical Relativism.* Polikarov rejects the Lakatosian kind of rational reconstructions of history of science. In his view, there is no general logic of the historical dynamics of science which can be extracted from the empirical history of science. (In his criticism of the idea of rational reconstruction Polikarov appears to be closely related to Holton-like structuralist analysis of basic episodes in the historical development of science.) But he has never given up the need for a general philosophical model of the history of science. Over the year he has claimed that normative-epistemological reconstruction of science is to be left behind in favor of an heuristic interaction between philosophical methodology and empirical historiography of science. In the context of this interaction, his main concern is the problematic of the proliferation and synthesis of scientific theories as historical processes. The study of these processes allows us to get a general picture of the historical dynamics of science. Closely related to this study is Polikarov's taxonomy of conceptual changes in the development of science, which provides interesting suggestions for avoiding the pitfall of historical relativism.

5. *Ideological Neutrality Without Affirmative Dogma.* Philosophical methodology must be developed not only as a logical and historical analysis of science's cognitive structure but as a critique of science as well. However, the critique of science is to be implemented without any prior commitment to a given system of ideological values. Polikarov has been a strong opponent of those Marxist philosophers who advocate an 'ideologization of philosophy of science'. In his view, the 'principle of neutrality' must not be given up in developing a philosophical methodology of science. Polikarov bases this neutrality on the non-meta-scientific character of his philosophical methodology. His proposal for a philosophical critique of science is not a kind of political critique of science (or better, a critique of science as a 'hidden politics'), as this view is spelled out by Latour, Woolgar, and many followers of Foucault. Moreover, Polikarov's philosophical critique of science does not entail the need of developing a normative-critical science of the sort of Horkheimer's 'critical theory', or of Habermas's 'reconstructive science'.

This thesis is to be attached to Marxism as well. In Polikarov's view, Marxism is a self-reflective theoretical program for revealing crisis-situations in the functioning of economic and political systems of class-structured societies. Because these situations shape the 'ideological consciousnesses' of different social groups, an important aspect of the program is the 'critique of ideology'. Considered as such a program, Marxism is neither a normative theory of 'human emancipation', nor a post-Hegelian meta-scientific point of view for analysing the forms of reification in modern societies. In Polikarov's reading, Marxism is simply one of the non-naturalist (i.e. heuristic) scientific programs. To be a Marxist in social sciences is the same as to be an adherent of an externalist approach in cognitive sciences, or a follower of the Copenhagen interpretation in quantum mechanics.

Polikarov's philosophical methodology has a critical potential not because of any hidden commitment to the values of a certain 'ideological meta-narrative' but because of its non-naturalist orientation. It is the 'Cartesian project of science' that is the target of this critique. And it is the promotion of non-Cartesian research programs in science that should be the positive outcome. (Here we see another argument against naturalizing epistemologyand philosophy of science. If one advocates the universality of the 'naturalist theoretical attitude' in modern science, and tries to apply the same attitude by doing philosophy of science, then one is unable to take a critical distance from science. Each attempt to reflect critically upon science will collapse into an affirmative naturalist study of science. If so, then the very 'naturalist theoretical attitude' will take the form of new universal ideology. To combat the transformation of naturalism into such an ideology is a task of prime importance in Polikarov's philosophical methodology.) Thus, *ex hypothesi* the search for new heuristic (non-naturalist) forms of scientific knowledge is a critique of science's self-understanding as a universal naturalist viewpoint. The proliferation of heuristic research programs is a critique of science.

Yet, in criticizing naturalism in science, Polikarov has not admitted a kind of philosophical critique of science which will claim a 'transcendental access' to the 'human minds in the world'. He has rejected all programs of developing a transcendental 'system of reference' for explaining and criticizing science, such as Husserl's phenomenology of the life-world, Heidegger's existential conception of science, or Apel's transcendental pragmatics. This philosophical critique of science must be developed between the Scylla of violating the principle of ideological neutrality and the Charybdis of transcendental idealism. To put the point in Rorty's terms, Polikarov is not trying to find a discourse which "combines the advantages of normality with those of abnormality" – the meta-scientific security of a transcendental position with the innovative and critical character of the heuristic research programs. Rather, Polikarov's aim is to intensify the abnormality – to promote the proliferation of heuristic research programs in all the different fields, including epistemology and philosophy of science.

JOSEPH AGASSI

WHO NEEDS ARISTOTLE?

It has been said repeatedly that Aristotle is the most common sense philosopher who has ever lived. This allegation seems to amount to two claims; first, there is such a thing as the peak of common sense, and second, that Aristotle came closest to it. This peak of common sense, or ideal common sense, or arch-common sense, or core, or epitome, or distillation of common sense – choose any metaphor you like – is what a philosopher will call the essence of commonsense. That it exists is the claim of Plato and Aristotle. In other words, the high praise of Aristotle comes from his own stable. Recently it has become a popular trend to find new ways to defend the idea of essentialism as strong common sense.[1] This, I fear, is not only objectionable on the ground that common sense is not the supreme court of reason; it is also objectionable on account of its circularity. For, in frequently observed and reported fact, there are varieties of common sense. We may, of course, appeal to the best common sense, to wit that sense which is common to myself and my peers, which is very appealing if I am Oxford or M.I.T.; but not if I am London School of Economics or Boston University. Even without belonging to the elite, I may still appeal to the best common sense: I may try to appeal to the universal essence of all common sense. (I shall talk of the universal later on.) It stands to reason, then, that in order to be able to appeal to the essence of common sense, I should postulate essentialism. But then, putting essentialism itself on the basis of such common sense will be rather circular.

The beginning of philosophy, says Aristotle, is in wonderment, is having a problem to solve. This makes my criticism of the new Aristotelian trend is, I feel, brief, unanswerable, and devastating. I say, members of this trendy school do not pose any problem; the essentialism which they offer thus solves no problem, nor is it meant to; rather, they offer revised versions of classical essentialism which, they argue, does not suffer from criticisms levelled against the classical versions of essentialism, and this surely is the proper mode of arguing. But the time-honored canons of dialectics require this to be the second step. The first step is to show that one's proposed theory is strong enough so as to solve the problems that its predecessors have come to solve. If the new essentialism would a) solve the initial problems, and b) be impervious to criticisms levelled in the past against past solutions, then my task as a commentator would be c) to attempt – only attempt – to look for newer criticisms. As it turns out, there is no need to take the commentator's obvious third step as long as the first step in the discussion is missing. Let me, then, raise the problem which the classical versions of essentialism come to solve, and examine whether the new version solves this problem, in order to show that the new version solves no problem at all and gains its plausibility from

D. Ginev and R. S. Cohen (eds.), Issues and Images in the Philosophy of Science, 1–11.

being too weak to be criticizable while its consequent inability to solve any problem is concealed by its adherents' claim that it is immune to the criticism of the old version. To repeat, this is true by the mere weakness of the new version.

I

Arch philosopher Thales wondered, we are told in Aristotle's *Metaphysics* Alpha, what is it that makes Tom or Dick or Harry the same man from one day to the next in spite of his being different from one day to the next? Let us call this problem The Problem of Identity. The identity of Tom as an individual may be contrasted with the identity of this table as a table. In this case, and it's historically important, the problem of identity splits into the problem of individuation and the problem of universals. There is no doubt that common sense supports this problem, or this pair of problems, by the way, and it is reflected in all literature repeatedly: even Wittgenstein in his *Tractatus Logico-Philosophicus* alluded to a German fairy tale relating to the problem of universals. In the United States the Problem of Identity was reflected in the folk tale of Rip-Van-Winkle who is the same, when old, as his younger self, yet his younger self is so much more like his son than like his older self. Common sense postulates a theory of continuity – Rip's identity must reside in a continuous space-time region and hence he can never become his son. The continuity theory is so strong that some proponents of the theory of the transmigration of the soul try to modify it so as to make transmigration violate continuity minimally: the new Dalai Lama is born at the very same instant as the old one has died, and also nearby – but not at the nearest vicinity. The theory that people possess individual souls itself already offers a solution to the problem of identity: Rip has a soul different from that of his son; the theory of the soul is in a sense a generalization of the theory of space-time continuity, then. But let us examine this a bit.

The soul theory and the continuity theory are both fairly common sense. They both came to solve the same problem. And so at the very best one of them is redundant. At the very best means here, if they are compatible. Descartes' theory of the soul, for example, or Leibniz's, are not compatible with continuity. Kant, who was a Leibnizian, was forced to change his mind by the criticism which Euler had launched against it: there is no conflict between Leibniz' theory and the assumption that my soul resides in the body of some African rhinoceros. But there are versions – two, as far as I know – of the theory of the soul, which seem compatible with, and so may be generalizations of, the theory of continuity. First, the claim that the soul is born at birth or at conception and dies with the body's natural death. Unlike Descartes, today's dualists – who assume the ghost to be in the machine – accept this assumption. Second, the soul resides in the body and comes and goes into it and out of it on proper continuous world-lines, like in Hollywood movies. The corollary from this second theory is that Tom's body is never identical with Tom and that when we speak of Tom we mean his soul, never his body.

But I fear the theories of continuity and of the soul concerning personal identity are in contradiction just because they are different solutions to the same problem. The point is that an explanation (*explicans*), far from entailing the explained (*explicandum*), actually contradicts it. I shall not discuss this point in detail, since it is presented in Popper's "The Aims of Science", Lakatos's "Proofs and Refutations", and my "Sensationalism". In brief, the situation is this. Consider any crucial experiment between two theories, say, between Einstein's and Newton's theories of mechanics or of gravity; it is possible if and only if the two theories are in a conflict. This everybody admits, those who say a crucial experiment is possible and those who say a crucial experiment is not possible. Also, everybody admits, a conflict is the outcome of having two answers to the same question. Now the crucial experiment is possible only because the two theories present observable facts somewhat differently. In other words, scientific explanations (*explicans*) modify the observable facts they come to explain (*explicandum*). What this amounts to is that a new explanation does not entail the old observation report taken verbatim, and hence, obviously, the celebrated deductive model of explanation is false, as claimed by Feyerabend and by myself, but as vehemently denied by Popper and by Lakatos. I think they abdicate logic here. The contradiction depicted in the present discussion of identity, incidentally, is not empirical as in the Einstein-Newton example; it is conceptual, as in the cases discussed so masterly by Lakatos. For, the theory of continuity, but not the theory of the soul, will declare the identity of a thing retained through all gradual changes; including the gradual change of each and every property and aspect of a thing: the theory of continuity makes spacio-temporal continuity a necessary but not a sufficient condition; the theory of the soul demands that the soul retains spacio-temporal continuity and on top of this ascribes to the soul its identity quite unconditionally.

II

To return to good old Thales, he postulated the existence of the unchangeable Tom within the changeable Tom in order to solve the Problem of Identity – he postulated, that is, the existence of the unchangeable small man within, of the soul. How this works in psychology we do not know, nor even whether the psychological example was offered by Thales or by later writers, perhaps even Aristotle himself. Thales' view, all is water, is understood by Aristotle to be a metaphysics, a grand theory of the physical or material world, and thus he considered Thales to be the father of the *physiologoi*, of the Greek physicists. His disciples, still according to Aristotle, agreed about the existence of the unchangeable, but denied that it is water. This goes for Aristotle as well.

In brief, Thales postulated a theory, which I wish to call Thales' essentialism, which says that the world comprises a complex and changeable appearances and a simple, unchangeable reality.

Thales' essentialism, alas, does not solve the Problem of Identity at all. For, it identifies matter but not a single material thing, be it a chemical substance or a cathedral – except perhaps water. Hence the force of Parmenides' criticism and his conclusion that only reality exists, not appearances (which are but lies); namely his conclusion that there is only one entity which has true identity, The One; The Unchangeable. Democritus postulated the identity of every atom to be that of an essence and concluded that atoms are unchangeable, eternal. He allowed, Aristotle tells us, to identify a thing, say a chemical or a building, as the variety of shapes and orders of atoms. This opened the way to Plato's version of metaphysics according to which a thing is matter and form. According to Democritus the atom, including its shape, is unalterable; according to Plato only the shape is unalterable. This is an advantage of simplification (since the shapes of all congruent atoms collapse into one shape) and the disadvantage of the added assumption of a Platonic realm of pure shapes (since the shape of an atom is no longer attached to that atom as the atom shares its shape).

Plato himself noticed that this is a solution to the Problem of Identity but an objectionable one: two things partaking in the same form are identical, whereas the problem of identity arises from our recognition of the difference between rip and his son though they have the same form. Leibniz was disturbed by this and postulated his principle of identity of indiscernibles in order to get out of trouble: Rip's son cannot entirely resemble Rip; no two entities, he said, can partake in the same forms exactly. He had two proofs of the principle of identity of indiscernibles.[2] The one invoked the richness of God's imagination. The other employed the theory of continuity referred to above: if Rip-Van-Winkle were identical with his son they should once occupy the same space-time region.

If we reject Leibniz' principle of identity of indiscernibles as the *deus-ex-machina* that it is, we shall admit the criticism of all past efforts to solve the problem of identification: Norbert Wiener has declared it logically true that any characterization of an individual that may be deemed a complete specification should be one that offers us a recipe for the reconstitution of the individual from its elements. Hence, by any theory of individuation which is at all satisfactory there is no such thing as an individual.[3] But the Problem of Identity arises from the puzzling fact that, however changed Tom is, we still recognize him as the same individual! The idea that there are no individuals is strongly supported today by many disciples of Wiener who are cyberneticists and information theorists of all sorts, on the ground of their claim that in principle we can reproduce anyone. And by now this idea is popular enough to appear on T.V. The beams in *Star Trek* are information beams (loaded with energy, of course); one episode of that series of science fiction was based on accidental transmissions of double beams, another on machines building robots duplicating individual humans, etc. This shows dangerous moral implications – ones which Norbert Wiener was the first to warn us against, and in each of his books Wiener claimed that we can only reproduce artificially what we can specify, but there is the unspecifiable. Apply this to the Problem of Identity and you may hear Wiener say, the Problem of Identity is inherently insoluble for irreproducible individuals like Tom or Dick, whereas a reproducible table or chair is hardly an individual. Where a Rembrandt painting stands is an open question as yet, all protestations of information theorists to the contrary notwithstanding.

III

We have thus arrived at the deadlock which R. G. Collingwood declared, in his *Speculum Mentis* of 1924, to be the one which constantly besets religion, science, and art, and which only art has thus far managed to evade to any extent whatsoever. To notice the severity of the deadlock we must notice the facets of the problem at hand. What makes Tom one and the same, yet different? The very problem is rooted in the observed fact that we have changed in the midst of constancy. The Parmenideans, ancient or modern, deny the problem by denying the existence of change. The Heracliteans, ancient and modern, deny the problem by denying the existence of constancy. Ludwig Wittgenstein's *Tractatus* is Heraclitean, and in his *Investigations* he admitted this to be false by recognizing the existence and validity of an intuition about the identity through change of, say, a game: when we change the rules of a game far enough, it becomes a new game and our intuitions tell us so. Hence, adds Wittgenstein, we need not worry about identity, as our intuition tells us fairly reliably what is constant-through-change and what is not. Hence, he concludes, we may ignore our problem.

I dare say Wittgenstein is fairly right about our intuitions. If we ever mistake Rip for his son then we are corrected without the aid of philosophers and without the use of explicit criteria. Of course, our intuitions can be stretched to their limits and get confused. Remember *Star Trek*. And, of course, Wittgenstein will protest against the stretching. Indeed, he blamed illegitimate concept-stretching to be the source of all philosophy. Some of us, however, wish to enlarge our intuitions and – perhaps quite intuitively – in agreement with the view of Émile Meyerson, R. G. Collingwood, and others: all past progress, artistic or intellectual, comes from worrying about the Problem of Identity, and by subsequent attempts to stretch our intuitions. We may even agree with Collingwood that to do so we may try to articulate our intuitive criteria in order to be able to apply them more consciously, to criticize them, to improve them. We can ask not only how long is Tom still Tom rather than Dick. We can ask, is the sound produced according to John Cage's specifications still music? Still in the tradition of Bach and Beethoven? Here Wittgenstein is of no help and his reliance on intuition is plainly poor.

IV

The above discussion has slid into another theory of essences, the theory of whatness so-called, of methodological essentialism,[4] as Popper calls it, which Aristotle ascribes to Socrates who "fixed thought for the first time on definitions" (*Met.* 987 b; he admitted some priority to Democritus and Pythagoras, *ibid.*, 1078 b). This is the starting point of methodological essentialism, of "seeking the essence" and of "inductive arguments and universal definitions [of essences], both of which are concerned with the starting point of science". Aristotle, again, ascribes the idea to Socrates (*loc. cit.*), and again with some hesitation. We ask what is X, when X is a familiar object; namely, we perform the task of searching for those properties which make X an X, whose absence will deprive X of its X-ness. Here we have used a technique: instead of varying in time, we vary by shifting from one member of a set to another. For example, what is man? You cannot say he is a biped, as there may easily be one-legged man; he is rational animal, says Aristotle. He is rational, since without reason he is a donkey; and he is an animal, since without flesh he is an angel. Ergo, Balaam's ass is human. If you accept Aristotle's definition you must accept Balaam's ass, as well as Aesop's cock and bull, as human. This, I submit, has both its great attractions and its great discomfort. If these examples are too unscientific for you, consider Mr. Spock of *Star Trek* fame, or any other extra-terrestrial rational animals of your choice. Methodological essentialists must debate the question, are they human? I will not.

Let me stress that one can be an ontological essentialist without being a methodological essentialist: Galileo and Bacon, for example, but not Descartes.

If we reject methodological essentialism, yet admit the intuitions of identity – i.e. of constancy in the midst of change – we may still hope to be able to *explain* (rather than explicate) these intuitions, test these explanations, etc. If we succeed in catching the constant in the midst of change, shall we call it the essence of the thing? Do things possess essences? Is not the very preoccupation with the Problem of Identity the admission of the existence of essence?

For my part, I will say, no. And thus I plainly reject what I have labelled Thales' essentialism; his claim that the world is comprised of reality and appearances. In contrast with this, I claim that there are things more ephemeral and things less ephemeral; that there are levels or degrees of reality (Popper, 'Three Views Concerning Human Knowledge'). Take the essence of man, and take Tom who essentially partakes in it. What is the essence of Tom? Not only that he is man, or else he will be essentially the same as Dick. And so we come again to the individual soul of Tom. Is it unique? Can it be copied? I submit that here is a philosophical morass that we are stuck in for twenty-five centuries. These questions shake methodological essentialism by attacking all ontological essentialism, namely by attacking the very polarization of everything to appearance and reality, the very dichotomy between the two. This dichotomy was first rejected, I think, by Franz Brentano and by Edmund

Husserl. (Husserl's very investigation into the essence of the appearances puzzled so many precisely because, according to the dichotomy, appearances cannot have essences but things have both appearances and essences.) Anyway, we can no longer ignore the fact that the dichotomy is false, and so all essentialism prior to the 20th century being based on Thales' essentialism, on Thales' dichotomy, must go.

V

On this issue, I feel, Aristotle's work is the worst and most confused philosophical discussion ever. First, there is the question, are Aristotle's essential definitions verbal or ontological: is man a rational animal because otherwise we shall not call him a human being, but by some other name, or is it that when he loses his reason he loses his very humanity? I shall remind you that already ancient commentators have noticed this slippery quality of Aristotle's argumentations (see W. and M. Kneale's *The Development of Logic*). Indeed, in *Metaphysics* Zeta, 1030a, Aristotle himself feels uneasy and tries to limit his theory of essential definitions expressly in order to prevent too much arbitrariness. He confines essences to cases of a species of a genus only. So he solves the Problem of Identity at most only for humanity as a whole, not for Tom or for Rip-Van-Winkle. Further, he identified essence with cause, and he had a theory of the four causes and of the substratum, in addition of the theory of essence. How do they all go together is itself unclear: I do not even know how clear is his theory of the substratum – that of things as being and becoming, as potential and actual.

Briefly, Aristotle's substance as potentiality and actuality is this. The essence of an acorn is its potentiality, its ability to become an oak, its very oakness. Essences can thus be hidden, or occult, or potential; or else it is manifest, or actual. Essences are the very potentials, becoming actual and fading away, generating and corrupting. Essences can move from the backstage to the limelight and back to the backstage. In this theory, substance is the stage of blooming and wilting essences. In it there is alas! little room for accidents. Also, as the existentialist Raymond Polin says, too little room for freedom. Also, may I add, though this picture of the world is not static, it is stationary; it does not offer much hope for progress – much contrary to Aristotle's grand theory of the unmoved mover which (supposedly) tops his grand metaphysics.

Thus, to sum up this point, it is never clear how integrated, or even consistent, are the various parts of Aristotle's views. This lack of clarity of Aristotle's leads to some astounding results. I shall mention the worst example, from *De generatione et corruptione* 328a, discussed in Sambursky's *Physics of the Stoics*, p. 12. The essence of wine is its ability to intoxicate us. Which is lost in dilution. Hence diluted wine is no wine; hence the essence of wine is (*inter alia*) its not being too dilute; whereas dilution is, we all know, a mere accident. This is absurd. The essence, concludes Aristotle, comes in pellets, in indivisible units, and so there is an essential limit to dilution and the absurdity is removed. After a whole volume of attacks on atomism, Aristotle endorses it – declaring essences indivisible – so as to overcome a rather marginal difficulty! Clearly, even in antiquity this was noticed as a major fault. It forced the Stoic physicists to make dilution central to physics and thus took them back to the Ionian school – to Thales' disciple, Anaximander, to be precise. I do not know how many more revolts against Aristotle we need before we see how out of step he always was.

VI

How do the neo-Aristotelians face the difficulties outlined here? I do not know if they do at all, though the literature includes some attempts in this direction.[5] For, in order to give an adequate neo-Aristotelian view one should somehow combine Aristotle's theories of substance, essence, the four causes, and the natural kinds, as well as the covering model of explanation; or else give up some of the ingredients in this list and explain how we can make do without what we give up. Moreover, the difficulty about the essence of wine should be handled too: what is the essence of wine, and what is the essence of methyl alcohol; how this is related to what is deemed the laws of nature concerning chemicals, intoxication; and more. I am utterly unable to decide whether any of the neo-Aristotelians would combine ontological essentialism with a methodological essentialism or not – I find clues going either way – because on the one hand the attraction of essentialism is methodological, yet on the other hand essential definitions are despised by mathematicians, analytic philosophers, and others. The problem here is of necessity, not necessarily of essences, though the two are connected, perhaps. We have necessity, by the laws of logic, by the meaning of words and such, called logical necessity or necessity *de dicto*; and we have the more significant necessity, that of the laws of nature, called natural necessity or necessity *de re*. If the laws of nature are the laws of essences, presentable by the definitions of essences, then definitions, not covering law models, describe natural necessity or pertain to necessity *de re*.

Possibly man is rational animal *de dicto*, i.e., by mere nominal definition, i.e., by our use of the word 'man'. Assuming man's rationality to be nominally necessary is not sufficient to enable us to decide whether you and I are (rational) human. Hence whether we are or not is undetermined as long as our rationality happens to be in doubt. Nominal necessity is quite different from the one based on the claim that all people, called by any name, but me and you included, are rational animals by some law of nature, by some natural necessity, which is necessity *de re*, or by essential definition. I submit that it is difficult to confuse necessity *de dicto* with necessity *de re*, yet the scholastics had to make great efforts to make this distinction since, to repeat, it was confused by Aristotle, as ancient commentators have observed. Clearing this confusion, as anyone trained in modern logic can do, only raises a problem. After the confusion is cleared, the claim that we are rational animals remains a hypothesis, not a necessarily true scientific proposition as Aristotle claimed essential definitions to be. That is to say, once Aristotelians admit all definition to be nominal, then he must view all putative essential definitions as mere hypotheses, and attempting to dodge this view they introduce the theory of natural kinds. Since natural kinds belong to Aristotle's philosophy of science as comprised of essential definitions, this amounts, precisely to Aristotle's confusion: by his theory, definitions are at times (when under attack) verbal and at times (otherwise) natural. The basic certitudes of informative theoretical scientific knowledge are, according to Aristotle, those concerning natural kinds; they have nothing to do with strictly nominal definitions. (All confusion of epistemology with essentialism, of course, is rooted in methodological essentialism, unless it is just any old confusion.)[6]

VII

Insofar as neo-Aristotelian essentialism is trendy, it also benefits from association with other trendy ideas. And, no doubt, the success of model theory and the claim that this theory does not square with traditional nominalism is conducive to essentialism (as long as we ignore methodological nominalism, that is; see note 4 above). And in particular, the essentialist fashion got boosted by the popularity of Saul Kripke's fashionable theory of possible worlds as the foundation of model theory, plus the fact that it has got snarled with the problem of naming, plus the fact that attempts are made to overcome the problem by postulating some sort of identity-through-possible-worlds, so-called. For the question this new situation naturally raises is, how is identity-through-possible-worlds preserved? And, admittedly, essentialism offers a solution to this new problem: essentialism entails identity through possible worlds: all rational animals in all conceivable worlds, including Balaam's ass and Aesop's cock and bull, not to mention Mr. Spock, are all human. This, of course, rests on the assumption that there is meaning to expressions or concepts like 'identity through possible worlds'. Can this be adjudicated? There is no viable theory of meaning to apply to the expression, and intuition may go in any direction. Perhaps intuition does raise a difficulty about possible worlds, which has to do with continuity: the continuity criterion of identity is intuitive and may easily contradict some versions of essentialism, e.g., the Cartesian, if not all versions of it (as I have argued).[7]

Of course, what one needs in order to apply essentialism to possible worlds is to use Aristotle's theory of meaning, and, indeed, originally essentialism encompassed a theory of meaning. The trouble with the idea of possible worlds, however, is that it is Platonic, not Aristotelian. Plato explained the common element between two different beds (which is the problem of universals), or between two time-slices of a given bed (which is our Problem of Identity) – he explained these as their sharing in the abstract idea of the bed or of bedness. Aristotle wanted bedness to be in the bed itself, not in a Platonic heaven.

Yet he wanted it to be universal. He could not have it both ways, as his scholastic commentators noted. Moreover, beds are not species of any genera and so it is hard to see how their essences reside in them, if they have essences at all. Yet he stuck to individual things since the very notion of possible worlds was too abstract for his taste though it is very palatable to a Platonist. This is why today, as in the Middle Ages, logicians who are not nominalists – Gödel, Quine – are Platonists, not Aristotelians. Aristotle simply falls between the two stools. (There are attempts to put a third stool in the middle, namely some versions of conceptualism. All conceptualism, however, is seriously entangled either in a confusion of necessity *de dicto* and *de re* or in frank identification of both: idealism.) It is therefore understandable that increasingly many Aristotelians undertake to examine the theory of possible worlds, nor should one undermine the insurmountable difficulty they face when undertaking this task: the very attempt to solve the problem posed by Kripke's work with the aid of Aristotelian (rather than Platonic) essentialism is inconsistent and doomed to failure.

And yet, somehow, Aristotle's theory sounds most convincing. Certain changes he says, do not affect identity, e.g., one's getting old; while certain changes do, e.g., death and putrefaction. This is repeatedly, I think, the strength of all essentialistic claims. All I can say to it is, it is true; but it is the *explicandum*, not the *explicans*; which is the source of the problem, not the solution. Neo-Aristotelian essentialism is a confusion of the *explicans* with the *explicandum*. We all agree that no matter how many wrinkles I gather, I am still myself, until I die, whereas a plum, after gathering sufficiently many wrinkles, ceases to be a plum and becomes a prune. Certainly this is true, and admitting this to be true seems to the admission of essentialism. We want, however, to know why. Aristotle says, there is an essence of me and there is an essence of a plum, the one has nothing to do with the wrinkles, the other a lot. This is a claim which can perhaps be made good by a fully fledged theory of the world which embeds a theory of essences, such that it agrees with our view that wrinkles make a prune but not a man. Aristotle's theory of essences would make good sense as a part of a theory of the universe which is a serious contender for the status of a true theory. This is too much to demand of Aristotle or of his modern followers, yet, the claim that wrinkles make prunes out of plums, is, to repeat, much too little. For, it is too little to say that while we change we may retain our identity, except when a change somehow makes us lose our identity. I would not call this small claim a theory, essentialism or otherwise. I would not call it essentialism, Aristotelian or otherwise. It is the reiteration of but a plain fact; not a solution to any problem, but the source of a problem to be attacked.

Aristotle's theory of essences sounds so very convincing, then, only out of context: when we confuse a problem with its solution. The classical seventeenth-century claim, then, as expressed by such diverse writers of that period, as Bacon, Molière, Spinoza, and Newton, is still the last word on this subject: Aristotelian essentialism is so facile and so hopelessly *ad hoc* as to be utterly unenlightening.

York University and Tel-Aviv University

NOTES

1 It is possible to view the Aristotelian school as a tradition continuous with the deep past, and all Thomist philosophers and all Aristotle scholars as the conveyors of that tradition. Yet I speak here of philosophers who come from the modern anti-Aristotelian tradition of modern logic, of modern science, and of modern enlightened liberal democracy. Of course, some giants, particularly Sir David Ross, stand out as belonging to both traditions. Friedrich Solmsen's book of 1960 on Aristotle's physics seems to me, however, a trail-blazer: despite Cherniss's harsh critique of Aristotle and despite Jaeger's attempt to offer a balanced view of Aristotle, Solmsen's muddled apologia for Aristotle was well-received (after Lane Cooper's thoughtful attempt to separate fact from fiction was ignored as too apologetic for Aristotle!).

2 There is also Leibniz's proof from causality: indiscernible causes will have indiscernible effects. This proof, however, is either a part of his metaphysical system of monadology, where every monad is in total isolation from the rest of the world, or it is the same proof as that from continuity. Certainly the proof from monadology is the stronger – indeed the only valid one – yet we must reject it along with Leibniz' whole system.

3 An element, say a chemical element, or an animal species, may still count as an individual without violating Wiener's thesis. But as long as there are individual samples of any element, this option must be excluded. Hence, the only option left by Wiener for individuation is the individual soul. Wiener himself tended to reject this option and so took refuge in nescience. This seems to me to be a cop out.

4 Karl Popper, *The Open Society*, Chapter 11, section II, is still the best restatement of methodological essentialism and objections to it. The same place also introduces methodological nominalism, in order to block the positivistic existential import (or rather export) which draws ontological conclusions from the nominalist method quite illegitimately. The commonsense of the positivistic reductionism as based on nominalism cum deductivism is counteracted by the seeming commonsense of the new Aristotelian essentialism: the two become intellectual poles as Lévi-Strauss-style myths that think for us. Of course both extremes stretch commonsense too mad, and only mixing them returns us to common sense. Hence we better reject both extremes and have no need to mix them to avoid their unpleasant extremism.

5 See, for example, Baruch Brody, 'Towards an Aristotelean Theory of Scientific Explanation', *Philosophy of Science* **39** (1972), pp. 20–31, reprinted in E. D. Klemke *et al.* (eds.) *Introductory Readings in the Philosophy of Science* (Prometheus Books, Buffalo, N.Y., 1980), pp. 112–23, where the difficulty is noted.

6 Much confusion arose from ignoring the fact that the question relating to unfamiliar words or objects are merely requests for information, translation, help, etc. When the request is for the "definition" of an unfamiliar word or to a word which refers to some unfamiliar object, the request has nothing to do with methodological essentialism. Hence Bacon's argument that questions concerning essences should properly come at the very end of a study, not, as Aristotle suggested, at its beginning. This amounts to the rejection of methodological essentialism or, as Bacon calls it, 'logic'. (Bacon's rejection of 'logic' is the rejection of the method of deductivism, or the method of anticipation as he called it; it is not a version of irrationalism.)

7 See my 'Naming and Necessity: A Second Look', *Iyyun* **44** (1995), pp. 243–72.

EVANDRO AGAZZI

NAIVE REALISM AND NAIVE ANTIREALISM

Today the expression "scientific realism" often designates epistemological conceptions which are quite different from each other; among these the following two are especially noteworthy:

a) science attempts to describe a "reality" which exists independently of it and according to which it is committed to measuring itself;
b) what science states and describes is an adequate image of this reality.

To admit the first conception is clearly less demanding than to admit the second: for we set ourselves the task of defending them both.

One can acknowledge that the question of realism in the sense set out above has almost uninterruptedly accompanied the path of modern experimental science. The last twenty years, however, have seen the diffusion of strong opposition to realism within that philosophy of science which can be considered in its own right, a development of empiricist-analytic epistemologies and of Popperian epistemology. This recent challenge to realism is no longer based, as was the case in the past, on the philosophy of knowledge, but rather on the philosophy of language. In other words, it is no longer a question of rejuvenating phenomenalist positions of a more or less directly Kantian flavour, insisting on the (real or presumed) impossibility for our knowledge, including scientific knowledge, to attain reality. It is a question instead of a further exploitation of the semantic thesis in which the meaning of terms depends totally on the whole context within which the terms are set. The first consequence of this, as is well known, is the thesis of the "incommensurability" of scientific theories. This first thesis immediately brings in another. The old realist claims of science were based on the more or less explicit presupposition that the observational terms established contact with reality, while it remained in doubt whether the same could be said of theoretical terms. Now, the possibility of neatly discriminating the first from the second having vanished (since it has to be said that all terms are in a certain measure theoretical), one cannot see what safe link with reality is guaranteed. Moreover, if it is true that the same term has a different meaning in two different theories, it is unavoidable that the hypothetical reality to which it might refer be different in the two cases. This leads to two equally paradoxical consequences; either one admits that each theory builds its own sphere of reality (which eliminates the idea of realism as the assertion of a reality which exists in itself independently of the science which discusses it), or one admits that realities can multiply indefinitely and become the object of different theories. The second would equally frustrate realist aspirations because it would not only conflict with the idea of the existence of one reality, but would anyway

D. Ginev and R. S. Cohen (eds.), Issues and Images in the Philosophy of Science, 13–26.

leave us with the impossibility of knowing "which" reality we are talking about at any one time.[1]

Realism and Referentiality

Within the new 'linguistic' perspective the question of scientific realism can be restated as follows: the realist position maintains that scientific discourse has an actual reference. As is known, at least since Frege's famous essay *Sinn und Bedeutung*, emphasis has been given to the distinction which exists between the sense of a term (the Fregean *Sinn*), which is a content of thought expressing "what is meant" by that term, and its *reference* or *denotation* (the Fregean *Bedeutung*), which is an object constituting "that about which" the sense in question is thought or expressed. Unfortunately, a similar distinction has been left unused just by those who, for a lengthy period of time, have occupied a pre-eminent position in elaborating theories of meaning, that is by mathematical logicians, who as far as the interpretation of formal calculi is concerned, quickly embraced an *extensionalist* semantics, according to which the sense of a term is precisely the sum of its references. The establishment of such an equivalency between sense and reference does not come of ignorance, but was made for practical reasons, supported moreover by the general 'philosophy' of logical formalism, according to which the symbols of a formal system have not and must not have any sense whatsoever.

Note in passing that this extensional semantics, which appears to be the semantics which is most concerned with quickly getting its hands on references without wasting time with the abstract world of concepts, shows all its weaknesses precisely when it is used in formalising empirical theories, i.e. those theories which intend to talk about some world 'external' to their language. The failures of extensional semantics in this field (which remain such despite the non-negligible quantity of articles which keep on being published in this sector in an effort to patch up this or that point) are a clear symptom of this essential fact: not only is it true that sense and reference cannot be made equivalent but also that neither of them can be eliminated, and that access to reference is guided by sense. Neither sense nor reference can be eliminated from discourse if all its fundamental characteristics are to remain intact. This may be otherwise stated as follows: by eliminating sense one would obtain a discourse which "does not say anything", by eliminating reference one would obtain a discourse which "does not talk about anything". A fully fledged discourse intends to say something about something.

At this point it may be clear why we have proposed the identification within the context of the philosophy of language, of the position of scientific realism with that which attributes reference to scientific language; on the one hand we must say that *without realism* one cannot give the *references* of a language, and this precisely because reference is defined as an extra-linguistic object which that specific language under consideration "refers to" as other than itself. In the case of scientific language, therefore, if one does not admit the existence of a reality different from the pure "language game" constituted by that given language, it is not possible to attribute to it the ability to refer to something (but only, at most, of proceeding according to the internal rules of the game). On the other hand if we interpret scientific language as purely and simply a language game, internally coherent and conducted according to rules accepted by a given community of speakers, but *without referential* purposes or possibilities, then we shall never be able to hold a *realist* position towards science because we shall already have accepted that it does not intend to talk about a reality distinct from its own language. The two theses, which imply each other, are therefore logically equivalent, and we are justified in saying that the thesis of the referentiality of scientific language is the translation of the thesis of scientific realism when moved from the gnosiological level to that of the philosophy of language.

Some might show dissatisfaction with these arguments of ours, making us remark that an authentic realist is not happy to maintain that scientific language 'refers to' something different from itself but pretends that this something is reality and not, for instance, a pure illusion, a mere intellectual construction, or even just the private world of one's sense perceptions. The objection carries weight especially because it invites us to specify what one can and cannot hope to establish about scientific realism by remaining within the philosophy of language. It is clear that within this philosophy one will never be able to say a lot about the 'kind of reality' belonging to the reference, and that for the good reason that this is not a linguistic problem. We have therefore no difficulty in acknowledging that the question of scientific realism is not *wholly* soluble within the philosophy of language; what is said above however, in support of the compatibility of linguistic reference and scientific realism, is tantamount to recognizing that the foundation of the referentiality of scientific language is a *necessary* condition for the establishment of the thesis of realism (indeed we shall see that it provides a great deal of essential ingredients for this foundation), and this fact justifies the interest that we now want to devote to it.

Symptoms of Referentiality

It is difficult to deny that the language of science intends to be referential because one needs only consider the attitude of all, or nearly all, practising scientists who share what has been called "spontaneous realism." On the basis of this, they *intend* first of all to devote themselves to a description and understanding of some sector of 'reality' (and not simply to the creation of mere intellectual constructions or to the development of some complex language game); second, they *believe* themselves to be doing something like that; finally, very many of them *believe* that science can succeed in this enterprise (others may be more sceptical and occupy intermediate positions).

It can therefore be accepted as uncontentious that science has a referential *intention*; it is however another thing to assert that it *succeeds* in constructing a referential discourse, and finally still another to clarify the *type of references* which scientific discourse may have.

We now want to touch briefly on the second point, contenting ourselves with a brief, elementary but fundamental remark; in empirical sciences it invariably occurs that *certain assertions cannot be accepted even though they are endowed with meaning*. This refers in particular to those which are disclaimed by empirical evidence. Given that these assertions are endowed with meaning, one cannot say that they are rejected because they do not correspond to the rules of the language game of that particular science in which they appear, but because there is a *non-linguistic* condition which prevents their acceptance.

One could object that in this case too the paradigm of the language game is not evaded because a rule common to all those language games which make up experimental sciences consists precisely in establishing that all those propositions which describe direct experimental results can, or even must, be accepted, while those propositions which are irreconcilable with propositions describing such results must be rejected. Despite appearances, the objection is very weak because it ignores the fact that a rule of this kind rests on a *non-linguistic* condition, such as that of taking into account operations, manipulations and observations of a concrete nature, which concern the sphere of 'doing something' rather than that of 'saying something'. Moreover, in the case of the rule of accepting propositions which describe experimental results and of rejecting propositions which contradict them, it would not be only naive but even misleading to ignore that this rule has been introduced in science because experimental results have always been attributed the role of providing the live view of 'reality' with which science sets out to concern itself. If we want to describe the situation as it is, we ought therefore to say that if there exists a reality which is endowed with its own structure, it is clear that it is not possible to say anything and everything of it because certain propositions which refer to it will turn out to be false as they state *what it is not*. The fact therefore that in experimental sciences there are propositions which turn out to be forbidden, because certain conditions of referentiality (experimental results) are opposed to them, is already an important symptom of the fact that these propositions speak of reality.

Semantic Logos and Apophantic Logos

The arguments so far presented centre around the distinction already made by Aristotle between the semantic logos and the apophantic or declaratory logos; the first consists in discourse which limits itself to "signifying" while the second is the discourse "which asserts and denies". The distinction might appear over-subtle but it is of the utmost importance. First it is advisable to realise that the establishment of the sense of terms does not imply asserting or denying in a literal sense, but rather a more general 'saying' to which, in particular, the dimension of truth and falsity is alien. It is however interesting to remark that the semantic logos, too, uses declaratory or descriptive sentences, as happens for instance in definitions; but why do we then say that definitions are not true or false, despite their being made of descriptive sentences? The question has raised a lot of discussion in the past and in particular it has added heat to the disputes on the difference between nominal definitions and real definitions. The only way to escape unambiguousity from misunderstandings appears to be this: the semantic logos is neither true nor false (and within it definitions, in particular, are of this kind) because *it is non-referential*. As soon as we give it a referential valency, it transforms itself into apophantic logos (this is the case of the so-called "real definitions", which actually must be considered as sentences which purport to be true of the real object which these definitions declare themselves to *define*, but to which they actually come to attribute some properties).

What has been argued allows us now to understand clearly what it would mean to deny referential import to empirical theories. It would mean reducing them to the level of the semantic logos, to pure instruments for establishing sense. Some might perhaps find this perspective acceptable, but it has the serious fault of not explaining the difference between the empirical sciences and purely formal sciences; if for these last it can be legitimate to say (even though with cautions on which we do not want to dwell) that they provide a contextual sense for their terms a major establishment of contextual sense for their terms, such a point of view is inadequate to characterise the empirical sciences because in them the presence of empirical data introduces at any rate something which spills over the pure and simple linguistic context. For this reason we must say that empirical sciences appear as discourses of apophantic or declaratory logos. The establishment of the apophantic logos is characterised by the fact that reference emerges, together with the sense, and furthermore does so in such a way as not to be independent from sense. In fact, as we have now repeated several times, the search for the reference requires a non-linguistic activity which in many cases (especially in that of sciences) is even of a clearly 'practical' type, such as operational manipulations with instruments, observation in suitably created conditions, and so on; this activity, therefore, consists in *exploring the world* and not in *exploring language*. However, it is not less true that this exploration of the world in search of reference takes place on the basis of sense. Otherwise, we would not be able to *recognise* the reference when we encounter it (here is the solution of the paradox already stated by Plato according to which one can only know what one knows already; the point really is that we all know a reference only because we recognise in it the characteristics expressed in the sense with which we had begun the search, but the reference was not already known to us before we met it). When the reference is traced in this way, several properties can be "asserted and denied" about it and in this way true or false sentences can be produced. The apophantic logos is therefore that in which the *notion of truth*, directly linked to that of reference, is established.

The Excessive Claims of Contextualism

For the sake of brevity the approach already mentioned, according to which each term takes up a sense which is totally determined by the context within which it is set, is here called "contextualism". From this it follows in particular that any two homonymous terms, if set in different scientific theories, consequently have different senses. The result, as we know, is the 'incommensurability' of scientific theories, the denial of true progress in science and the impossibility of referring theories to a common reality.

Faced with this position, it is more than legitimate to ask why the comparison of theories should take place on the basis of their sense and not instead on that of their references? After all, it was the traditional conviction of scientists and epistemologists that two rival theories, which speak about the same reality, may be compared in the sense that one may be found false and the other true about that reality, and this even if they assert *different* (but not necessarily contrary or contradictory) things about it. This remark is very important because it shows that a theory can be better than another even though it cannot be compared with it regarding sense if, on the basis of its criteria of experimental control, it can be presumed to be true, while the other, on the basis of its experimental controls, must be declared false. Why cannot a similar point of view continue to be adopted? One might say because it is a point of view founded on realism, which today we reject. But it is clear that this answer cannot serve as a reason for *justifying* the rejection of realism. Different reasons, therefore, have to be put forward which in particular should hit the nodal point of the discourse which we have called traditional, that is, the thesis that theories with different sense can deal with the *same* references.

Let us first remark that the relationship between determinacy of sense and identifiability of references is not as strict as it may at first sight appear. In the first place, some indeterminacy of sense is compatible with the possibility of identifying the references; it is sufficient, in fact, that some of the senses are determined, that is, those which a certain linguistic community has agreed to use to identify certain references. For instance, whales were once classified as fish and today as mammals, so that the sense of the term has undoubtedly altered. Should we then say that the references too have altered, that is, that we no longer call whales that kind of animal that used to be so-called? Not at all; in fact there exists a sufficient number of properties of whales (for instance those of their simple external morphological nature) which enable us to identify whales and which remain the same even today. Analogously, in different empirical theories groups of characteristics can exist which remain unchanged even within two different contexts and can be used to trace the references and to recognise that they are the same. From what has been said several times above, we can infer that these characteristics will be those linked to experimental results, or more precisely, those which are connected with operations of ascertainment and measurement which are materially the same in both theories. When this happens we are able to say that the references are the same and we may then proceed to the comparison, even without denying that all the terms of the two theories, because of the influence of different theoretical contexts, result in more or less different sense. In other words, a 'referential part' of the sense of certain terms exists which doesn't appear to be sensitive to contextual 'meaning variance' because it is linked with that *extra linguistic* and operational component which precisely characterises empirical sciences.

Let us therefore conclude that it is true that the sense of a term always depends on the context within which it is set, provided this 'depending on' be understood somehow as the 'result' of an addition, the addenda of which are not *all logically interconnected* to the point of being *interdefined*. In certain cases, therefore, those which one could call 'free addenda' or 'free components' can reappear in other contexts while continuing to remain free and can, thanks to their liberty, guide the search for common references. In other words, one can accept the 'contextualist' approach to meaning without reaching the extremist consequences to which this approach has led for reasons which have really nothing to do with its internal consistency. Ultimately, it is a question of not losing that degree of common sense which allows us to understand that the identity of "that of which one talks" does not require the identity of "that which is said about it", but only the compatibility of the two.

Reference and Reality

We have already had a chance to mention above the fact that realism is a necessary condition of referentiality, in that reference, introducing itself as something which is of an extra-linguistic nature, already has characteristics which are attributed to 'reality' when dealing with the philosophy of language. We have, however, also remarked that within this approach one cannot pretend to specify which 'type of extra linguistic reality' the reference has. Before going on we consider it useful to draw attention to the fact that it is not correct to challenge realism on the basis of a concept of reality which is arbitrarily inflated and pretentious. In other words, we claim that one has the right and the duty to declare that something is real if one is compelled to somehow admit that it is *different from nullity*. It is clear that from this viewpoint, even a dream, a mathematical calculation, something imaginary, a hallucination, must be considered 'real' because, despite being different from what we call concrete and material reality, they are also different from nullity, as witnesses the fact that we can describe them, and even describe them in a true or false way (I can assert that I have dreamt of a white horse while 'really' I dreamt of a black cat). We shall say therefore that these various types of things differ not in the *fact* of existing, but in the *way* of existing (a house exists in such a way as to be capable of being perceived with the senses and to be operationally used for living in, while its image exists as a mental entity at different levels and according to various modalities). It would therefore be totally arbitrary to maintain that only things which belong to a single and well-determined *kind* of reality (that is to the reality of material type) are real.

If on the one hand we should not be trapped in the pretence that reality is of one type only, neither can we use the distinction between the fact of existing and the way of existing as a kind of jemmy to reach the claim that one cannot at any rate avoid being a realist. How then can one avoid conflicting misunderstandings? The answer appears to us to be simple enough; we shall describe a discourse as realist if it *intends* to talk of a reality *of a certain type* and succeeds in this intention. Therefore, a discourse which intends to talk of physical reality is realist only if it can claim that it actually succeeds in talking about this reality and not, instead, only about intellectual 'images' of it. However, a discourse which proposes to deal with dreams or hallucinations will truly be 'realist' precisely if it succeeds in its aim, independently of the fact that dreams and hallucinations are not physical objects; on the other hand, this discourse would not be realist if it succeeded only in talking about physical situations which accompany the dream and hallucination, such as electrical or chemical states of the brain (physicalist reductionism is quite far, therefore, from being a guarantee of realism; it is rather its negation).

What has been said so far using the notion of "type of reality" or of "way of existing" can be better specified in terms of *quality* and of *criteria of referentiality*. To say that not all that exists has the same type of reality ultimately means to acknowledge that various entities possess different qualities and that we are able to 'refer' to different types of entities depending on the ability we have to accede to their properties used in this way as "criteria of referentiality". It is enough to look back on what was said where we spoke of sense as a 'guide' to trace the reference; it will be easy to realise that this discourse was nothing but a description of how properties serve to identify references since properties are things which are 'attributed' to references in an intentional act of the subject. They cannot fail – and this is the point – to involve the subject as well; however it is no less essential to acknowledge that they spring not only from the subject but from the meeting between it and reality. For instance, a toothache is as real as anything can be (it is enough to think about the profound difference between the being and the not-being of a toothache for the suffering subject); however, it does not have a colour, a mass, a localisation in space, a shape nor many other properties which would allow us to qualify it as a physical entity. In this case, the criterion of referentiality is only a subjective state of pain which is sufficient for us to state that it exists. Not even with all possible effort would we succeed in attributing to it a colour or other 'properties'. It is therefore not within the capacity of the subject to attribute properties to things at its leisure. On the contrary, we shall find it 'necessary' to attribute a colour (for instance, green), a mass, a shape, to a leaf, but we will not be able to say whether it is odd or even, monosyllabic or polysyllabic, introvert or extrovert, given that these are properties which serve to qualify and identify other 'types' of entities.

Our reasoning has so far been developed at the level of common speech but can be extended without difficulty (indeed, in an even simpler way) to the case of the discourses of sciences. As I have by now for a long time and in several of my writings tried to make clear, each scientific discipline presents itself as a discourse which has an intentional relationship to reality from a certain 'viewpoint', that is, it sets itself the task of investigating only certain aspects or qualities of reality; because of this it selects a certain circumscribed number of predicates and, for the purpose of succeeding in its referential effort, associates them with some standardised *operations*, which we can call indifferently "criteria of referentiality". These operations 'clip out' specific objects of a given science within the vast sphere of reality, and, precisely because these operations do not apply to nullity but to already identified references ('things' of daily experience as it is practised within a certain historically determined collectivity) and are furthermore subjected to empirical and not purely linguistic and intellectual manipulation, this gives rise to specific references which cannot avoid being *real* as well. One could wonder whether the *properties* attributed to these references are or are not real, but at this point the deep-seated naivety of this question should be clear, given that in any particular science only those entities which have such properties are acknowledged as objects, so that, for that given science, the object is nothing but the set of the properties which can be operationally attributed to it, precisely because they are operationally 'related' to it and not only 'thought' about it. This, evidently, does not exclude at all the possibility that that determined reference possesses other properties as well, which can be enquired into by other sciences or which can even be the object of non-scientific discourse.

Realism and the Possibility of Error

Against what we have been saying so far there appears to militate the fact that no scientific theory is ever certain of its own truth and that, moreover, the history of science attests to the continuous change of theories, which, it seems, one could interpret either as an uninterrupted series of "falsifications", or as the indication of a lack of reference.

In this accumulation of arguments several different aspects are present which it is absolutely indispensible to distinguish. First, it is necessary to specify that the falsity of a theory can in certain cases indicate that it is without reference, but in other cases it cannot. Second, it is a question of seeing whether the cases in which it is thought that a theory has been 'falsified' are really cases of falsification, or more simply cases of a change of reference.

Let us discuss the first point and see how, in certain cases, the falsity of a theory implies the acknowledgement of the non-existence of its reference. An example relating to a singular term is that of all the stories that for long centuries were narrated around Hermes Trismegistus, considered the author of the writings known as *Corpus Hermeticum*. Relatively recent criticism has shown that this figure, in whom very serious Renaissance scholars such as Marsilio Ficino still believed, never existed and that the *Corpus* was written during the age of Imperial Rome by philosophers of neo-Platonic inspiration who invented the existence of this mythical scholar, roughly co-eval with Moses, for the purpose of adding credence to their doctrines. The falsity of the theory therefore coincides, in this case, with the elimination of its presumed reference. A certain analogy exists with the phlogiston theory, once adopted as the basis of the incipient science of chemistry and today abandoned. In this case, too, one can say with strict rigour that the falsification consisted in discovering that the term "phlogiston" does not have a reference; however, one could be slightly more tolerant and maintain that we actually call by another name (for instance, hydrogen and oxygen) certain gaseous products which can be seen to emanate from certain chemical reactions and that once were covered by the single term "phlogiston".

However, it must not be at all taken for granted that the falsification of a theory implicitly denies the existence of its references; in the case of the Ptolemaic and Copernican theories, for instance, one can claim that the references remain the same (Earth, Sun, planets) and that the latter has shown that certain assertions of the first relating to the static position of the Earth, rather than of the Sun, in the planetary system, are false. We can say that this is, after all, the most common situation, which well corresponds to the fact that in general a discourse can be said to be false when it is referential but 'says' something about its references which they 'really are not'.

When one specifically concentrates on studying scientific theories, those cases acquire a much greater importance in which the purported falsifications must be interpreted neither as an elimination of their respective references nor as a discovery of false assertions made about these references, but as *a change* of references. We easily realise this possibility when we bear in mind that in reality the references of every scientific theory are 'clipped out' within 'things' (that is within the references of common sense) through standardised, precise and limited operations. From this it immediately follows that if the set of these operations changes, the operational meaning, that is, on account of what has already been said more than once, the 'referential meaning' of certain basic terms, alters and with it also alter the *objects* to which theories refer. For instance, one can read the transition from classical to quantum mechanics in this way. But it is then clear that two rival theories can both remain true, each obviously *about* its own objects, or references as one may prefer. In this way, we shall in fact have broadened the range of known truths, as the new ones take their place beside the old ones and do not displace them.

Note that in this perspective one can explain what Popper wants to claim by his unhappy theory of "verisimilitude", which claims that there exists *truth in itself* which is intrinsically unattainable, despite the fact that successive theories are more and more closely approximate to it in an endless, asymptotic process. The misunderstanding lies here in having reified truth, so that the cognitive enterprise is not thought of as a process which aims at 'knowing reality', but at 'knowing truth'. Now, while there is nothing absurd in stating that the enterprise of knowing reality can be an ideally infinite task, because each set of *true* knowledge about it only picks out partial aspects of it, it appears absurd to say that we are certain of approaching truth even if we have no possibility of taking this last as a term of comparison to assess whether we have really come closer to it.

These last considerations allow us to acknowledge in the realist position the most solid basis for talking of a cognitive progress of science. This progress may consist in the elimination of errors equivalent to showing the non-existence of purported references, or in the elimination of earlier erroneous, assertions about references which are preserved, in the technical sense in which the new theory retains the same 'objects' as the preceding one. One can think also of cases in which certain references of common sense are successively 'objectified' through operational predicates which are totally or partly different; we shall then say that the different theories permit one to increase true knowledge about these references by emphasising different 'aspects' of them. When the diversification of operational criteria is such as to leave doubts about the fact that the references of common sense are still the same we shall certainly talk about non-comparable (or "incommensurable") theories, but incommensurable on an empirical-operational basis and not only on a semantic-contextual basis, and scientific progress will consist in having brought to light new 'objects' of knowledge.[3]

In all these cases it is perfectly legitimate to talk about cognitive progress even in a 'cumulative' sense where either one knows more and better about the same references, or one knows more because new references have been discovered. Both truth and error contribute to this progress in the way sketched above, and this justifies the common conviction, which is also that of the scientific community, according to which human knowledge, even if fallible, proceeds nevertheless in the discovery of the true (that is, in the acquisition of true propositions) in as far as it has more and more success in describing and understanding the structure of reality.

Naive Realism or Naive Anti-realism?

It has been customary in the idealistic tradition to call realism "naive", and this qualification was intended to stress that the spontaneous commonsense conviction – that things exist 'independently of our thought', or 'outside our mind' – lacks totally the awareness that the very condition for expressing such claims is thinking, so that things allegedly existing 'outside our thought' are included in it in the moment we 'think' of them as being external. In a similar way things which are claimed to exist 'independently' of thought cannot be affirmed as existent otherwise than by an act of thinking. It may well be that realism is often naive in the sense just mentioned (a sense which was sharply criticized already by Berkeley), but this still does not imply that it is 'wrong'; in fact it could be right, in spite of its obliviousness to – or too little awareness of – the reasons for its being right, and the remedy of this situation would simply require making more precise the sense in which the 'independence' of things from thinking should be understood. In fact, idealism too may become naive (though in a different way) when its correct claim that no discrepancy can be affirmed between reality and thought is pushed to the extreme of claiming the total equivalence of reality and thought (i.e. to the claim that everything is nothing but thought). These two opposite naiveties have a common root, consisting in what one could call 'epistemological dualism', a philosophical commonplace originating in 'modern' philosophy. According to this view, what we know are our representations, our sense impressions, our 'ideas', and not things as they are 'in themselves', so that we have to look for a guarantee both of the real existence of things and of their being faithfully represented by our 'internal' ideas. The impossibility of solving this problem was definitely clarified by Kant, but the problem is in effect insoluble because it is a pseudo-problem. Indeed, there is no evidence nor argument for claiming that what we know are our ideas, rather than things themselves 'through' our ideas. It is furthermore because of this bizarre fantasy – that things are there 'behind' appearances, 'outside' our thoughts, and so on – that the impossible task of reaching them by transgressing the border of our thinking was proposed. What Berkeley already had attempted was accomplished by the 'classical' German 'transcendental' idealism of Fichte, Schelling and Hegel, and in the most consequent and strict way by the 'absolute' idealism of the Italian philosopher Giovanni Gentile. The result of their speculation was however twofold: on the one side they correctly eliminated any 'dualism', or 'heterogeneity', or *separation* between reality and thought; on the other side they denied any *distinction* between them, reducing the whole of reality to thought, and this was much less correct. Indeed, the price of this reduction was no less dogmatic and unsatisfactory than the 'dualistic presupposition'; in order to avoid the patently awkward claim that things exist only while I am perceiving (or thinking of) them (subjectivism and solipsism), they were obliged to claim that things are the content of a superindividual act of thinking, be it the thinking of God (Berkeley) or of the Spirit (transcendental and absolute idealism) and were obliged to propose very unplausible reasons in order to account for the 'illusion' everybody spontaneously cultivates regarding the existence of several subjects, and of things existing independently of our thinking of them.

Why do we say that a *separation* of reality and thinking is not legitimate? Because, if thinking were not thinking of reality (understood not in the misleading sense of something which is mysteriously unaccessible, but simply as something which is different from nullity), it would be 'thinking of nothing' and therefore no thinking at all. Moreover, if thought were itself different from reality, it would simply be reduced to nullity. However, this conclusion does not imply that reality is *identical* with thinking, or equivalently, that thinking exhausts the *whole* of reality. In other words, while it is absurd to deny that thinking always is of some kind of reality, it is gratuitous to say that thinking is *only* thinking of itself. In fact, one of the most serious difficulties which the idealists were never able to overcome convincingly, is that thinking is always 'thinking of' something, that it necessarily entails a bipolarity. It is true that this bipolarity is given *inside* thinking, but this does not mean at all that the two poles are two aspects or moments of thinking itself. The real situation is rather that reality is *present* to thinking, is *given* within it, this presence and giveness being the 'intentional presence' which may be also expressed as an 'intentional identity', but this does not mean 'ontological identity'. Rejecting this position even comes down to reintroducing surreptitiously the 'naive' epistemologic dualism, claiming that all we know are our ideas, but at the same time endowing these ideas with the characteristics of giveness which are the distinctive marks of reality. A symptom of this fact is that the problem inevitably arising out of epistemologic dualism, i.e. the problem of the 'origin of the ideas' is not eliminated at all in the idealistic philosophies; instead of saying that these ideas are put in us by God, these philosophies say, no less dogmatically, that they are created by the Spirit of which we are part, in a way which remains unconscious for us and no less mysterious than the different forms of innatism advocated in the history of philosophy.

These reflections concerning idealism and realism are no digression with respect to our topic. In fact, the total absorption of reference into meaning is the linguistic version of the idealistic attitude, since language is in many respects the contemporary counterpart of the idealistic transcendental subject. The thesis of the intranscendibility of thinking is the parallel of the thesis of the intranscendibility of language, and both have the same strong and weak points. It is in fact true that a referent may be present in a discourse only through a structured set of meanings, exactly as ideas are the way in which reality is present to thinking. Therefore there is no 'separation' between meaning and reference. A naive realist would be therefore someone pretending that the existence of reality may be claimed as something residing 'behind' meaning or 'outside' language. But an anti-realist would be equally naive if he were to claim that our language does not say anything about reality, either because reality remains 'hidden' (epistemological dualism, which would mean that language 'speaks of nothing'), or because language is supposed to exhaust the whole of reality, and this would be tantamount to claiming that language can only speak about language. Both naive realism and naive anti-realism in this way express a confusion between the 'referential identities' of language and reality and their 'ontological identities', which is as untenable as the idealistic confusion between the intentional and ontological identities of reality and thinking.

University of Fribourg, Switzerland

NOTES

[1] These theses, as is well known, have been slowly emerging within the discussion of "observational" and "theoretical" terms of empirical science, and have been corroborated within those doctrines which have supported the "theory-ladenness" of all scientific concepts indifferently. The most extreme thesis in this line of thought, can be traced especially to Feyerabend.

[2] I take the liberty of referring, for a more detailed discussion of these theses, to two works of mine: E. Agazzi, 'The Concept of Empirical Data. Proposals for an Intensional Semantics of Empirical Theories', in M. Przelecki *et al.* (eds.) *Formal Methods in the Methodology of Empirical Sciences* (Reidel, Dordrecht, 1976), pp. 143–57; and also *Commensurability, Incommensurability and Cumulativity in Scientific Knowledge*, *Erkenntnis* **22** (1985), pp. 51–77.

[3] For a detailed exposition of these various alternatives we refer to the article already quoted: *Commensurability, Incommensurability and Cumulativity in Scientific Knowledge.*

BABETTE E. BABICH

AGAINST POSTMODERNISM AND THE "NEW" PHILOSOPHY OF SCIENCE: NIETZSCHE'S IMAGE OF SCIENCE IN THE LIGHT OF ART

In what follows I offer a polemical – and inevitably elliptical – review of the current state of the philosophy of science and argue for a radically hermeneutic philosophy of science, following Nietzsche's recognition that although science represents the fulfilment of the modern project of a self-grounding ground, the problem of science as such cannot be posed on its own ground. Yet I am aware that some readers will find the following closer to a sociology of science and knowledge as an argument for the recognition of historical factors (though I focus on no specifically social categories and employ no sociological concepts) than to a critique of the philosophy of science. I have neither the competence for nor the intention of offering either such a sociology of knowledge or a history of science. What I will do is to challenge the kind of conceptualizing fetishism – an Aristotelian legacy – that would make such distinctions to the detriment of the critical scope proper to philosophic thought.

But by arguing Nietzsche's case against the foundations of modernity and modern science, I do not support what might be named a postmodern alternative in what follows. Rather, I criticize the conviction concerning the postmodern sophistications of science and technology as fundamentally deficient. This seductive vision of accomplished sophistication is suggested by Jean-François Lyotard, first and foremost, but also by Don Ihde, among other philosophers of technology, and by, for one example, Stephen Toulmin among philosophers of science, and, among other scientists, the late David Bohm. The idea of postmodern sophistication represents the ultimate conversion of the modern in architecture, music, art, as well as theory and public consciousness, and is perhaps best exemplified in the virtual image of cyberspace. With earnest levity and declared irony, postmodern sophistication promises a solution to the limitations of the modern progress ideal in the commercialistic, corporate, quasi-industrial guise of what is quite soberly or 'technically' called 'playfulness'. It is not that such a resolution to the perceived problem of our age would or could be *wrong* as such, but only that it is more elusive and much harder than and not at all as light as its advocates seem to imagine.

Indeed, play requires something like what might be called, speaking with Gillian Rose, hell in one's heart – and all hope and expectation besides.[1] It is this that brings one close to what is real, and therewith to the problem of play itself. As an image beginning not only with Heraclitus (as the child's game of cosmic eternity), articulated in Kant and further elaborated in the German Idealist as in the Romantic perspective, and more recently invoked by Nietzsche and Heidegger, play has a complex and vast literary and philosophical history.

D. Ginev and R. S. Cohen (eds.), Issues and Images in the Philosophy of Science, 27–45.

Conversely, the regulative ideal of 'playfulness', in turn and beyond play as such, might be said to have begun with the structuralist move (via Benjamin or else Derrida or Lacan) that effected a literal rendering of metaphor and metaphorical transformation of description. This innovation enables all manner of loose associations, not least among which is deconstructive practice.[2] But the original problem of play must be understood as the quintessential concept of tragic antiquity, not of the optimistically modern or indeed the coyly post-modern era.

The postmodern salvific project of playfulness underestimates the innocence of invention or creativity and fails to prise the supreme elegance, poise, and utter absorption without remainder, ironic or otherwise, that is play. This is, all apologies to football fans, not a matter of the challenges of sport or the daring risk of speculative calculation. In all its aspects, play must be thought closer to a kitten's fierce fight with a piece of cotton wool than to modern competition in any of its many masculinist varieties – whatever the contrary claim in the gaming examples proposed by Wittgenstein or Gadamer or MacIntyre. As distant from real hell as from real hope, the game is easy to describe and prescribe, be it a language game, a ball game, a chess game, a game of dice, or the unreal game of postmodern playfulness. With the concentration of hell, the innocence of play lacks any awareness beyond itself. Unfathomably real, everything and nothing at all is at stake, as a child, lost in the toss, spins a top or a rhyming line or a kitten plays with string.

A case might be made that Derrida's idea of "free play" is related to this tragic notion of play as it is also Nietzsche's understanding of the real universe and the necessary, interactive possibilities of human perspectival understanding.[3] The same case may not be made for the scientist, despite popular and postmodern claims for such virtuosity. The scientist's version of "free play", or the freewheeling affirmations of a philosopher of science such as the late and great Paul Feyerabend (the cry of "Anything goes!" imputed to his anarchic epistemology), is less an affirmation of child's play or chance than the fortuitous exploitation of this creative possibility (as it is what science is) in the precise interval of prediction and control.[4]

Against the secure conviction of modern advance, in a reverse dialectic of influence and interest, a hermeneutic questioning of science that might be counted as a Nietzschean philosophy of science reminds us that it is the rudely historical case and social matter of political fact that alternatives of whatever kind are routinely co-opted or absorbed within the ambit of that against which they were posed as alternatives in the first place. The defeat of critical or alternative approaches in this way is pernicious and seemingly ineluctable. It is the triumph not of pure reason but its image, and this is driven by the interest-laden dynamics of oppression and institutional political, academic, and market practice.

It is and remains as urgent as ever for critical reason to speak not only in its own name but also in the name of the voiceless, of what is unintelligible to reason because it is oppressed, the awful suffering perforce muted so that unspeakable deeds may be done and may escalate as muffled evil always escalates. The resources of critical reason, materialist or idealist or otherwise, have failed and continue to fail to do this. To see why, one needs a hermeneutics of power and its project. This is a hermeneutic of suspicion exactly as Nietzsche teaches it in the twists of interest precisely to be found in the name of moral reckoning and the call for equal and human rights. Thus Nietzsche writes that we are unknown to ourselves and that we do not know what speaks in us in our desire for power, what Nietzsche calls our will to power, as such a will dominates moral reasoning and our demands for justice. Nietzsche raises the self-same question concerning the philosophic or scientific solicitation of truth. Seeking truth, as men of science or philosophical labourers, Nietzsche asks what we wish to gain? What in us, Nietzsche asks, *really* wants truth; what is the nature of our desire for truth, and what the object? These are the questions Nietzsche asks us to pose, the suspicions or doubts he raises concerning the value of truth and the value of the search for truth.

Nietzsche reminds us, just as Kant does at the start of his critique of foundational morality, that we can never be sure if the very heart of our purest, most selfless desire does not include, as desire necessarily includes, self-interest or advantage. But simply by virtue of acknowledging the necessity of such questioning, knowing that we must always second-guess ourselves and our own purest notions of truth, justice and the American way (or, indeed, the ideals of a united Europe), does not, Nietzsche emphasizes, eliminate the problem. This is the anti-dialectical, critically rigorous turn in Nietzsche and it is this which offers a chance for a radically philosophic perspective on science, conceived on the basis of art (and of art in life).

The tendency to what I have named co-option or complicity and ultimate devaluation must be recognized. Recognition alone can alter nothing. That is the point. But this point makes it obvious that the philosophic program of questioning is more than an exercise in hyperbole, like Descartes's project, doubting for the sake of security and conviction, but for the sake of a resolute disruption. Not to found but to shake the foundations of what we think we know and how we know it. And if this looks like the old account of revolution, like the traditional narrative of scientific discovery and insight, this is no accident, but a paradox that will be justified if it proves productive.

THE CASE FOR CONTINENTAL CURRENTS IN AN ANALYTIC DISCIPLINE

Until recently, the nomological-empiricist approach, often described as the "Received View", characterized the analytic philosophy of Anglo-American philosophy of science. Since the writings of Thomas Kuhn and Paul Feyerabend, the adequacy of this analytic scheme has been challenged as an unreal or idealizing or inaccurate account of the scientific enterprise. In a bald, if simplistic statement: the method of science is not the so-called Scientific

Method, be it verificationist or be it, following Karl Popper (as so many theorists of science do), falsificationist. Yet confessed as inadequate to the task of describing, let alone analyzing science, this axiomatization-preoccupied, essentially science-approbative approach remains the ideal of today's more sophisticated expressions of the philosophy of science.[5] In other words, this is to say that in spite of what some would call the Kuhnian "revolution" in analytic philosophy of science, the idealization (that is sometimes adulation) of science within the philosophy of science and scientific culture continues apace.

Steve Fuller appears to express a contrary statement, speaking of a state of crisis internal to the philosophy of science by adverting to this crisis directly in a section title of his book, "The Mythical Status of the Internal History of Science, or Why the Philosophy of Science Is Suffering an Identity Crisis". Fuller, an advocate of the social perspective so often excluded on disciplinary grounds from the compass of the philosophy of science,[6] is not the only scholar, analytic or otherwise, to bruit about today's state of chassis in the philosophy of science. An eminent and important philosopher of science who has since taken his energies to the broader essayistic fields of popular science studies colonized by the likes of Oliver Sacks, Ian Hacking speaks of a "crisis of confidence". "Philosophers," according to Hacking, "long made a mummy of science". It was this mummification (the axiomatization or calcification of theory-making and its conflict with the history of science), which, late in the horror-show, could not but predispose philosophers of science to crisis: "When they finally unwrapped the cadaver and saw the remnants of an historical process of becoming and discovering, they created for themselves a crisis of rationality".[7] Alasdair MacIntyre, employing an altogether different analytic tack, categorizes this same crisis as "epistemological", resolving it in terms cohering with the traditional perspective represented in *After Virtue*. MacIntyre understands this last traditional scheme as history (and *not* historiography as Kuhn understands it): "An understanding of the concept of the superiority of one physical theory to another requires a prior understanding of the concept of the superiority of one historical narrative to another. The theory of scientific rationality has to be embedded in a philosophy of history".[8] For MacIntyre, this philosophy of history must be conceived as closer to the research programmes of Lakatos than to the sociologistic and economic canons favoured by Fuller or the Foucauldian perspective ruling the accounts offered by Joseph Rouse and Gary Gutting.[9] MacIntyre continues to note that when "the connection between narrative and tradition on the one hand, and theory and method on the other, is lost sight of, the philosophy of science is set insoluble problems".[10] This dichotomy is none other than that between so-called continental (or in this context, historically sensitive or traditional) philosophy and analytic or Anglo-American style philosophy and MacIntyre in a recent book adverts to the chasm between incommensurable thought-styles as unbridgeable.[11] The problem lies in the stylistic self-limitation of the prescriptive definition of the philosophy of science itself conceived exactly in the light of science, i.e., towards the ideal of a "science of science", and never in the broader Nietzschean hermeneutic context, which illuminates science through art and art itself in the light of the life it serves.

What is at issue institutes and preserves at a stroke the conceptual, stylistic divide between the current world practices of philosophy, a division we may trace to its roots in the conceptual bifurcation that stemmed from Frege's formal interest and Husserl's phenomenological theory and is today to be seen in the differences between analytic and continental styles of philosophy. But it is important to add that beyond the question of philosophical pluralism, this same "identity crisis" recurs in the conceptual opposition between *Verstehen* and *Erklären* in the field of the social sciences. It is also, conceived from an even broader perspective, the problem of rationality and its other, which Nietzsche names with his other co-masters of suspicion, Marx and Freud in the same contextual gambit, the problem of the subject (be this the subject of philosophy, the unconscious, or capital).

In the context of this essay, the problem of the division between analytic and hermeneutic approaches is constitutive of the (analytic) philosophy of science. Within the analytic tradition of the philosophy of science, what the above philosophers have named a "crisis" (of epistemology, confidence, identity, or rationality) is not a matter of a conceptual crisis but a practical, even pragmatic crisis, and thus, very plainly, as I have named it above (after Joyce): a state of chassis. It begins with the realization of the ineluctable relevance of scientific (social) practice and scientific history in opposition to, rather than in the wake of, what Karl Popper tried to mint as his own singular thought-invention as the contextual distinction of scientific discovery and scientific (and aspiring philosophic) justification.[12]

Deceptively persuasive, Popper's falsificationist strategy sidesteps the epistemic question of theory confirmation (truth) or verification for the falsificationist strategy that won him enormous influence, despite its theoretical and practical limitations. This influence is at least in part due to the fact that Popper is an advocate of the full progress ideal of enlightenment modernity and because the practice of scientific research that he explains in the various contexts of discovery and justification do not revise popular or academic conceptions of the practice and progress of science. If verification is (logically) impossible, science nonetheless advances toward truth by the indirect method of *directly* eliminating alternatives. Popper's falsificationism thus has the advantage of leaving the traditional view of both science *and* the philosophy of science untouched while explaining the progressive sense and logic of experiment. Popper's views have survived and doubtless will continue to survive sociological analyses and historical studies of science because Popper's theory of science as falsified is itself, as Nietzsche would say, *interpretation*, and is therefore flexible enough to reflect any practice. If science tends in practice or in the minds of its practitioners towards theory-confirmation, if scientists do not in fact or effectively abandon theories in practice even after they have been "falsified", if once-falsified theories are later revived, one world-inverting revolution after another, Popper is vindicated even so. The problem of verification is conceptual not practical or historical.[13]

Against Popper in a stylistic fashion reminiscent of Ernst Mach, N. R. Hanson can be read as arguing that the philosophers of science must consider the details of science in practice, whether through direct experience or as provided by historians and sociologists just because philosophers of science fail to consider the actual research or experimental practice of science (represented by the complications of quantum mechanics or "microphysics") in their accounts of science. For Hanson, the "background information, the 'set' that makes an explanation stand out, derives as much from what is obvious in a situation as from discursive knowledge gained through training".[14] This inculcated setting-up of the observational context of the scientist's trade activity, the experiment as such, offers the first expression of what would later gain such central importance in Kuhn's concept of the (research) paradigm. Any understanding of the physicist's activity requires a recognition of the paradigmatic project of physics. Commenting on Duhem's description of laboratory life, Hanson notes that the physicist's absorption in physics permits him to be a physicist, seeing an X-ray tube where the child sees only a complicated light bulb.[15] For Hanson, "the paradigm observer is not the man who sees and reports what all normal observers see and report, but the man who sees in familiar objects what no one else has seen before".[16] It is a short, but quite typically unattributed, step from Hanson's original "paradigm observer" to Kuhn's later notion of paradigm in the historiographical distinction between the activity of so-called "normal" science and so-called revolutionary science.[17] And whether we speak of context or paradigm, the sociological turn introduces the distinction that later separates philosophers of science into the respective camps of the historian/sociologist (in the range from Kuhn to Latour and preceded by Hanson and Duhem, among others) and the programme/methodologists (Lakatos-Feyerabend in the line of Popper, and so on.)

THE MONOTONY OF POSTMODERN PLURALISM AND THE TRAGIC GROWTH OF KNOWLEDGE

To take stock here, it seems obvious that if the philosophy of science is incomplete or in need of alternative approaches, the problems in the philosophy of science must be patent and clear to all concerned; that where there are plain problems, there are equally plain solutions, ready for action and contending for implementation, proffered and arbitrated from within the tradition itself.[18]

In recent accounts of the philosophy of science both within and outside of the profession (read: analytic, always analytic) philosophy of science, as indeed in the present context: putting the philosophy of science in question, we are always asked to count science as an advocate, since, we once again speak, after Galileo, of the "new" science. Where once one might have sought alternative approaches to science and scientific culture, science has become alternative to itself.

This critical point concentrating the giveness of a "new" science, or more pointedly, of a "new philosophy of science", reflects the times themselves and the changes in philosophy as a whole. After all, in these postmodern, post-coldwar, post-pluralist times, it's obvious that analytic philosophy is no longer quite the conceptual behemoth it once was. Philosophy is now a kind of post-philosophy, modified, modest, but above all: *new and improved.* A postmodern vision or condition rules. Why am I here issuing a call for revolution *after* the revolution has succeeded?

I have been arguing that there has been no revolution but only that opposition has been declared unnecessary. And this is old hat and a very routine operation of capitalist-cum-marketing ideology: this is the functioning of *co-option* in practice and effect. And if such a conversion of opposition is subtle, it is also extremely effective. Simone de Beauvoir, speaking of the dialectical nuances of the biological-historical-social oppression of woman invoked the *complicity* of the oppressed to describe the co-optive process. Nietzsche, who called it *ressentiment*, claimed that such a dynamic provided the very motor of western culture. This co-option, complicity, *ressentiment* effects the very considerable preservation of the same, the old, in the guise of the new: the weak represented as strong, the "new" but no longer oppressed femininity. Note that the process of co-option, complicity, *ressentiment* is not fundamentally subversive. Ideals are not converted against themselves, they are simply, in Orwell speak, declared fulfilled. De Beauvoir emphasizes that bad faith is part of the mechanism of complicity, as of co-option, and indeed on both sides. Nietzsche's analysis of *ressentiment* takes its point of departure from the coincidence. For Nietzsche, the will to power of the weak seeks the representation of weakness as strength. This is a routine ideological (now, in Adorno's sense) mechanism for preserving the status quo. To recognize a challenge, conceding its claim as viable, as reasonable, etc., defeats the challenge and this means that nothing needs to be changed. A challenge can only work in a climate of opposition. Co-option nicely defuses opposition by declaring its work at end, its goal accomplished.

Indeed, a sense of accomplishment or closure is part of the contemporary condition of knowledge and culture, that is also the postmodern *episteme* described by Foucault, Baudrillard, and, of course, Lyotard. This postmodern *episteme* cannot be understood apart from the modern and even the *avant-garde episteme* characteristic of the early twentieth century. The negative watchwords of the latter are familiar to us: alienation or anomie, nihilism, Nazism, anti-humanism, fascism, Dadism, structuralism, existentialism. The antidote to all these negative ideologies or simple bad *-isms* may be traced in the intellectual trajectory of the word pluralism. Pluralism of course has recently been renamed as a conceptual attitude or social stance – the party line of what is "politically correct". Beyond the front pages of popular news and culture journals, the ideas of pluralism like that of the politically "correct" seem to have had their day. The point of the past tense here must be drawn out for it emblematizes the political ideology essential to the not-yet-exhausted project of being politically correct and it dramatizes the troubles that beset good ideas such as pluralism. For one might think that the ideal enfranchisement of variety as such (pluralistic or multicultural values) would be complex in the extreme and the development of a plan or programme for reflecting that variety would take time, require a series of reviews and compromises, all the very messy details that are part of execution and implementation. But, lo! no sooner is the notion of pluralism breeched in a democratic society than it wins the day and is so forth and already the plan of business. And why ever not? Who, in America and today, in the entire world, would oppose the idea of an open society?

But the assertion of success, already consummate or already fulfilled, is part of the mechanism for maintaining the status quo, the standard schema in the guise, the sheep's clothing, of the new. In the same way, philosophy of science remains as little pluralistic as mainline philosophy.

For my reading of the "postmodern condition" of knowledge and the philosophy of science, *any* alternative approach to the philosophy of science conceived on the basis of "continental" philosophic perspectives requires considerable defense. Thus the project of any continental approach to the philosophy of science remains inherently (ant)agonistic. Yet here again the problem of co-option recurs.

Historically, conflict has had a nice and not divergent heritage in traditional philosophy and in this the tradition has ways of co-opting the eristic, not ironically, but ingenuously. We are not in the wake of revolution (post-revolutionary philosophy and science) but in a co-opted revolution. Thus it is that Anglo-American philosophy is dominated by the same analytic perspectives as ever but now offered (in analytic style) by analytically trained and analytically read philosophers hitting all the analytic high notes in a popular motet riffed with the low bass of continental names (Foucault, Habermas, even Heidegger, etc.). In this way, it is exactly significant that one learns that it is not necessary to go all the way to a continental point of view, where, as Richard Rorty tirelessly repeats, John Dewey's American pragmatism embodies what's best in Derrida, Foucault, Heidegger *and* Nietzsche – that is, *all* of French and German thought – all rolled into one really readable philosopher. In a more robust version of the same edifying perspective, the politically correct stance represents the conviction of the success of revolution: racism, sexism, ageism, lookism, all bad things, but no longer to be lamented as they have long since been corrected, revised, overcome, fixed.

In the philosophy of science this same sophisticated appreciation of the so-called continental turn translates into the contention that there is nothing really or ultimately "alternative" about continental approaches. Mainline or traditional (whatever about analyticity) philosophy of science has, in its own way, managed to overcome its own limitations, inventing in an elegant instance of convergent evolution the wheel, not again but independently: self-sufficiently. This self-sufficiency means it is not in need of advice from those who have spent years or even a lifetime in formerly alternative traditions. All, all say the same thing. To borrow Rorty's thunder once more: Heidegger? Dewey? What's the difference? Thus, as Putnam has suggested, Wittgenstein, Austin, *and* Husserl, voice the same conclusion in the end.[19] Ergo, there is no need for the *articulation* of an alternative approach to the philosophy of science, especially one which would require a rethinking of the philosophic function of the philosophy of science, a rethinking of philosophy, of science and technological practice (experiment, perception, reason, logic, intuition, and, even, engineering and the mechanical arts) and so on. Once again, there is no need to rethink the philosophy of science in light of alternative philosophical perspectives because a *new philosophy of science* is already underway.

I've already noted the wide agreement that renovation has already been effected, with adherents representing both continental and analytic formations. Frederick Suppe on the analytic side argues that there is no need for a consideration of the opposition as such except, of course, for the purposes of clarification. Thus towards the end of his recent book on *The Semantic Conception of Theories and Scientific Realism*, by way of a few tidying details, Suppe introduces his consideration of Kuhn's and Feyerabend's views on scientific knowledge (their respective "relativisms") by noting (as a *fait accompli*) that neither the straw man's perspective of logical positivism nor the opposing "highly relativistic 'subjective' views of Kuhn, Feyerabend, and others" should be said to enjoy widespread acceptance today.[20] For Suppe the democratic ideal has been achieved: the best insights of both styles of thinking are absorbed and the deficiencies rejected. This parallels the functioning of modern science and the rationality patterned on this same modern scientific ideal. Thus "a central focus of contemporary philosophy of science seems to be on developing an account of scientific knowledge which accommodates the sound insights of both perspectives while avoiding the known defects of either".[21] The effects are revolutionary in the best scientific sense. The so-called "Received View" was discredited in the 1960's, calling "into question the entire positivistic portrait of science" so that today the "dominant thrust of contemporary work in the philosophy of science is the development of new views of science which proceed from, and tend to depend heavily on, close examination of actual scientific practice and products (e.g., theories, explanations)".[22] But although Suppe seems to be saying that the historical and sociological practice of science has become increasingly important for the philosophy of science, this in fact is not his claim. Thus the focus of *The Semantic Conception of Theories* is not the history of scientific practice but rather on scientific knowledge. Suppe's interest is "an improved philosophical understanding of the nature and growth of scientific knowledge".[23]

For Suppe, both Kuhn/Feyerabend and (note the magnanimity of Suppe's conception) logical positivism are "influential *relics* of the history of the philosophy of science". I have italicized the term "relics" because Suppe himself places this emphatic tag squarely on the side of the epistemic relativism of

Kuhn and Feyerabend fitting their epistemologies in the reductive line of 18th century British empiricism, and naming them "the last gasp of a moribund epistemology, which if one shares my philosophical instincts about what an adequate epistemology should be like, mercifully should be put to sleep".[24]

One continental thinker, Don Ihde, in a text written at about the same time as Suppe's own, offers a perspective that could only encourage Suppe's tender mercies. Ihde argues that what he calls the "New Philosophy of Science" emerges out of the anathematic (that is, as regarded from the perspective of the "Old" philosophy of science) "witch's brew" of New Scholarship, including "postwar phenomenology and hermeneutics, structuralism, revivals of Neo-Marxian and Neo-Freudian methods, and later, feminist and deconstructive techniques as well".[25]

What is intriguing in Ihde's as in Jean-François Lyotard's similarly framed account of the liberating potential of science on the level of the 'little' as opposed to the 'grand' narrative, is the effective mythologization of the demythologizing of science.[26] Lyotard, in what is doubtless an unintentional reprise of Herbert Marcuse's equally well-intentioned but not less short-sighted vision of the *politically* liberating potential of radio and the *humanely or creatively* liberating potential of recorded music (for Marcuse, and note the clash with today's standard technologies: reel to reel tapes and LP's), argues that a mythic transformation of the information era (on-line services, world-wide accessible ATMs, the Internet, etc.) can be accomplished by one simple, idealistic move: granting "equal", and that means free and public, access to computer information systems. Who said socialism lacked a human face? Although such informational anarchism is decried by computer trade journalists as "Internet mentality" – a something-for-nothing, download-it-for-free style of thinking[27] – it is also clear, to take Lyotard's side, that those who make their living in the computer trade are necessarily devoted to perpetuating the extraordinarily robust fulfilment of the wildest fantasies of waste-capitalism represented by the industry. In the course of a didactic and pedagogically minded argument, Lyotard demythologizes nothing less than the original master of the grandly humanist narrative, at least in the academy, Wilhelm von Humboldt himself. Demythologizing Humboldt's paedagogic ideal, Lyotard declares: "a professor is no more competent than memory bank networks in transmitting established knowledge, no more competent than interdisciplinary teams in imagining new moves or new game".[28] It is important to note that, according to Fritz Ringer, Humboldt's ideal of *Bildung* could only yield a mandarinic tradition at the core of the university and the grand narrative of culture or Western civilization, and one necessarily predisposed to decadence.[29] Demythologizing *Bildung* and the ideal of pure, elite knowledge, however, only works by mythologizing the transparency of CD-ROM access to cultural resources and the vitality of virtual art in the on-line gallery or museum tours or other multimedia sites on the World-Wide-Web. So far from training the perfect 'mandarin' professor (who would seek "knowledge for its own sake"), Lyotard argues for the perfect microprocessor and the specificity of performative excellence. This will work, so Lyotard suggests, as a liberating program given the student's access to training and to information banks ("data"). This "pure research" mentality, itself so characteristic of academe as to be an industry cliché, offers the potential for exploiting untapped creativity, where for Lyotard, such innovative "performativity" consists not in having access to databanks as such but in "arranging the data in a new way".[30] Thus "it is possible to conceive the world of postmodern knowledge as governed by a game of perfect information, in the sense that the data is in principle accessible to any expert: there is no scientific secret".[31] In Lyotard's fraternalistic republican vision, all men would be free to make new moves or to "change the rules of the game".[32] New moves, according to the performative ideal, qua achieved, assure legitimacy. But the cult of the new (mousetrap or computer program) is the earmark of the modern progress ideal. Only, in Lyotard's vision, rather than the romantic genius or the human visionary, progress is a matter of access to the databanks that "transcend the capacity of each of their users",[33] and, of course, a question of teamwork. The end remains the same, the postmodern engine is different. And we go, in Nietzsche's words, "straight into – the *old* ideal".[34] Ihde's new science similarly debunks the old in an apotheosis of the new: "the new philosophy of science, Davidian in its initial hermeneutic, had – if not slain – at least wounded the older giant".[35]

Now, against Ihde's optimistic conviction, and as any Nietzschean knows the tactics whereby the weak inevitably overcome the strong, David triumphs through his own cleverness but much more by surreptitiously shifting the terms of the battle: changing the 'rules' of the game. Thus the ascetic ideal triumphs over the clumsiness of the giant, the master narrative, and this is the image of the lumbering vulnerability of Big/Old Science. But who is to be David here? In any Theogony, the young gods battle the old titans, who for their part won the world from their fathers before them. It's always David against David and the father must be child to the man, or as the poet Hölderlin mused against Schiller, gently chiding, Jupiter, the son of time: the once deposed titan must be restored to honour in the end.

It is significant that a recent *New York Times Sunday Magazine* interview with the biologist Lynn Margulis concludes with Margulis's repetition of Kuhn's (or as we have seen: Hanson's earlier) word on the persistence of scientific paradigms. For Margulis (the biologist does not think to name either Kuhn or Hanson), it is only when the adherents of a particular scientific worldview die "and differently behaving people take their place",[36] that a new scientific theory can hope for a scientific life.

The modern ideal of continuous scientific progress employs a basically biological metaphor of evolution and development, presupposing a natural and softer procession along the scientific tree of theoretical truth. In what may be called the tragic view of theoretical succession in practice, viewed by Margulis, the scientist, and Kuhn and Hanson, historian and philosopher of science respectively, scientific transformation or innovation is possible only after the reigning scientific worldview plays itself out. There is a kind of natural selection and a version of punctuated revolution and equilibrium among theories in this view, but it is of a meteorological or catastrophic order. What is important in this tragic view is the inadequacy of the genealogical or evolutionary metaphor of the growth of scientific knowledge, of scientific progress and development. For such a worldview admits no alternatives in its domain and those that it does descry it names anathema. And, as Margulis ruefully emphasizes, given the theoretical blinders which constitute a scientific dynasty, only when its last member has perished may a new vision of science come to stand.

Much more "realistic" and less poetic or tragic an account of the state of science and the philosophy of science may be found in the assessment offered by Feyerabend, a thinker hardly less fond of "science" than Suppe, no less committed to the history of ideas, and although far less familiar with continental thought than Ihde, withal a philosopher (despite his own protests), and considerably more aware than either Suppe or Ihde of the nexus of the limits of theory, abstraction, reason, and the political fate of the "open society".[37]

For Feyerabend the philosophy of science and science itself are concerned (more often than not) with two different things. Not by accident, not because the philosophy of science is on the wrong track but necessarily because of what science is.[38] For Feyerabend (and I understand the project of the philosophy of science to date likewise) a "theory of science that devises standards and structural elements of all scientific activities and authorizes them by reference to some rationality-theory may impress outsiders – but it is much too crude an instrument for the people on the spot, that is for scientists facing some concrete research problem".[39] There is no "new" philosophy of science for Feyerabend because the nature of science and of reason are not yet recognized. The fantasy of objectivity keeps thinkers from recognizing that "scientific theories are human creations and that science is one tradition among many".[40]

The issue is not to endorse pluralism but rather to highlight the technical function of scientific or paradigmatic focus. For Feyerabend, what is key in noting the liberation from the particular is that it betrays the very interest, the traditional working of modern (as, of course, of pre- and postmodern) science: "if we pay attention to some facts – the general features selected by science – we can disregard others".[41] Thus we continually ignore the multifarious, horizonal complexity of the world to focus on the world constituted from a given scientific perspective. Nietzsche would note that we have no choice.[42]

QUESTIONING THE PHILOSOPHY OF SCIENCE

I have argued that beyond traditional, analytic approaches to the philosophy of science (historically) traditional continental approaches take a critical stance beyond "scientificity". The reflexive advantage maintained by "continental" philosophy compared with analytic approaches to philosophy may be explained by noting the difference between what may be called the "Copernican turn" (the heritage of Kant in the hermeneutic/phenomenological tradition) and the linguistic turn (in the Wittgensteinian heritage of analytic philosophy). I have also noted the difficulties implicit in launching an argument for the value of approaching the philosophy of science from a hermeneutic, phenomenological perspective.

Beyond the political project of excluding alternative approaches, this means that ordinary analytic philosophy of science does not question the providence of scientific inquiry. As such a belief system characterizes the philosophy of science (as a kind of metaphysical faith in or 'church' of science), the philosophy of science is non-philosophical. It stands in relation to science as a believing people stands in relation to its God.

Analytic philosophy of science assumes the viability of science and, subscribing to its ideal of rationality, seeks to emulate this ideal as a "science of science". To say this is only to say that scientific rationality is the paradigmatic ideal of rationality. Classically, we recall, rationality as such names the metaphysically guaranteed correspondence or proportionality of human intellect and object. A practical orientation expressed in a metaphysical framework marks the commonality between science and the philosophy of science. Although science has been opposed to metaphysics in the tradition of modern philosophy, it is erroneous to think that this opposition is more than superficial. And although science does not seek answers to the questions traditionally posed as metaphysical, as antinomically illustrated by Kant, for example, it nevertheless locates the answers to the questions it does pose on the same level. The efficacy of theoretical insight proves the empirical potency of metaphysical rationality. This is what Nietzsche means when he writes in *The Gay Science*, in a passage he repeats in a concluding section of *The Genealogy of Morals*, "No doubt, those who are truthful in that audacious and ultimate sense that is presupposed by the faith in *science thus affirm another world* than the world of life, nature, and history. . . . It is still a *metaphysical faith* upon which our faith in science rests".[43]

Elsewhere I show how this metaphysical faith works to deny the world of life, nature, and historical context.[44] Here, I must be content to note that unlike its magical preludes, scientific inquiry effectively, empirically – and not ideally or transcendentally, in any typically metaphysical sense – does uncover the secrets of nature and that it thus advances the possibilities of technological control, which technological advance confirms the scientific project as true (and the empirical or material attains to the metaphysical). Likewise, philosophic inquiry into the functioning of science concerns the secrets or effective technique of that functioning and confirms the adequacy of its revelations in the domain of viable, productive applicability. The metaphysical faith ideal of rational progress reveals the effective or empirical domination of scientific reason or knowledge.

The paths of Explanation and Understanding are the diverging paths of science and philosophy described by Dilthey in a typology appropriated by Weber and Adorno but also C. P. Snow and von Wright. These paths do not share common ground, nor are they parallels, traversing lower and higher roads. Rather, the topology of alienated mutuality describes a single path. That is the path of objective science; the philosophy of science follows anxiously behind. Eager to keep up, eager to move in the same by-ways, eager, most of all, to be recognized as a proud companion. But if scientific rationality is the only conduit for reflection, the distinction between explanation and understanding is the differentiation of expression and attending elaboration. This deficiency has been acknowledged, and recent efforts in the philosophy of science seek to overcome the limitations of this perspective. Yet the efforts of such a reform are not yet accomplished and it is necessary to review the continuing *temptation* in the philosophy of science to approximate the paradigmatic ideal of a "science of science". The elaborative, science-sycophantic or calculative, reductive way of understanding (and the denigration of discursive or narrative reflection implicit in this ordering) yields, in effect, a single way of explanation. The power of scientific explanation outstrips the illumination of philosophic (here to be allied with poetic and other) understanding in productive efficacy. The explanations of science remain just as Kant ruefully observed more than 200 years ago, working explanations. It is an index of this superiority that the effort of the philosophy of science is directed towards an understanding of scientific explanation.

I have said that the philosophy of science takes its objective focus and subjective definition from science. This is problematic as it reduces philosophy to science, and this reduction is not a patent advance. Nietzsche rejects the reduction of philosophy to science because one abandons the project of philosophy to that of science and because the problem of science cannot be adjudicated on its own fundament. Critical philosophy must be recalled to its original and challenging mandate. This is imperative because the repetition of science in philosophy moves in a circle that is more than vicious and radically non-hermeneutic but crucially vapid.

Science's analytic or critical perspective is conducted from within its own smallest circle – avowed as Ockham's principle, which approximates what Nietzsche literally calls the "greatest possible stupidity" ("*grösst-möglichen Dummheit*")[45] – and operates by discounting the interpretive perspectival circumstance of creative knowing, hence taking itself and its project too seriously. This scientific "gravity" is the cardinal sin from Nietzsche's perspective. Gravity of this kind is not offensive because the scientist could never appreciate the real weight of things, but rather because the scientist lacks distance, proportion, and style. When science, or its philosophy, takes itself too ponderously (as potentially or progressively absolute or true) it winds up taking itself to be the only valid valuation, and that means that it deceives itself. We need Nietzsche's recommendation: we need to look at science as he reminds us, not from the basis of science and scientific reason, but seeing science as a version (and thus on the ground) of art. Because art is, if anything is, even more of a problem for philosophy than science, we further need to look at art, and it is at this juncture that Nietzsche refers us to life.

To propose to reflect science in the light of art and art in the light of life is to propose nothing like a perspicuous conclusion. Talk of science from the perspective of art (or indeed life) seems talk of the creative potential in science. This looks easily enough like yet another modernistic paean to the inventive openness of the new scientific project, replete with information frenzy and abundance (another way to regard Lyotard's ideal of free public-access to "data" or, better for today's sensibilities: to the Internet), chaos creativity, and thoroughly holistic sensitivity (gender-free, non-agist, non-racist). But it is not such an optimistic endorsement of the redemptive potential of science. Such optimism too is still science, regarded on the basis of science rather than, as Nietzsche would suggest, recognizing that science is brought into being from the same urgency as art, drawing upon the same resources at art's disposal.

One needs to remember that the greatest imperative of and the greatest challenge to critique is that of rigorous thought. Like Zarathustra's animals (or like so many of Nietzsche's interpreters) it is all too easy to reduce critical teaching (for example Nietzsche's teaching of Eternal Recurrence) to a cliché, a commonplace, a hurdy-gurdy song. The routine destiny or fate of politically correct doctrines like racial or gender equality, like enfranchisement of the economically disadvantaged or the handicapped, is that such ideals, however manifestly right and true, become, at worst, trivial and at best, laughable. Nor is the problem limited to an arcane doctrine (everything flows, everything returns, eternal is the wheel of time). This is the fate of multiplicity, of pluralism, of alternative philosophies (as indeed of alternative music and lifestyles).

Although I have here come to the end of this essay, I do not claim to have solved the radically hermeneutic problem of critique. For the problem of rigorous thought is the hermeneutic problem of the co-option of critique. This is the *ressentiment* interior to radical doubt or the question that vanishes in a complicit reply. The problem of rigorous thought, which Heidegger simply named thinking, persists beyond the philosophy of science, beyond the modern and the postmodern images of science and its achievements in our era and its values and beliefs.

Fordham University at Lincoln Center

NOTES

[1] The image is inspired by the epigraph to Gillian Rose's autobiographical record of living and dying, *Love's Work: A Reckoning with Life* (Schocken Books, New York, 1995): "Keep your mind in hell, and despair not".

[2] In the structuralist context, Derrida uses free play in a positive sense as a game or the field of finite language: "a field of infinite substitutions or the closure of a finite ensemble". From 'Structure, Sign, and Play in the Discourse of the Human Sciences', in Richard Macksey and Eugenio Donato (eds.), *The Structuralist Controversy: The Language of Criticism and the Science of Man*, (Johns Hopkins University Press, Baltimore, 1970), p. 260.

[3] Thus Derrida writes of "... Nietzschean affirmation – the joyous affirmation of the free play of the world and without truth, without origin, offered to an active interpretation ..." as an affirmation which "determines the non-center otherwise than as loss of the center. And it plays the game without security", *op. cit.*, p. 264. For a comprehensive articulation of the modern German idealist understanding of "play" as an aesthetic category, expressed from a critical Derridean literary perspective, see M. Spariosu, *Dionysus Reborn: Play and the Aesthetic Dimension in Modern Philosophical and Scientific Discourse* (Cornell University Press, Ithaca, 1989). While claiming a more-than-phenomenological perspective, Spariosu relies extensively on Ingeborg Heidemann's *Der Begriff des Spiels und das aesthetische Weltbild in der Philosophie der Gegenwart* (De Gruyter, Berlin, 1968), from which last Spariosu nonetheless distances himself by criticizing it as "onto-phenomenological". By employing the term "historical description", Spariosu is able to suggest that his reading exceeds phenomenological provinciality while temporarily forgetting the meaning (contamination?) of hermeneutics.

[4] It is worth noting the rigor of Feyerabend's thinking here, for it is similar to the spirit of Nietzsche's philosophic rigor. But the image of scientific playfulness remains more a matter of idealization or even public relations (invented mythology) than not. Hence the popular science paperback, *Surely You're Joking. Mr. Feynman* may well appeal to readers as a demonstration of such playfulness (Hey! Right there in the title!) but it can also be read as an account of a preciousness taken to be and preached as if it were playfulness. Without wishing to malign the late popular science author's essayistic motives, this so-called playfulness should be considered as the essential *méconnaisance* of the scientific project, brought perhaps to perfect pitch in the postmodern condition. I suggest that we will need more than Mr. Feynman's jokes to catch the point of the levity, the "light feet" of a Nietzschean challenge to scholarly gravity.

[5] If science is defined in terms of theoretical and technical (practical) progress, philosophic analyses of science are (ideally) taken to offer principles for further scientific advance.

[6] Steve Fuller, *Philosophy of Science and Its Discontents* (Westview Press, Boulder, Co., 1989), p. 7. [The 'chassis' appears in Sean O'Casey's play *Juno and the Paycock* (1924: "The whole worl's in a state o' chassis". – RSC, ed.]

[7] Ian Hacking, *Representing and Intervening* (Cambridge University Press, Cambridge, 1983) p. 1. Hacking mentions both Duhem and Hanson. This does not stop him from giving the ballpark dating of this crisis as happening "around 1960". Thomas Kuhn's *The Structure of Scientific Revolutions* (University of Chicago Press, Chicago, 1970; 1962) is the 'paradigm' temporal locus in all taxonomies and histories of the crisis in the philosophy of science. As will be further emphasized below, there is some justification for moving this date back a few years: to 1958, the date of N. R. Hanson's *Patterns of Discovery* (Cambridge University Press, Cambridge, 1958; reprinted, 1961).

[8] Alasdair MacIntyre, 'Epistemological Crises, Dramatic Narrative, and the philosophy of Science', in Gary Gutting (ed.), *Paradigms and Revolutions: Appraisals and applications of Thomas Kuhn's Philosophy of Science* (University of Notre Dame Press, Notre Dame, 1980), p. 71.

[9] See Joseph Rouse, *Knowledge and Power: Toward a Political Philosophy of Science* (Cornel University Press, Ithaca, 1987) and Gary Gutting, *Michel Foucualt's Archaeology of Scientific Reason* (Cambridge University Press, Cambridge, 1989). See also Todd May's recent illustration of a Foucauldian critique of social scientific knowledge, discussing authors ranging from Gilles Deleuze to Wilfrid Sellars: *Between Genealogy and Epistemology: Psychology, Politics, and Knowledge in the Thought of Michel Foucault* (Penn State Press, University Park, 1993). MacIntyre, to be sure, has in recent years adopted a position closer to these perspectives on Foucault.

[10] MacIntyre, 'Epistemological Crises, Dramatic Narrative, and the Philosophy of Science', p. 73.

[11] Talk of incommensurability in this context indicates that the philosophical stage of crisis interior to the philosophy of science precedes Fuller's conception precisely as precedes Joseph's rouse's related political expression, or Gary Gutting's, or Hacking's and even MacIntyre's broader

contribution as well as the efforts of the needless-to-mention names (Saints Feyerabend and Kuhn). For MacIntyre's signal articulation of this communicative quandary, see his *Three Rival Versions of Moral Inquiry: Encyclopaedia, Genealogy, Tradition* (Notre Dame University Press, Notre Dame, 1990).

[12] See Steven Shapin, *A Social History of Truth: Civility and Science in Seventeenth-Century England* (University of Chicago Press, Chicago, 1994 and Steven Shapin and Simon Schaffer, *Leviathan and the Air Pump: Hobbes, Boyle, and the Experimental Life* (Princeton University Press, Princeton, 1985).

[13] And Popper's putatively neo-Kantian solution sidesteps the problem of empirical extension to infinity by way of an exploitation of the power of the counter-example. For those who reduce scientific theories to claims on the order of "All swans are white", and employ an inductive method which would seek, as falsificationism does not, to examine every swan (the idea of absolute induction is based on an impossible tabular collocation of individuals) cannot succeed. In contrast, the method of counterexamples or "falsifications" offers compelling advantages. But the method has its limitations for those taxonomers who consider melanism in swans not an alternative or opposite quality but a phenotypical modification of white feathers.

[14] Hanson, *Patterns of Discovery*, p. 62.

[15] Hanson, pp. 15–7.

[16] Hanson, p. 30.

[17] The analytic tactic of non-citation, usually reserved for ignoring the precedent of tradition in philosophy as well as for consigning one's continentally-oriented colleagues to the dustbin of ignorability, was recently exploded in a charge of, of all things, plagiarism (but this is Lockean, intellectual-property conscious America, land of copyright). Thus, Saul Kripke, once lionized by the New York Times itself as the genius of the century, has been accused of appropriating the philosophic insights of Ruth Barcan Marcus, a Yale philosopher less celebrated than disliked. The case, as documented by Quentin Smith and reported by Jim Holt in a recent issue of *Lingua Franca*, appears more than plausible. As thoughtlessly but as little blamelessly as Kuhn himself read and forgot having once read Ludwik Fleck, it would seem that a young Kripke heard and engaged with, but forgot having heard and engaged, Ruth Barcan Marcus in the presence of a witness for the analytically controversial theory of sense and meaning: W. V. O. Quine. As influence is more than a matter of copyright, both Kuhn's and Kripke's work may be judged centaurs incorporating the reflected value of inspiration. The phenomenon, is as Quentin Smith casually notes, standard stuff in historical interpretations of art and literature.

[18] As the hallmark of progressive, modern scientific thinking, resolutions are always available. In a context proper to Nietzsche's own thinking, in the struggle against the reign of morals and the ascetic ideal, after two long essays in search of a critical ally or advocate to make common cause with him, in the third essay of *On the Genealogy of Morals*, Nietzsche observes that he ought not have to look far for a champion "they tell me it is not lacking". Characteristically, the purported hero is science itself. Nietzsche, *On the Genealogy of Morals*, trans. Walter Kaufmann and R. J. Hollingdale (Random House/Vintage, New York, 1968) III:23, p. 146.

[19] Hilary Putnam, 'Is the Causal Structure of the Physical Itself Something Physical', in Peter French, T. Uehling, Jr., H. Wettstein, *Midwest Studies in Philosophy IX, 1984: Causation and Causal Theories*, p. 10.

[20] Frederick Suppe, *The Semantic Conception of Theories and Scientific Realism* (University of Illinois Press, Urbana, 1989) p. 301.

[21] *Ibid.*

[22] Suppe, p. 415. Suppe refers to his earlier 'Afterword' to *The Structure of Scientific Theories* (University of Illinois Press, Urbana, published in the second edition, 1974), pp. 650–716. This is not the place to more than note the rhetorical reticence evident in the "heavy" reliance of contemporary philosophy of science on actual scientific practice. The nature of Suppe's own loyalties and interests is betrayed by his listing of the "products" of science "(e.g., theories, explanations).

[23] Suppe, p. 418.

[24] Suppe, p. 331.

[25] Don Ihde, *Technology and the Lifeworld: From Garden to Earth* (Indiana University Press, Bloomington, 1990), p. 202.

[26] Jean-François Lyotard, *The Postmodern Condition: A Report on Knowledge*, trans. Geoff Bennington and Brian Massumi (Minnesota University Press, Minneapolis, 1984), p. 31. French publication in 1979.

[27] John C. Dvorak, 'Internet Insanity', *Popular Computing Magazine*, February 6, 1996, p. 85. Owing to such deranged assumptions, "anybody who tries to make money on a site is hacked by a group of anarchists". This "anarchist" mentality, assuming "free e-mail, free information", is "insane" because, as everyone knows, following Madison Avenue's version of post-depression American history: 'There is no free lunch'. But Dvorak chides the popular computing reader, a free-access mentality is "a doomed way of thinking". Capitalism is as capitalism does.

[28] Lyotard, p. 53. One American professor of English has taken Lyotard's hint as a veritable program for revising the style of graduate-level research. David Damrosch, *We Scholars: Changing the Culture of the University* (Harvard University Press, Cambridge, MA, 1995). Like the sciences, Damrosch proposes, aspiring doctoral candidates should learn to network with other experts and to produce work in concert with other scholars. Such a cooperative strategy, according to this Columbia University professor (the New York pinnacle of the American academy) would have the beneficial result of increasing the number of English PhD's (an endangered species). Other, less ivy-minded reformers, have more pedestrian goals. They see the promise of computer-assisted learning and they see the ideal as the complete replacement of the teacher, particularly at primary school levels, particularly in the inner cities, where it is easier to raise money for high-tech equipment than to guarantee low-tech teacher's salaries or attract such teachers to teach in such areas. A computer, just as "competent" as a professor, in Lyotard's anti-Humboldtian judgment, and by the same logic as good as a teacher at any level, would be handicapped by a fear of neither social stigma nor personal violence.

[29] Fritz Ringer, *Decline of the German Mandarins: The German Academic Community, 1890–1933* (Harvard University Press, Cambridge, MA, 1969).

[30] Lyotard, p. 51.

[31] Lyotard, p. 52.

[32] *Ibid.*

[33] Lyotard, p. 51

[34] Nietzsche, *The Genealogy of Morals*, III, 25, p. 155. Translation modified for context.

[35] Ihde, p. 202.

[36] E. Royte, 'Attack of the Microbiologists', p. 23, *New York Times Magazine*, January 14, 1996, pp. 21–3.

[37] To illustrate Feyerabend's hermeneutic emphasis, 'anything goes' is conceived more as trammel than as license. As Feyerabend writes, in the ultimate essay of *Farewell to Reason*, such a slogan makes life *harder* not easier for both researchers and philosophers of science. " 'anything goes' is not a 'principle' I defend, it is a 'principle' forced on a rationalist who loves principles but also takes history seriously". *Farewell to Reason* (Verso, London, 1987) p. 289. Here, as always and as scrupulously as any other analytic philosopher, Feyerabend avoids referring to the continental tradition.

[38] Feyerabend likes to think that there are occasions when the philosophy of science and science come together, as for his taste, in the thinking of Ernst Mach where what is at stake is a matter of reading a rule of thumb (this explains Feyerabend's affection for Aristotle's *Nicomachean Ethics*) and he has to concede an attack which fails to grasp – it is Feyerabend's view that this cannot be helped – the intuitive reference. Feyerabend, *Farewell to Reason*. In particular see p. 189. See also Chapter Seven, 'Mach's Theory of Research and its Relation to Einstein', as well as pp. 15–6 and also 280–96.

[39] Feyerabend, p. 281.

[40] Feyerabend, p. 122.

[41] *Ibid.* Feyerabend, p. 122.

[42] The ineluctability of focus (Nietzsche's perspectivalism) was not lost on Feyerabend but Feyerabend underscores the annihilatingly exclusive nature of perspective (and this too is also a Nietzschean insight). To be sure and to be fair, coincidences between Feyerabend and Nietzsche should not be overplayed. For one thing, Feyerabend remained an analytic philosopher all his life. For another, despite routine references to such correspondences between Feyerabend and Nietzsche, from critics and friends alike, Feyerabend himself claimed to have been underwhelmed by Nietzsche, finding "much to agree with, but nothing to write home about". Feyerabend, personal letter.

[43] Nietzsche, *The Gay Science*, trans. Walter Kaufmann (Random/Vintage, New York, 1974) 344, pp. 282–3. Cf. *The Genealogy of Morals*, Part III, Section 24.

[44] See my *Nietzsche's Philosophy of Science: Reflecting Science on the Ground of Art and Life* (State University of New York Press, Albany, 1994).

[45] See Nietzsche's *The Birth of Tragedy*, Section 14 and *The Will to Power*, Section 617. In this context, Alister Moles's reading of Nietzsche's so-called "positivism" is worth noting. See Moles, *Nietzsche's Philosophy of Nature and Cosmology* (Peter Lang, New York, 1990). If the author did not mean to deliberately misread Nietzsche here, it is because Moles takes Nietzsche's position on science's "greatest possible stupidity" as approbative, that is: Moles holds with the view that Nietzsche was a positivist in the best conscience. This view is hardly uncommon and it is offered from a position of considerably greater sophistication by Rüdiger Safranski in his recent, *Wieviel Wahrheit Braucht der Mensch? Über das Denkbare und das Lebbare* (Hanser, Munich, Vienna, 1990) For Safranski, "*Nietzsche bleibt auf verhängnisvolle Weise ein Kind seines wissenschaftsgläubigen Zeitalters*" (p. 71) and fell victim to the appeal of biologism and naturalism, rendering his thought susceptible to the misappropriation of Nazi theorists.

EFTICHIOS BITSAKIS

ON THE VALIDITY OF VON NEUMANN'S THEOREM*

The present paper is an attempt at a critical analysis of von Neumann's argument, concerning the impossibility of a more complete specification of quantum states, by means of the introduction of supplementary (hidden) variables. The "impossibility proof" of von Neumann (1932) has been considered as a negative and definite answer to the question of hidden variables. However, the validity of the theorem of von Neumann was challenged, especially in the fifties by L. de Broglie, his collaborators, and many other physicists of the realist school. The debate concerning the possibility of a local and causal description of quantum phenomena, received a new impetus after the formulation of the famous inequalities of J. S. Bell (1964). However, the question of the hidden variables remains always open. The same holds for the validity of von Neumann's theorem.

THE TOPICAL CHARACTER OF VON NEUMANN'S ARGUMENT

The German original of the *Mathematical Foundations of Quantum Mechanics* was published in 1932, while the English translation appeared much later, in 1995.[1] One of the major contributions of this book lay in its conclusion, stating that the introduction of hidden variables in quantum mechanics is impossible, and that – consequently – nature does not respect the principle of causality. This "impossibility proof" constituted for a long time one of the fundamental arguments of the Copenhagen Interpretation (CI). Von Neumann's formal proof, on the other side, was "complementary" to the epistemological analyses of the matter, propounded by Bohr, Heisenberg, Pauli, and other members of the Copenhagen School.

But, as is well known, the validity of this proof was never accepted by Einstein, Schrödinger and other outstanding personalities who adhered to the thesis of realism. It is also well known that since 1952, the validity of von Neumann's theorem was questioned by many physicists. Thus according to de Broglie there is a kind of "vicious circle" in von Neumann's argument. The same objection was later formulated by A. Landé, namely, that von Neumann's concluding argument suffers from the logical defect of circularity. According to D. Bohm, on the other side, von Neumann in his proof has implicitly restricted himself to an excessively narrow class of hidden parameters and, in this way, has left out of consideration precisely those types of hidden parameters which Bohm himself proposed in his classical paper. Moreover, Bohm and Bub maintained that von Neumann's conclusion concerning the impossibility of hidden variables was tacitly assumed in the premises on which the argument was based. According to S. P. Gudder, on

D. Ginev and R. S. Cohen (eds.), Issues and Images in the Philosophy of Science, 47–62.

the other hand, the impossibility proof is "irrelevant", while for J. S. Bell, the essential arguments of von Neumann are "unreasonable". Many other writers, finally, challenged in the following years the validity or the generality of the demonstration of von Neumann.[2]

Von Neumann's proof has lost, since 1952, its incontestable prestige. But as I will try to show, the circularity is not the unique weakness of this argument. There are further crucial problems to be discussed, such as those concerning the definition and the transformation of the quantum states, and evidently, the general question of the validity of the principle of causality in quantum mechanics. I will try to formulate and to discuss these points, which are directly involved in the actual debate concerning the interpretation of quantum mechanics.

The main conclusions of this paper are the following:

- Von Neumann's conception of quantum states is a static one.
- Consequently, and because he confuses causality and determinism, his theorem is very restrictive (it concerns only the possibility of a deterministic description).
- As a result, von Neumann excludes the possibility of probabilistic hidden variables theories which are, in principle, possible.

STATISTICS *VERSUS* CAUSALITY

According to von Neumann, the object of his book "is to present the new quantum mechanics in a unified representation which, so far as it is possible and useful, is mathematically rigorous".[3] The question of the validity of causality is fundamental in this book.

Classical mechanics, von Neumann states, is a causal discipline. If we know $q_1, \ldots q_k, p_1 \ldots p_k$ at $t = t_0$, we can calculate the state at any later time t. Of course, a statistical method is also being used in classical mechanics, but this is an independent addition: if we don't know all the variables defining the state, but only some of them, we make statistical statements. However the use of statistical methods in classical mechanics does not exclude, in principle, the possibility of a *complete* knowledge of the state, obtainable by the introduction of supplementary variables. As opposed to this, the statistical statements in quantum mechanics have a different character.[4]

Von Neumann emphasizes the well-known fact that the initial state of a quantum system determines the state at a subsequent time t, on the basis of the time-dependent Schrödinger equation. However, the statements of the values of physical quantities are of a statistical character. And the statistical character of quantum mechanics is, according to von Neumann, incompatible with the validity of the principle of causality. For him a relation is either causal or statistical. In this way von Neumann identified the principle of *causality* (existence of causes) with the principle of *determinism* (determination of the effect, and mode of determination), thus establishing a formal antithesis between statistical and dynamical laws.[5]

The laws of quantum mechanics, von Neumann says, are statistical, *non-causal*: The statement P is statistical, non-causal, because it does not tell us what values the variables get in the state Φ, but only with what probability they assume all possible values.[6] However, what is questioned here is not the existence of causes (some of them, at least, are known) but the classical determination (probability equal to 1) of the result by its causes. Because of the interaction with the measuring apparatus, the statistical ensemble N realizes its potentialities. Consequently, a new form of determination corresponds to the (generally) multiple potentialities of N. We will discuss this later.

So, let us return now to the argument of von Neumann. According to his conception the statistical interpretation abandons the principle of causality in microphysics. Chance in classical mechanics is non-essential; it is of a subjective character, while the probabilities in quantum mechanics are *essential*. Thus, the statistical distribution is *non-reducible* to a dynamical description. The causality of the microscopic world, von Neumann says, can in any event be simulated by the levelling action which is manifest whenever many elementary processes operate simultaneously, i.e. by the "law of large numbers".[7]

Von Neumann establishes a formal antithesis between dynamical and statistical laws. The latter, for him, are not causal. Nevertheless we can maintain that the statistical laws too, are causal, because the causes of the statistical distribution exist and are generally known. The real problem is the *mode* of determination of the effect by the causes, and the *completeness* of the theory.

Von Neumann distinguishes two types of evolution of the state vector:

1. The *"causal" evolution*, characteristic of the movement of a free particle. In this case it is possible to calculate Φ_t causally from Φ_0, because the Schrödinger's time-dependent equation

$$\frac{h}{2\pi i}\frac{\partial \Phi t}{\partial t} = -\mathrm{H}\Phi_t$$

determines uniquely the whole path of the particle:

$$\Phi_t = e^{-2\pi_i/h\,(t\,-\,t_0)\mathrm{H}}\Phi_0$$

Consequently determinism is not contested in this idealized situation.

2. In the above case there is no external perturbation and the system conserves its identity. But if we make a great number of measurements, we obtain a statistical distribution of the possible values of a given variable. The measurement R, von Neumann says, has the consequence of changing each state Φ into one of the states $\Phi_1, \Phi_2, \ldots$, which are connected with the respective results of measurement, $\lambda_1, \lambda_2, \ldots$ This "discontinuous transition" is not of the type described by the time – dependent Schrödinger equation (continuous causal change of Φ). This second kind of evolution is, according to von Neumann, *non-causal*.

At this point, von Neumann raises the question of the possibility of reducing statistical relations to "causal" (i.e. to deterministic) ones: "If we want to explain the non-causal character of the connection between Φ and the values of the physical quantities following the pattern of classical mechanics, then the interpretation is clearly the proper one. In reality Φ does not determine the state exactly. In order to know this state absolutely, additional numerical data are necessary. That is, the system has other characteristics or coordinates in addition to Φ".[8]

So, Φ does not determine the state exactly. With Φ alone, only statistical predictions are possible. The question of a more complete specification of the state is then posed.

It is customary, von Neumann writes, to call these hypothetical additional coordinates, "hidden parameters". In classical physics the explanation by means of the hidden parameters has reduced many apparently statistical relations to the causal foundations of mechanics. The view that an explanation of this type is possible in quantum mechanic, "has at present prominent representatives" and if this view were correct, the present form of the theory would be a provisional one. However, the introduction of hidden variables is not possible, according to von Neumann, without a basic change in the present theory.[9]

Von Neumann's impossibility proof is based on the argument that there exist no dispersion-free ensembles in microphysics. His conclusion is the following: "The decision is made and it is against causality; because all ensembles, even the homogeneous, that is to say, even the pure states, have dispersions".[10] However, what is the meaning von Neumann attributes to the concept of homogeneous (or pure) ensembles and of mixtures? And what is the value of his statement, that even homogeneous ensembles are not dispersion-free? Let us examine first the definitions of von Neumann.

THE DEFINITION OF THE QUANTUM ENSEMBLES

Let us take the state Φ. The physical quantity R has an expectation value

$$\rho = (R\Phi, \Phi)$$

and a dispersion ε^2. *The statistical dispersion of R exists, according to von Neumann, even though Φ is an individual state.* But the statistical character becomes even more prominent, von Neumann says, if many states Φ_1, Φ_2, . . . with respective probabilities w_1, w_2, . . ., constitute the description.

However, what is the meaning of the concept of individual state? And which is the nature of the dispersions of such states? And how is the initial state related to the statistical distribution of the states in the final ensemble?

Von Neumann uses the concept of the homogeneous ensemble: "Ensembles we are accustomed to consider to be composed of individual particles, all identical and all determined causally".[11] The expression "all identical" is evidently used to define the pure states. In fact, von Neumann identifies in other contexts of his book the concept of homogeneity with that of pure ensemble.[12] Yet, what is the meaning of this definition?

According to von Neumann, the homogeneous (or pure) ensembles are characterised by

$$U = P_{[\Phi]}(||\Phi|| = 1)$$

and since these are actual states of the system S, i.e. not capable of further resolution, he calls them *states*. This definition gives the impression that the homogeneous (or pure) ensembles correspond to *sharp* states, i.e. to states with only one possible eigenstate (states described in a one-dimensional Hilbert space). The elements of reality of such states can be actual or potential. In both cases the result of a measurement is unique and no further resolution of such ensembles is possible. Consequently, the only statistical dispersions of such ensembles are those described by the Heisenberg inequalities. This is not clear in the argument of von Neumann.

Von Neumann examines afterwards the case in which U has a pure discrete spectrum, perhaps with the eigenvalues $w_1, w_2, \ldots$ and the eigenfunctions $\Phi_1, \Phi_2 \ldots$, which form an orthonormal set. We have in that case (we use the notations of von Neumann):

$$U = \sum w_n P[\Phi_n]$$

Such ensembles represent, evidently, a superposition of states, "which are chosen to be pairwise orthogonal". Von Neumann calls these "general ensembles", mixtures.[13] *But a superposition is not a mixture.* The initial ensemble is a pure state, because all its elements are described by the same state vector. *The final state, on the contrary, is a mixture*, created by the interaction S + A (by the realisation of the potential states of the ensemble). To call it a mixture means that the different states $\{\Phi_n\}$ pre-exit and that the same holds for the statistical dispersions of the quantum ensemble. However, before the measurement the dispersions are those defined by the inequalities of Heisenberg. After the measurement, on the contrary, the dispersions are those of a mixture of different states.

This point of view, which does not take into consideration the processes of transformation occurring during the measurement, processes transforming the initial pure state into a mixture, constitutes a fundamental assumption in von Neumann's impossibility proof. Thus, to quote L. de Broglie: "von Neumann's demonstration presupposes that the distribution of probabilities accepted by quantum mechanics exist before the measurement. This assumption is contrary to the facts, because at least some of the probability distributions are created by the measurement". And de Broglie concludes, "There is a vicious circle in von Neumann's theorem".[14]

We can, in fact, distinguish the following cases of measurement:

- A pure ensemble in the narrow sense (sharp state) is transformed into a pure ensemble. The elements of reality characterizing the state exist before the measurement (real elements) or are created during the interaction S + A (potential elements) by the transformation of other, pre-existing elements of reality. The only dispersions existing in that case are those statistical dispersions described by the Heisenberg inequalities.

- An initial mixture of states $\{\Psi_i\}$ is separated into a number of pure sub-ensembles. In that case the statistical dispersions pre-exist and there is no "non-causal" transformation.
- The case of superposition (pure ensemble transformed into a mixture) poses the question of the deterministic or non-deterministic character of the processes of transformation involved. In that case the states $\{\Psi_i\}$ do not pre-exist. They are *potential states* and they are actualized during the measurement.

Consequently the statistical dispersions in that case correspond to the existence of a spectrum of (different) eigenvalues. This is the specific characteristic of the new ensemble, and not the dispersions of the pure states, corresponding to the Heisenberg's inequalities. By identifying the superposition (pure state) with a mixture, von Neumann's account fails to understand the dynamical character of the interaction S + A. At the same time, he puts the question of determinism in a very restricted manner.

THE IMPOSSIBILITY PROOF

We saw that von Neumann identifies the superposition with a mixture. At the same time he seems to consider the superposition as a pure state. In fact, he defines the wave function Φ as a point of a Hilbert space (this definition is valid for "sharp" states, as well, as for a superposition of states). Then he writes: "Although we believe that after having specified (Φ) we know the state completely, nevertheless, only statistical statements can be made on the values of the physical quantities involved".[15] It is evident that this assertion concerns the superposition of states and not a mixture.

Von Neumann attempts to explain the statistical dispersions *via* the hypothesis of the hidden variables. For statistical ensembles, he says, it is not surprising that a physical quantity has no sharp value a_0, and that a positive dispersion exists. Von Neumann envisages, in the following, two possible ways of explaining the existence of such dispersions:

I. The individual systems $S_1 \ldots, S_N$ are in different states, so that the ensembles $[S_1 \ldots, S_N]$ are defined by their relative frequencies. In that case, the fact that we do not obtain sharp values, is due to our lack of information. This case is well known.

II. "The individual systems $S_1 \ldots, S_N$ are in the same state, but the laws of nature are not causal. Then the cause of dispersions is not our lack of information, but is nature itself, which had disregarded the "principle of the sufficient cause". This case II, "is important and new".[16]

The interesting case is consequently that of a pure state (superposition) which is transformed into a mixture.

One might object, von Neumann says, that nature cannot violate the "principle of sufficient cause," because it is a principle of identity: Two systems S which are in the same state will remain identical in all conceivable interactions. So, two systems that could react differently to the same interaction cannot be called identical. Therefore, in an ensemble which has dispersions relative to a quantity R, the individual systems $S_1 \ldots, S_N$, cannot all be in the same state. These systems are not equal to one another, and the description by the wave function is not complete. So, other variables must exist – the "hidden parameters".[17]

It is evident that *von Neumann's conception of the hidden variables is a static one*: it presupposes the possibility of subdividing the initial ensemble in *a permanent way*, into a number of "homogeneous" ensembles (sharp states) each of which would react differently to the external perturbations. He then uses this formal assumption to refute the initial hypothesis.

The statements contained in the last sentences, he writes, "are incorrect, because we did not consider the fact that the measurement changes the measured system". Each step, von Neumann says, destroys the result of the preceding one "and no further repetition of successive measurements can bring order in this confusion. In the atom we are at the boundary of the physical world, where each measurement is an interference of the same order of magnitude as the object measured, and therefore affects it basically. Thus the uncertainty relations are at the root of these difficulties".[18] Some remarks are needed here: First of all, the essential in the measurement is not the destruction and the "confusion" but the creation of an ensemble of eigenstates with definite and (generally) predictable probabilities. Also, the root of the difficulties is not the existence of the inequalities of Heisenberg but the unknown, non-linear processes of transformation of the initial pure ensemble into a mixture.

In order to decide on the hidden variable hypothesis, von Neumann poses then the question of the possibility of dividing the statistical ensemble into a number of sub-ensembles constituting sharp states. In this way and contrary to his previous thesis, he advances the assumption that the eigenstates exist before the measurement, identifying once more the superposition with a mixture of states. He then takes into consideration the fact that the measurement changes the initial state, to refute this assumption. However, the initial pure state is formed by identical particles in the same state; von Neumann's argumentation is therefore based on a false assumption.

Given the above assumptions, von Neumann's conclusion that the hidden variables do not exist is logically unavoidable: We have no method, he says, which would make it always possible to resolve further the dispersing ensembles (without a change of their elements) or to penetrate into those homogeneous ensembles which no longer have dispersions. But, continues von Neumann, we could attempt to maintain the fiction that each dispersing ensemble can be divided into two (or more) parts different from each other and from itself, without any change in its elements.

The static conception of the state is once more formulated: the states exist before the measurement and it would be possible to separate them. In this way it would be possible to obtain a number of pure sub-ensembles. Should these ensembles be dispersion free?

Each dispersion-free state, von Neumann says, should be pure. "But our question at the moment is the converse one: Is each pure Exp (R) function dispersion-free?" After a long discussion von Neumann concludes that "*there are no ensembles which are free from dispersions*". All these ensembles have dispersions, "even the homogeneous". By this way, "the decision is made and it is against causality". But there would still be the question of "hidden parameters" to be discussed, i.e. "the question as to whether the dispersions of the homogeneous ensembles characterized by the wave functions Φ are not due to the fact that those are not the real states, but only mixtures of several states". But this is impossible, because the dispersion free states which would have to correspond to the "actual" states, do not exist.[19]

We have already noted that von Neumann identifies the homogeneous ensembles with the "sharp" states of quantum mechanics and that these states do not manifest other dispersions than the statistical ones predicted by the Heisenberg inequalities. *But these dispersions are irrelevant to the problem of hidden variables.* The problem arises only in the case of the superposition, with the creation of an ensemble of different states $\{\Psi_i\}$ from one and the same initial state Ψ.

However, in order to establish the possibility of hidden variable theories, is not necessary, as von Neumann claims, to analyze every pure state into a number of sub-ensembles without dispersions. The states $\{\Psi_i\}$ do not pre-exist. They are potential states. So, the Hilbert space to which the state vector

$$\Psi = \sum_i c_i \Psi_i$$

belongs, *is not the space of actual states, but the space of potential states of the quantum ensemble under given external conditions*. Von Neumann's conception is a static one. As opposed to that, the hidden variable theories presuppose a dynamical conception of the state: the existence of a fine structure of dynamical, stochastic character, and the possibility of a transformation of the state under given external conditions.

Von Neumann rejects the possibility of a more complete quantum mechanical description by means of the introduction of supplementary variables. However, his negative conclusion is not an absolute one: "The introduction of hidden parameters is certainly not possible without a basic change in the present theory". In fact, the existing theories of hidden variables are, epistemologically judged, different form conventional quantum mechanics. Also, in another point of his book, von Neumann asserts that "the present system of quantum mechanics would have to be objectively false, in order that another description of the elementary processes than the statistical one be possible".[20] However, it is not certain whether this second statement of von Neumann is correct. In fact, the specialists are today divided on this crucial question. Thus it is well known that according to Bell's inequalities, the validity of causality and locality in quantum mechanics would lead, in certain special cases, to statistical predictions different from those of the actual quantum theory. But according to another point of view, the hidden variable theories must reproduce the statistical predictions of the conventional quantum mechanics (de Broglie and others). Thus, from this point of view, the present form of quantum mechanics would not have to be necessarily "objectively false" in order that a more complete description of the elementary processes be possible.

FURTHER REMARKS CONCERNING THE "IMPOSSIBILITY PROOF"

Some further remarks seem now necessary, concerning the argument that hidden variable theories are impossible, on the ground that dispersion-free quantum ensembles do not exist.

As already noted, we can distinguish three types of quantum ensembles and, correspondingly, three different cases of measurement.

1. Pure states in the narrow sense (*sharp* states). One and only one state is possible after the measurement (one-dimensional Hilbert space). In that case we have two possibilities.

1a. The state is actual. No change in the state vector occurs. It is only the apparatus that passes from its initial state into an eigenstate:

$$\Phi_0\Psi_\omega \xrightarrow{\text{interaction}} \Phi_\omega\Psi_\omega$$

(Φ denotes the apparatus states. Ψ denotes the system states).

1b. The state is potential and is realized during the interaction of the system with the apparatus:

$$\Phi_0\Psi \xrightarrow{\text{interaction}} \Phi_\omega\Psi_\omega$$

In this case, new elements of reality are created during the process of measurement.

Thus, in the case pure state → pure state, the only dispersions present are the dispersions taken into account by the Heisenberg inequalities. The final state is uniquely determined in conformity with the dynamical form of causal determination. Thus the validity of the principle of causality is verified in that special case, as well as the principle of a dynamical form of determinism.

2. Mixtures constituted by a ensemble of states $\{\Psi_i\}$. The initial ensemble can be sub-divided into a number of pure sub-ensembles. The probability of the state Ψ_i is:

$$pi = \frac{n_i}{N}, \quad N = \sum_i n_i$$

In that case, too, there is no question of determinism. The result of each measurement is uniquely determined, because of the fact that the distribution of probabilities exists before the measurement.

3. The most interesting and difficult case is that of a pure state (superposition) transformed by the measurement into a mixture. In that case we have:

$$\Psi = \sum_i c_i \Psi_i \quad \text{and} \quad \sum_i c_i \Psi_i \xrightarrow{\text{interaction}} \{\Psi_i\} \text{ with } p_i = |c_i|^2$$

How is this transformation realized? It is well known that within the conceptual framework of the CI, this transformation is impossible. The consciousness of the observer is the *deus ex machina* necessary for the (non-causal) "reduction of the wave packet". Schrödinger's cat is a very well-known illustration of the impasse of the orthodox interpretation.[21]

One can consider from the point of view of the question of determinism, the cases (1) and (3) *as identical.* It is true that the two cases are represented by a unit state vector in a Hilbert space. It is also true that both of them are characterised by a density matrix.

$$M_{mm} = \langle m|\Psi\rangle \langle\Psi|m\rangle$$

Nevertheless this formal identity does not preclude an essential difference from the physical point of view. In case (1) we have only one possible state (potential or real). In that case the question of determinism, as already emphasized, does not arise. In case (3) on the contrary, the states are not real before the measurement. So the density matrix is the measure of the multiple potentialities of the statistical ensemble. The same is true concerning the physical meaning of the state vector. It is the dispersions corresponding to this "superposition" of states that posit the question of the status of determinism in microphysics and not the dispersions corresponding to the inequalities of Heisenberg.

Von Neumann maintains that this third type of measurement corresponds to "a non-causal change in which each of the states Φ_1, Φ_2 . . . can result, and in fact does result with the respective probabilities". But von Neumann considers the initial state as a mixture. The mixture U goes into the mixture U′:

$$U \rightarrow U' = \sum_i (U\Phi_{n'} \Phi_n) P_{[\Phi n]}$$

and "since the states go over into a mixture, the process is not a causal one".[22]

But what is the meaning of the expression "a non-causal change," if the initial ensemble was already a mixture? The same thesis was advanced 30 years later by E. Wigner, who writes that "in order to obtain a mixture of states as a result of the interaction, the initial state must have been a mixture already". This, according to Wigner, "follows from the general theorem that the characteristic values of the density matrix are constants of motion".[23]

This formal conception does not take into consideration the external perturbation (interaction of S with A) which results in the creation of new elements of reality and (in consequence) in the realization of the potentialities of the quantum ensemble. The states $\{\Phi_i\}$ do not pre-exist. The initial ensemble was a pure state and the final states were potential states before the interactions S + A. *Consequently, von Neumann's impossibility proof is irrelevant: The hidden variable theories do not presuppose the possibility of analyzing the initial ensemble into a number of dispersion free sub-ensembles.* More than that: the conception of von Neumann is very restrictive. It presupposes a classical form of determinism for any type of hidden variables theory. However, as we shall see, statistical hidden variables theories are also possible. This form was not envisaged by von Neumann and, consequently, is not excluded by his impossibility proof.

The initial state was a pure one. The interaction S + A results in the realisation of the potentialities of the statistical ensemble, transforming it into a *mixture*. The ensemble is now characterized by the statistical operator:

$$W = \sum \frac{n_i}{N} |\Psi_i\rangle \langle\Psi_i|$$

This formal difference is an expression of the fact that the interaction resulted in the creation of new elements of reality and, correspondingly, in the creation of new states.

However, the case (3) raises a real problem: How is it that from the initial state Ψ many different states $\{\Psi_i\}$ are possible, under identical conditions? Is this fact not a proof that nature does not respect the principle of determinism? Let us examine this question.

1. The above situation does not contradict the principle of causality. The causes of the phenomenon (or, at least, some of them) are known. They are the different forms of interaction known in contemporary physics.

2. It would be possible, however, to maintain, that this situation contradicts the principle of classical determinism (the unique determination of the effect by the causes). Nevertheless, the distribution of the probabilities is *determined* by the nature of the quantum ensemble and its conditions. An eventual modification of the conditions results in a modification of the statistical distribution. More than that: it is possible to calculate in advance the probability of each possible state, on the basis of the nature of the system and the conditions. The quantum-mechanical predictions had always been verified. This fact is, therefore, an additional proof of the validity of the principle of determinism in microphysics. The conservation laws and the corresponding selection rules are the formal expression of the deterministic character of the "reduction of the wave packet", that is to say, of the transformation of the quantum systems. However, in the case of quantum mechanics, we can transcend the classical conception of determinism (unique determination of the effect by the causes) in favour of a more flexible form of determination, corresponding to the multiple potentialities of the quantum ensembles.

3. One further objection would be the following: Why is it so that in every individual measurement the system realises one or another of the possible states? Is not this fact an indication that nature makes an arbitrary choice, among all possible states?

The hidden variable-hypothesis is a possible response to that question. The quantum-mechanical description may be considered incomplete and the introduction of a number of supplementary variables would allow for a dynamical description of the movement and of the transformation of quantum systems. But the eventual introduction of such variables does not mean that the initial pure ensemble should be divided into a number of sub-ensembles representing different pure (sharp) states. It is more realistic to think of a fine structure of the state, of a variable, stochastic character, namely of a fine dynamical structure not affecting the statistical distribution of the possible states, *which is determined by the already known essential variables*. In this way it is not necessary to claim that the states pre-exist, i.e. that the initial ensemble is a mixture. We can maintain that in quantum physics a new mode of causal determination operates, different from the mechanistic and the dynamical one, the *quantum statistical determinism*.[24]

The strategic aim of hidden-variable theories is not to restore the Laplacian (mechanistic) form of determinism in microphysics. The real question concerns the possibility of a *dynamical* or, even a *statistical*, more complete description of certain microphysical processes. The history of physics is, from a certain point of view, the history of discovery of "hidden variables". The atom was believed, in the last century, to be a "hidden variable". The same is true of the elementary particles, before the discovery of the proton, the neutron, etc. Spin and other quantum numbers were "hidden variables" which have now been discovered and so became manifest.[25] Consequently, why should we exclude the possibility of a deeper understanding of quantum processes?

In fact, we cannot exclude, *a priori*, the possibility of the hidden variable theories, to reduce some actual stochastic processes to dynamical ones, establishing a concrete dialectic between chance and necessity. But already, the notion of *quantum statistical determinism* establishes this dialectic between two concepts, mutually irreducible in the frame of formal logic.[26] Nevertheless a hidden-variable theory does not necessarily give a dynamical description of quantum-mechanical phenomena. *Statistical hidden variable theories are also possible.*

As we have seen, von Neumann considers and excludes only the possibility of a dynamical description of quantum-mechanical phenomena on the basis that non-dispersive states do not exist. On the contrary, it is quite plausible to consider the possibility of dispersive hidden-variable theories. Von Neumann does not consider the possibility of such theories. The interest of statistical hidden-variable theories lies in the fact that they refute in practice the CI argument, concerning the completeness of quantum-mechanical description, giving at the same time a more profound understanding of the microphysical reality.

The probabilistic type is the most general case of physical law. The dynamical law is an idealization of the real state of affairs and this is true even for classical mechanics. So, the search of probabilistic hidden variable theories[27] opens up a new domain of investigation, which may give a solution to the crisis created by the violation of the inequalities of J. S. Bell.

BEYOND THE LIMITS OF THE CLASSICAL FORMS OF DETERMINISM

The debate concerning the possibility of a more complete specification of the quantum states has not come to an end. However, a hidden variables theory exist since 1927 (*Théorie de la double solution* by L. de Broglie).[28] Also, D. Bohm elaborated a non-local, deterministic hidden variables theory, twenty five year later (1952).[29] Since then, as is well known, many other hidden-variable models have been proposed.

For Einstein a hidden variables theory ought to be local and deterministic. At the 7th Solvay Congress, he formulated a well-known paradox, in order to demonstrate the *non-statistical* and *non-local* character of the CI. The Copenhagen Interpretation implies, according to Einstein, that "un même processus élémentaire produisît une action en deux on plusieurs endroits de l'écran". This non-locality presupposes "un mécanisme d'action à distance tout particulier, quí empêche que l'onde continument répartie dans l'espace, produise une action en *deux* endroits de l'écran".[30]

Einstein never accepted the single system and non-local interpretation of Q.M. Also, L. de Broglie formulated another paradox demonstrating the non-local and non-deterministic character of the CI.[31] With his theories, over and above, he gave a *de facto* refutation of the "impossibility proof" of von Neumann. "Les théories de la double solution et de l'onde pilote," he emphasized, "si elles ne peuvent être considerées comme prouvées *existent* cependant et l'on peut se demander comment leur existence est conciliable avec le théorème de von Neumann".[32]

One of the more serious defects of the argument of von Neumann was, according to de Broglie, the fact that he considered that "toutes les répartitions de probabilités admises par la Mecanique Ondulatoire ont une existence physique avant même que l'on ait fait l'expérience qui fait entrer en jeu l'une de ces répartitions".[33] The defect was also emphasized by J. Bub: "According to von Neumann W_{AB} represents a mixture which does not in any way depend on the initial quantum state of the photon. But the initial state is *required* to be a plane wave (and not, say, a mixture of plane waves) in order to obtain interference effects."[34]

The thesis that the initial state must be a mixture was once more formulated explicitly by von Neumann in 1938: "Il est donc clair que, si une théorie statistique n'est pas purement statistique, c'est-à-dire si elle peut être dérivée d'une théorie causale par la 'méthode des paramètres cachés', tout ensemble non causal doit être mixte-puisque il doit inclure l'établissement de la moyenne des différentes valeurs des coordonnés supplémentaires ('paramètres cachés')."[35]

I analyzed in the preceding sections this defect of the "impossibility proof" of von Neumann. This is one of the aspects of his static conception of the quantum ensembles. And this conception is coherent with his narrow and restrictive understanding of the hidden variable theories – with the fact that for him every such theory ought to be deterministic. Probabilistic hidden variables theories are excluded from the narrow horizon of the theorem of von Neumann.

It is well known that a number of deterministic models were formulated by many physicists in spite of their "impossibility". One general idea of these theories is that the probabilities are the result of a random distribution of the hidden variables, a distribution not affecting the probabilities calculated on the basis of the actual formalism of quantum mechanics. In particular, the so-called collapse of the wave packet is considered as an objective, irreversible and dissipative deterministic process.[36]

As I maintained, von Neumann (as the majority of physicists) confuses causality and determinism. Consequently, he does not consider the possibility of dispersive hidden variables theories. However, such theories are possible and, moreover, they are not ruled out on the basis of the theorem of von Neumann. Such theories transcend the horizon of the theorem of von Neumann, unaffected as they are by his restrictive conditions. As Th. Brody puts it, "deterministic" or dispersionless hidden-variables, but only these, must be ruled out. Such theories, Brody states, must be statistical in nature and possess non-negligible dispersions. Consequently, von Neumann's argument, "so far from constituting an impossibility proof should therefore be seen as giving a strong hint about the structure of hidden variables theories to be developed for quantum mechanics".[37] The basic assumption of von Neumann (assumption V) is in general false, according to D. Bohm and J. Bub. It is valid only in a certain special case which includes the ensembles of quantum mechanics. On the contrary, sub-ensembles are possible which have less dispersions than the "normal" ones and even, in the limiting case, dispersion free ensembles, in which they are all precisely defined.[38]

One can accept that hidden variables constitute another level of organization, and that they are in random fluctuation. The essential variables of the quantum level determine the spectrum of the potentialities of the quantum ensemble. Consequently, the hidden variables do not modify the statistical predictions of quantum mechanics. Statistical hidden variable theories are, in principle, possible. And, as Th. Brody puts it, if hidden variables are to be introduced in a statistical theory in order to obtain a more detailed description, they must themselves be dispersive.[39]

Dynamical laws are often the macroscopic (and formal) expression of ensembles of random effects. Probabilistic laws express more real situations.

They are the more general form of physical law and they are compatible with causality – they are the dialectical negation of the deterministic laws and of the inexorable necessity associated with them. Probabilistic hidden variables theories do not presuppose dispersion-free states. Such theories are causal without being deterministic in the classical understanding of the term. The following definition constitutes the general frame for such theories.

DEFINITION: If, during a first kind of measurement, it is possible to predict with a probability P_i the actualisation of a state Ψ_i, characterized by an element of reality π_i, then we can postulate that this element has been actualized during the measurement *via* the transformation to one or more elements of reality $\{\lambda_i\}$ defining a state Ψ_0, capable of generating Ψ under the given conditions. In this case, we postulate the existence of more than one ensemble $\{\lambda_i\}$ generating different states Ψ_i with probabilities P_i. The so-called superposition: $\Psi = \sum_i c_i \Psi_i$, for example, is the quantitative expression of the potentialities of a pure ensemble which is transformed into a mixture.[40]

Probabilistic hidden variables theories are not excluded from the logical point of view also. For example, analyzing the "impossibility proof", S. Gudder writes: "Strangely enough, there have also appeared explicit examples of hidden variables theories. Clearly, there is something wrong there. One cannot have hidden variables theories if it is impossible". The opponents of such theories, Gudder notes, advocate a return to classical mechanics or, at least, an embedding of quantum mechanics into a classical mechanical framework. However, this is not so. Consequently, "although von Neumann's general idea is correct, his mathematical definition is too restrictive. We contend that the dispersion-free states need not be defined on the entire proposition system".[41]

Only a partial embedding is needed (and possible) for dispersion-free states, which allow predictions for single measurements. Probabilistic hidden variables theories need even less restrictive conditions.

University of Athens, Greece

NOTES

* The present paper is a more elaborated and developed version of an article published in *Determinism in Physics*, E. Bitsakis and N. Tambakis (eds.), Athens, 1985.

1 John von Neumann, *Mathematical Foundations of Quantum Mechanics* (Princeton University Press, 1955).

2 See: L. de Broglie, *La Physique Quantique restera-t-elle indéterministe?* (Gauthier-Villars, Paris, 1953); A. Landé, in *Determinism and Freedom*, S. Hook (ed.) (N.Y. University Press, 1958); D. Bohm, *Phys. Rev.* **85** (1958), p. 166; Bohm and Bub, *Rev. Mod. Phys.* **38** (1966), p. 453; S. P. Gudder, *J. Math. Phys.* **11** (1870), p. 431; J. S. Bell, *Rev. Modl. Phys.* **38** (1966), p. 447; Th. Brody, *The Philosophy Behind Physics* (Springer-Verlag, 1993); Th. Angelidis, *J. Math. Phys.* **34** (1993), p. 1635.

See also: J. J. Bellinfante, *A Survey of Hidden-Variables Theories* (Pergamon Press); M. Jammer, *The Philosophy of Quantum Mechanics* (John Wiley, N.Y., 1974).

[3] J. Von Neumann, *op. cit. Preface.*
[4] *Ibid.*, pp. 206–7.
[5] *Ibid.*, pp. 196, 206 and elsewhere.
[6] *Ibid.*, p. 206.
[7] *Ibid.*, p. 210.
[8] *Ibid.*, p. 209.
[9] *Ibid.*, pp. 209–10.
[10] *Ibid.*, p. 323.
[11] *Ibid.*, p. 305.
[12] See, *Ibid.*, pp. 307, 312.
[13] *Ibid.*, pp. 328–9.
[14] L. de Broglie, *op. cit.*, pp. 17–8.
[15] J. von Neumann, *op. cit.*, p. 207.
[16] *Ibid.*, pp. 301–2.
[17] *Ibid.*, pp. 302–3.
[18] *Ibid.*, pp. 304–5.
[19] *Ibid.*, pp. 308–25.
[20] *Ibid.*, p. 325.
[21] See E. Bitsakis. *Ann. Fond. L. de Broglie* **5** (1980), p. 263.
[22] J. von Neumann, *op. cit.* chapters *V* and *VI.*
[23] E. Wigner, *Am. J. Phys.* **31** (1963), p. 6.
[24] Concerning the concept of quantum statistical determinism, see: E. Bitsakis, *Found. of Phys.* **18** (1988), p. 331.
[25] See, K. Popper, 'Realism and Quantum Theory', in *Determinism in Physics*, E. Bitsakis and N. Tambakis (eds.) (Gutenberg, Athens, 1985).
[26] For a further discussion see, E. Bitsakis, *Le problème déterminisme en Physique*, Thèse, Paris 1976. *Id.* Physique et Matérialisme (Editions Sociales, Paris, 1983).
[27] See, for example, F. Selleri, *Found. Phys.* **12** (1982), p. 645 and F. Selleri and G. Tarozzi, *Lett. Nuovo Cim.* **22** (1980), p. 533.
[28] See, for example: L. de Broglie, *Le Physique quantique restera-t-elle indéterministe? op. cit. Id, The Current Interpretation of Wawe Mechanics* (Elsevier, 1964). See also the *Proceedings* of the fifth Solvay Congress: *Electrons et Photons* (Gauthier-Villars, Paris, 1928).
[29] D. Bohm, *Phys. Rev.* **85** (1952), p. 166 and **85** (1952), p. 180.
[30] A. Einstein, in *Electrons et Photons, op. cit.*
[31] L. de Broglie, *J. Phys. Rad.* **20** (1954), p. 963.
[32] L. de Broglie, *La Physique Quantique restera-t-elle indéterministe?, op. cit.*, p. 17.
[33] L. de Broglie, *Ibid.*
[34] J. Bub, *Jour. Phil. Logic.* **6** (1967), p. 381.
[35] J. von Neumann. *In, Les Nouvelles Théories en Physique* (Warsaw, 1938).
[36] See, for example: L. de Broglie, *Journal de Physique*, 1927 (théorie de la double solution). D. Bohm, *Phys. Rev.* **85** (1952), p. 166 and **85** (1952), p. 180. D. Bohn and J. Bub, *Rev. Mod. Phys.* **38** (1966), p. 453. N. Gisin and C. Piron, *Lett. Math. Phys.* **5** (1981), p. 279. A. O. Barut and M. Bozic, *An. Fond. L. de Broglie* **15** (1990), p. 67. N. Gisin, *Synthese* **89** (1991), p. 287. A. O. Barut, in *Bell's Theorem and the Foundations of Modern Physics*, A van der Merwe *et al.* (eds.) (World Scientific, 1992). Th. Angelidis, *Jour. Math. Phys.* **34** (1993), p. 1635.
[37] Th. Brody, *The Philosophy behind Physics*, Chapter 15 (Springer-Verlag, 1993).
[38] D. Bohm, J. Bub. *Rev. Mod. Phys.* **38** (1966), p. 453.
[39] Th. Brody, *The Philosophy behind Physics, op. cit.*, chapter 15.
[40] See, E. Bitsakis, *Physics Essays* **4** (1991), p. 124.
[41] S. P. Gudder, *J. Math. Phys.* **11** (1970), p. 431.

RONALD N. GIERE

EXPLAINING SCIENTIFIC REVOLUTIONS

SCIENCE AND THE PHILOSOPHY OF SCIENCE

It is difficult to exaggerate the importance of science and technology in contemporary culture. The popular media are full of reports on new developments in the bio-medical sciences, in information and communication technologies, and on the role of science and technology in maintaining an economy that will be competitive in world markets. And these are only a few of the most prominent current areas of general interest. This widespread interest generates a need to understand science as a cultural phenomenon. What kind of activity is it? How does it work? How does it interact with other aspects of contemporary culture?

One might reasonably have expected philosophers of science to provide some enlightenment on these important issues. But if one turns to the recent literature within the philosophy of science, one may be disappointed. Many philosophers of science focus their energies on the logical or conceptual structure of particular scientific theories, such as relativity theory (Friedman 1983) or the theory of natural selection (Sober 1984). Even those philosophers of science whose interests transcend particular scientific theories tend to be preoccupied with the peculiarly philosophical concern to develop *normative* criteria for such activities as constructing scientific theories (van Fraassen 1980), formulating scientific explanations (Salmon 1984), or determining which of several rival hypotheses should be preferred (Levi 1980). The source of these concerns is to be found not so much in the sciences as in the history of philosophy itself.

The Greek founders of Western philosophy were much concerned with the nature and sources of human knowledge. Later, in the Christian era, this concern was focused on knowledge of the Divine Will. Most philosophers were theologians, and they took it as their task to elaborate and defend the principles of Christian theology against heretics and other non-believers. As the impact of the reformation took hold and the scientific revolution began to unfold, some philosophers took it upon themselves to defend the emerging, new sciences, often against the skepticism of theologians. But their conception of the task, and forms of argumentation, remained similar to those of the theologians they sought to displace. The assumed task was to provide an *autonomous*, philosophical, justification of the new science. Philosophical proofs of the justifiability of scientific knowledge have a similar overall structure to proofs of the existence of god. They begin with premises regarded as self-evident and proceed by logical demonstration to the desired conclusion.[1]

D. Ginev and R. S. Cohen (eds.), Issues and Images in the Philosophy of Science, 63–86.

In the seventeenth century it was reasonable to assume both that science needed philosophical justification and that philosophy had within itself the resources to provide an autonomous, philosophical justification. As we near the end of the twentieth century, neither of these assumptions remain plausible. The obvious success of science, particularly in the twentieth century, eliminates any serious worry that there might be something fundamentally unsound about the way science is done or about its results. This does not mean that we understand why science has been so successful, or even that we are clear about the connection between its success and the veracity of the pictures of the world it has produced. But science now needs no wholesale philosophical legitimization. Indeed, the soundness of science as a whole is now more secure than any philosophical proof of its legitimacy could possibly be.[2]

As for the resources of philosophy to provide the kind of autonomous justification traditionally sought, these seem to have run out. If one denies oneself any appeal to empirical claims of the kind scientists might investigate, where is one to turn? To logic? To intuition? To ordinary linguistic usage? These have all been tried without lasting success. That fact does not, of course, constitute a proof that no such demonstration is possible. But it does provide a powerful incentive to come down to earth and inquire, *within* a scientific framework, how science works and why it is successful. That, I suggest, is the real challenge to contemporary philosophy of science.

In what follows I will begin by reviewing post-World War II attempts to construct a general account of the nature of science. This review will include work by historians and sociologists of science, as well as by philosophers of science. One of the major features of more recent philosophy of science is its increasing involvement with these other fields, an involvement that has been both complementary and antagonistic. This review will provide the springboard for my own suggestions as to the directions future investigations should take.

PHILOSOPHY AND SOCIOLOGY OF SCIENCE AFTER WORLD WAR II

At the end of World War II there existed two major theoretical approaches to the study of science as a human activity, one philosophical and one sociological. For the most part these approaches were *complementary.*

Logical Empiricism

Between 1945 and 1960 philosophical thinking about science was dominated by Logical Empiricism. This philosophical movement was well named. It was *empiricist* in the classical sense that the ultimate appeal for all scientific claims was said to be direct sensory experience and nothing else. It was *logical* in its methods, which derived from Russell and Whitehead's work on logic and the foundations of mathematics (1912–14), and from Russell's philosophical writings, particularly those on metaphysics and epistemology (1914). Two aspects of the Logical Empiricist analysis of science are relevant here.

First, it was epistemological and highly "theory centric."[3] That is, focused on the analysis of scientific knowledge, and it assumed this knowledge to be encapsulated in scientific theories. Moreover, it insisted that scientific theories are to be understood as axiomatic systems to which the methods of logical analysis could be applied.

Second, it insisted that the epistemological relationship between sense experience and theory is itself a "logical" relationship. Thus epistemology becomes a kind of applied logic. Of course it early became apparent that the "logic" could not be the deductive logic of mathematical reasoning. So the program was to develop an enriched *inductive* logic, of which probability logics were the most popular (Carnap 1950). The program of developing a probability logic for science continues to this day (Jeffrey 1985).

It follows that science has the properties of being both *representational* and *rational*. It is representational in the sense that theories are the kind of thing that can be *true* of the world. The probability of a theory is the probability that it is true. And science is rational because the probability of a theory provided by inductive logic was interpreted as expressing the rational degree of belief in the truth of that theory relative to existing evidence.

It is a consequence of the Logical Empiricist analysis of science that there is an unbridgeable gap between the content and methods of science and all other aspects, such as the psychology of scientists or their social organization. The former are fully analyzable using the *a priori* methods of modern mathematical logic. The latter are the subject of empirical social science. This gap was codified in the Logical Empiricists' rejection of "psychologism" and in their distinction between the "context of discovery" and the "context of justification."

Functionalist Sociology of Science

Between 1945 and 1960 the reigning sociology of science was the structural-functional account of Robert K. Merton (1973). In this framework, the job of the sociologist was to exhibit the ways in which the *social structure* of an institution, as defined by its social norms, promotes its *functions*. Here Merton was applying a general approach that was at the time widely used by sociologists in investigating a wide variety of institutions within society.

A nice example of structural-functional analysis is Merton's account of the role of priority disputes in science. They serve the function, he argued, of reinforcing the norm that original discoveries should be rewarded by suitable recognition. This is functional because the desire for recognition is an important motivating factor for scientists to pursue the goal of producing new knowledge.

Merton took it for granted that the function of science is to produce "certified knowledge." He explicitly renounced any role for the sociologist in analyzing the *content* of this knowledge or the *methods* by which it became certified. For Merton, then, the role of the sociologist of science is complementary to that of the Logical Empiricist philosopher of science. The philosopher is concerned with the logic of science, the sociologist with its social structure. These are distinct enterprises.

THE STRUCTURE OF SCIENTIFIC REVOLUTIONS

During the past twenty-five years, Thomas Kuhn's *Structure of Scientific Revolutions* (1962) has emerged as the single most influential work on the nature of science to be published since World War II. The magnitude of its success seems to have surprised even its author. Part of its success was surely due to the timeliness and originality of its analysis. But there were fortuitous components as well. Like Darwin, whose *Origin of Species* was a quickly produced "outline" of a projected but unpublished larger work, Kuhn was also forced by circumstances to publish a condensed version of a projected longer work.[4] It is doubtful the larger work would have been so accessible to so many. In addition, by the middle 1960s talk of "revolution" was very much in the air. Kuhn became a hero of the "cultural revolution" in spite of himself.

Kuhn's Stage Theory

On the surface, Kuhn developed a "stage theory" of science. The stages in the life of a scientific field unfold chapter by chapter throughout the book. Indeed, the stages – pre-paradigm science, normal science, crisis, revolution, and new normal science – provide the main organizing theme of the whole work. Kuhn is at his best describing the characteristic activities of each stage. He is somewhat less successful at pinpointing the *mechanisms* that drive the process from one stage to the next. I will begin by following the order of Kuhn's own presentation, but my conclusion will be that, for the purpose of developing a general theory of science, the stages are misleading. It is the mechanisms that matter.

Pre-paradigm Science

Kuhn's major examples of pre-paradigm sciences are optics before Newton and electricity before Franklin. In such periods there is a roughly defined subject matter and two or more schools of thought about how the subject should be conceived. Typically these schools are social as well as intellectual entities, with disciplines and students following the teachings of a single teacher. The different schools may coexist for a considerable length of time with little real contact among them and slow progress only within competing schools.

Even in Kuhn's own framework, "pre-paradigm" science is better described as *multi-paradigm* science. By his own characterizations of what constitutes "having a paradigm," each of the competing schools has a paradigm. What is lacking at this stage are not paradigms, but a *dominant* paradigm that can guide the energies of the vast majority of practitioners concerned with the same general subject matter. Without a single, dominant paradigm, energy is wasted in fruitless debates over fundamentals. Only when such debate is past can all the practitioners focus their energies on new, and often quite esoteric, issues with some confidence that they are at least on the right track.

Thinking of this early stage as a multi-paradigm period actually strengthens the case for Kuhn's cyclical picture of science. It makes this early stage very much like the later "revolutionary" stage which is also characterized by competition among fundamentally different approaches. And it reduces the question how a multi-paradigm situation resolves itself into a context with a single paradigm to the question of how a revolution gets resolved. The process should be fundamentally the same in both cases.

Normal Science: The Meaning of "Paradigm"

The stage of normal science is characterized by a single, dominant approach to the subject matter. There is little questioning of the fundamentals of the approach. Rather, energy is devoted to working out its implications and extending its application to new phenomena. A phenomenon apparently at odds with the dominant approach is not taken as evidence that the approach might be mistaken. It merely creates a "puzzle" to be solved. How can this phenomenon best be incorporated into the existing framework? Failure reflects poorly not on the approach, but on the competence of the scientist.

The Structure of Scientific Revolutions is notorious for the ambiguities in its key analytical notion, "paradigm." In later writings (1977), Kuhn himself distinguished two quite different senses. In one sense, a paradigm is a very general "world view" that includes specific theories, instrumentation, and even metaphysical presuppositions. It is this very general sense of a paradigm as a "theory plus" that the initial philosophical commentators (Shapere 1964; Scheffler 1967) took as fundamental. This made it possible to assimilate Kuhn's problem of how revolutions get resolved to the standard philosophical problem of why one theory is to be preferred over another. Despite many of Kuhn's own comments, however, this was not the sense of "paradigm" that was in fact central to Kuhn's theory of science.

The sense that *was* central appeared in the early pages of *The Structure of Scientific Revolutions* where Kuhn emphasized the importance of "the concrete scientific achievement" (p. 11) which provides "models from which spring particular coherent traditions of scientific research" (p. 10). Thus, Newton's account of the motion of a planet or Franklin's account of the Leyden jar provided "paradigms" for the practice of mechanics and the science of electricity respectively. In later writings Kuhn refers to these "concrete models" as "exemplars."

Contrary to a common view of scientific activity, normal science, for Kuhn, is *not* a process of "applying" general laws to new cases. Rather, solutions to new problems are developed by modeling them on the exemplary solutions that underlie the general approach. When Kuhn later talks about "the priority of paradigms" (Ch. 5), he means that the practice of normal science is guided and sustained not by general theory or method, but by these exemplary solutions to earlier problems.

Crisis

At any point in a stage of normal science there will always be *anomalies*, that is, phenomena which no one has yet been able to explain with reference to standard exemplars. The tides, for example, were an anomaly for Newtonian mechanics for half a century. Since there are always some anomalies in any normal science research tradition, their mere existence provides no basis for dissatisfaction with the existing tradition. Nevertheless, there comes a time when practitioners in a tradition begin to lose faith that the anomalies can be resolved using the resources of their tradition. The result is a stage of *crisis*.

Just why scientists lose faith in their tradition is, as Kuhn admits (p. 82), very difficult to explain in general terms. It is clear that no simple answer in terms of the *number* of anomalies, or other, qualitative features, can do justice to the historical facts. Nor, Kuhn argues, can it be explained by pressures from outside the immediate scientific community. So here we have a clear instance in which, as Kuhn himself suggests (pp. 85–6), that the stage theory needs to be supplemented by deeper investigation of things like the psychological reactions, and social responses, of people when accepted ways of operating no longer seem to be paying off as before.

Revolution

Whatever its detailed causes, crisis leads to the proliferation of new approaches and a situation resembling "pre-paradigm" research. There is once again discussion of fundamentals and even of general methodological and philosophical principles. Eventually a new set of exemplars is created which gives rise to a new normal science tradition.

What leads a scientific community to abandon the old tradition for the new, to put aside one set of exemplars and take up another? Here Kuhn appeals to factors that are psychological, sociological, and institutional. Individuals, he says, experience something like a "gestalt switch." They come to see things differently, sometimes quite suddenly. Of course, not everyone experiences a similar "conversion." However, those who do not convert tend to be members of the older generation who were educated in the old tradition. Eventually they die off, leaving the field to a new generation that was educated in the new tradition. Finally, new textbooks are written from the standpoint of the new tradition. Members of the old tradition are thus literally written out of the field.

Incommensurability

Why does Kuhn explain the resolution of revolutions by appealing to psychological, social, and institutional factors? Why can't scientists simply compare two approaches and realize that one is objectively better? The reason is that he believes rival traditions to be "incommensurable". What does that mean?

In a great many passages Kuhn explains incommensurability in *linguistic* categories. Adherents of different traditions, he says, talk past one another. They use the same words with different meanings. Moreover, Kuhn had a theoretical basis for this way of describing rival traditions. Wittgenstein taught that the meaning of words is a function of their use in actual life. Kuhn applied this idea to the scientific life.[5] And it fit in well with his emphasis on the importance of concrete exemplars of scientific achievement for defining a normal science research tradition.

Kuhn's insistence on the incommensurability of research traditions attracted considerable attention from philosophers. Prominent philosophers such as Kripke (1972) and Putnam (1975) invented new theories of meaning and reference partly in response to Kuhn's apparent challenge to the objectivity of science. Nevertheless, in spite of Kuhn's own presentation, and its widespread acceptance by others, the linguistic interpretation of incommensurability seems to me mistaken. The important incommensurability is not one of meaning but of *standards* and *authority*.[6]

By Kuhn's own account, adherents of rival research traditions appeal to different exemplars to authenticate their solutions to new problems. Moreover, there are, according to Kuhn, no higher standards to which one might appeal. This is because the laws, theories, and methodological principles of a research tradition are themselves ultimately grounded in the exemplars. Exemplars are primary. Thus one need not invoke a dubious theory of linguistic meaning to conclude that a principled resolution of a conflict between rival research traditions is impossible. An incommensurability of standards is sufficient.

This understanding of incommensurability is very much in keeping with the fundamental *political* metaphor underlying talk of scientific "revolutions." The breakdown in a political revolution is not primarily linguistic. People may not listen to one another, but not because they cannot understand what is being said. The breakdown is one of *authority*. There being no court of ultimate appeal, one uses persuasion, manipulation, and, finally, force.

Cognitive and Evolutionary Aspects of Kuhn's Account

Although overtly a stage theory, there are indications in Kuhn's own work that the stages are not so fundamental as it appears. One indication already noted is that Kuhn appeals to cognitive mechanisms, both psychological and social, to account for the dynamics of scientific development. For example, he claims that scientists do not learn or understand their subject matter in terms of general laws or theories. Nor do they proceed by following general methodological principles. Rather, they learn by example and appeal to these examples to validate solutions to new problems. Similarly, crisis occurs not because a general theory has been refuted, but because satisfactory solutions to new problems cannot be found. The crisis ends not because scientists believe some new theory to be true, but when sufficiently many practitioners in the field become convinced that some new solutions provide the most fruitful exemplars for future research.

From this more cognitive perspective, the stages appear as almost tautologous consequences of the simple fact that members of a scientific community do switch from one approach to another that is markedly different; for example, from a particulate to a wave theory of light. The main content of the stage theory, then, is the denial that scientific fields exhibit a cumulative development, and the contrary assertion that "radical" change does occur.

Most commentators on Kuhn's work have been aware that there is a cognitive theory underlying the stage theory. What is less often remarked is that there is also an *evolutionary* theory. It appears in the last few pages of *The Structure of Scientific Revolutions* where Kuhn attempts to explain how there can be scientific progress without there being any "truth" toward which science moves. His answer invokes an analogy with organic evolution. According to evolutionary theory, species do become better adapted to their environment, but there is no final form toward which they are evolving. Progress is measured relative to the previous state, not in terms of movement toward a final goal. Scientific progress, he claims, should be similarly conceived.

Kuhn warns his readers that "the analogy that relates the evolution of organisms to the evolution of scientific ideas can easily be pushed too far" (p. 171). And well he might, because to develop the evolutionary analogy would be to undermine the stage theory that informs everything that went before. The idea of stages itself has its roots in a *biological* analogy, that of the development of individual organisms – birth, growth, maturity, decline, death. But even when applied to other sorts of *individual* development, the analogy is weak, as in Piaget's (1954) stages of cognitive development or Kohlberg's (1973) stages of moral development. Applied to collective, social phenomena, it is even worse, as in the stages of economic and political development proposed both by Marx and later, capitalist, thinkers. For social phenomena, *evolutionary* models may be much better, but then one loses any neat progression or cycle of "stages." Later I will suggest that we retain the cognitive and evolutionary aspects of Kuhn's account, and abandon the stage theory altogether.

POST-KUHNIAN PHILOSOPHY AND SOCIOLOGY OF SCIENCE

The reaction to Kuhn's account, particularly by philosophers (Shapere 1964; Scheffler 1967), was initially very negative. The reason is not far to seek. Kuhn's account was neither representational nor rational. Moreover, he refused to recognize the barrier philosophers and sociologists of science had erected between the content and methods of science and its social structure.

For Kuhn, science is not a search for a true representation of the world. It is a puzzle-solving activity which results in something better characterized as an "interpretation" of the world. Furthermore, in appealing to psychological and sociological factors to explain the historical development of science, he implicitly rejected any search for rational principles of theory choice. Philosophers, in particular, saw Kuhn as opening the door to a dangerous *relativism.* If the choice of a research tradition is a matter not of logic but of individual judgment and social solidarity, what "objective" basis could there be for claiming that a new tradition is in any significant sense "better" than the old?

It is hardly surprising, therefore, that many philosophers and sociologists of science simply rejected Kuhn's account and went about their business as before. But others in both disciplines were more deeply influenced.

Progress and Rationality

Following publication of Kuhn's book there arose within the philosophy of science a "historical school" which took up the problem of accounting for the development of science. But by and large these philosophers retained the earlier philosophical goal of showing how the development of science could be, if not progressing toward the truth, at least objectively "rational."[7]

Laudan (1977, 1984) provides a good example of a philosopher both deeply influenced by Kuhn's account, yet deeply critical of it. Like most philosophers, he focuses on the more global interpretation of "paradigm," reinterpreting Kuhn's "research tradition" as a series of theories, each understood as a set of laws, pretty much as portrayed by Logical Empiricists. Following Kuhn, he takes it as the business of a research tradition to "solve problems." But, unlike Kuhn, he thinks that the number and importance of problems solved within a tradition can be used as an objective measure of the "problem-solving effectiveness" of that tradition, and thus of the *acceptability* of the tradition and of the theories which comprise it. He even suggests that the *rate* of problem-solving effectiveness can provide a measure of the pursuit-worthiness of a tradition. Objective *progress* is measured by problem-solving effectiveness. Scientific *rationality* is then characterized as accepting the more progressive of rival traditions while pursuing the more promising. The particular judgments of individual scientists play no role in the account. Nor is there any role for social or institutional variables.

In the end, therefore, Laudan claims to recover much of what philosophers and sociologists of science thought they had achieved before Kuhn came on the scene. He portrays science as a rational activity in which the content and methods of science are distinct from its social structure. The price he pays is giving up the picture of science as a representational activity in favor of Kuhn's account of science as a problem-solving activity.

Note that on Laudan's account there is no incommensurability among research traditions. Problem-solving effectiveness can be calculated within each tradition and the tallies compared. Thus, there are no "scientific revolutions" in Kuhn's sense. There is just the rational choice by scientists to pursue and accept the objectively more promising and more acceptable traditions. As a result some traditions fade away while others prosper.

None of the philosophical reactions to Kuhn's account has gained widespread support. In Laudan's case, for example, there are serious questions about its adequacy both as a *descriptive* account of how science has been done and as a *normative* account of how it should be done. Few historians or philosophers of science, including Laudan, have even attempted to show in detail that scientists in historically important cases actually made their choices of pursuit or acceptance in accord with standards based on problem-solving effectiveness.[8]

Nor is it clear that the supposed standards could be applied as Laudan intends. If judgments of problem-solving effectiveness were to be made by members of the rival traditions, it is pretty clear that members of each tradition would give their own tradition higher rankings. In particular, each would regard its exemplary problem solutions as being very weighty indeed. But if the scientists themselves cannot be trusted to produce an objective ranking, who else is to hand out the blue ribbons? And what difference would it make to the scientists themselves? Why should they pay attention to "outsiders"?

Science as a Social Construct

During the last decade or so, European sociologists of science have developed an alternative to the functional approach of the Mertonian school. Their approach emphasizes Kuhn's notion of a "paradigm" as an *exemplar* for guiding further research. And, like Kuhn, they insist on the indispensability of judgments by scientists, both individually and collectively. The result is an account of science which, like Kuhn's, portrays science as being non-representational, non-rational, and lacking any fundamental separation between content and methods, and social structure.

British sociologists, particularly the group at Edinburgh, have pursued a conclusion that Kuhn resisted. If it requires individual judgment to determine which are the exemplary solutions and what counts as a successful application to a new case, then there is room for what even Kuhn would regard as "non-scientific" values and interests to enter into the scientific process at the most fundamental level. These interests might be personal, professional, social, or political – or a combination. Historians and sociologists associated with the Edinburgh school have produced an imposing set of historical cases which they claim illustrates the role of various kinds of interests in the development of science.[9]

Other "new wave" sociologists of science are even more radical. It is not just that non-scientific interests "influence" scientists' judgments about theories. Rather, science is totally *constituted* by human interests and interactions. Science is simply a social construct, like morals or the law. For these sociologists, then, moral relativism and scientific relativism are of a piece. Just as there is no objective basis for preferring our Western cultural values to those of more primitive peoples, so also there is no objective basis for preferring our Western scientific belief system to theirs.[10]

Summing Up

One does not have to believe in any extra-scientific basis for scientific rationality to be quite sure that, in some sense, the pursuit of science during the last three hundred years has produced a greater understanding of the physical world than was available earlier. And indeed, progress in the past fifty or one hundred years has been dramatic. Yet, over twenty five years after publication of *The Structure of Scientific Revolutions*, we are left with at least two very different approaches to understanding how science works. Philosophers, focusing on one sense of "paradigm," have been seeking objective criteria for rational scientific progress. Sociologists, focusing on another sense of "paradigm," argue that scientific progress is no different from political or social progress.

One wonders whether there is not a middle way. I think there is. It may be characterized by a different combination of the traditional scientific virtues than that embraced by either historically inclined philosophers of science or recent sociologists of science. The winning combination, I suggest, is one that gives up the search for criteria of scientific rationality, abandons the attempt to separate the content and methods of science from psychological and sociological reality, but preserves the view of science as a representational activity.[11] A framework for such an account may be found, I suggest, in the cognitive sciences and in evolutionary models of scientific fields. As noted earlier, these components of an account of science were already present in Kuhn's original work, but were largely ignored by his philosophical and sociological heirs.

EVOLUTIONARY MODELS OF SCIENCE

There are several important negative lessons to be learned from the past twenty-five years of debate about "scientific revolutions." One is that we should abandon the *political* metaphor now embodied in the very term "revolution."[12] Thinking of scientific changes this way makes them seem more social, and more arbitrary, than they in fact are. It unnecessarily encourages the idea that scientific facts might be just like social facts. In science there is not only interaction among individuals and social groups, there is also causal interaction with the world. Although it cannot be claimed that this causal interaction uniquely determines the scientists' picture of the world, it does play a major role in the story. The task is to explain that role, not to ignore it or deny its existence.

A second negative lesson is finally to abandon the *developmental* metaphor that underlies both Kuhn's stage theory of science and its recent philosophical rivals. Given the generally low repute accorded stage theories of other social phenomena, it is surprising that Kuhn's stage theory of science has not been more severely criticized on these grounds alone. Stage theories take a biological concept that applies primarily to the development of *individual* organisms and attempts to apply it to social groups. Taking the evolutionary analogy more seriously provides an immediate explanation of why developmental models do not work very well at the collective level. The variety of possibilities for evolution in a population is very great. The actual path taken in the evolution of a biological population depends not only on variations in the gene pool, but also on myriads of details in the local environment. Much is left to chance. So there simply cannot be a single pattern of evolution for populations. The circumstances are far too variable.

In fact, we can now see why Kuhn's political metaphor and his stage theory are at odds with each other. Political revolutions, like social phenomena generally, are high variable. Any stage theory of political change is immediately subject to many counterexamples which can be avoided only by rendering the theory so vague as to empty it of useful content. This "internal contradiction" in Kuhn's views has been obscured because "revolution" was only one stage in his theory. But unless one grossly exaggerates the differences between "normal" and "revolutionary" science, the political analogy must be applied to the whole theory, or none of it. The somewhat paradoxical conclusion is that to explain scientific "revolutions" one should first stop thinking of them as revolutions in the full-fledged political sense. Rather than trying to understand "scientific revolutions," we should be thinking in terms of "the evolution of science."[13]

The Evolution of Scientific Fields

There is currently no consensus on how best to proceed to develop an evolutionary model of science.[14] Here I can only identify the major alternatives and indicate the directions that now seem to me most promising for further research.

Evolutionary theory applies to *populations* of individuals. What, then, is the population to which an evolutionary model of science might apply? The literature contains two sorts of answer to this most fundamental of questions. The answer favored by philosophers and intellectual historians of science, such as Toulmin (1972), is a population of something conceptual, such as theories or concepts. On this approach one investigates the evolution of scientific ideas. The answer favored by psychologists or sociologists, such as Campbell (1960, 1974), is populations of *scientists*. That is, what evolve are not theories but scientific communities. On this view, then, theories as such do not change; rather, scientists change the theories they espouse. The apparent evolution of theories is an artifact.

Of these two approaches I would urge pursuit of the latter. In general, the former view seems a holdover of the old intellectualist "theory centric" orientation that characterized Logical Empiricist philosophy of science. On the latter, agent-centered, view, theories do not have a life of their own. Nor do they exist in a timeless, Platonic heaven. They must be embodied in people who create and use them.

Beyond such generalities there are more specific reasons for taking scientists, not theories, to be the basic individuals in a theory of science. One is that a scientific theory of science should itself be a *causal* theory. Concepts themselves don't have any causal powers; they can't *do* anything. But people can. Of course having a particular concept, or believing a particular theory to be correct, is often a causal factor in a scientist's decision to choose one course of action rather than another. But in this ideas are no different than physical skills, say in operating some experimental apparatus, or personality traits, like ambition or risk-aversion. These traits also cannot operate on their own, but only as part of a functioning human being.

The model of biological evolution itself provides another reason for making scientists the basic individuals of an evolutionary account of science. To take a standard example, we say that the neck of the giraffe evolved from shorter to longer. But that way of talking is recognized to be mere shorthand. Individual necks did not get longer. Rather, interaction with the environment made it possible for giraffes with longer necks to have more successful offspring. So the average length of giraffes' necks increased from generation to generation. The operative causal interactions, however, were between individual giraffes and their environment, and with each other. Having a longer neck was just one causal factor, among many, that made some giraffes reproductively fitter than others.

It sometimes happens, as with Marie and Pierre Curie, that literal reproduction between scientists yields new scientists. But even in these rare cases, the child, like other pupils, had to be taught the parents' views. Being a holder of a particular theory is always an *acquired* trait. This is one important respect in which the analogy between biological evolution and the evolution of scientific fields fails to be complete. Populations of scientists sharing a theory grow largely through teaching, indoctrination, and professionalization. They also grow (or decline) through the in- (or out-) migration of people who, apart from their original training, individually choose whether or not to accept the theory in question.

On a traditional Darwinian account of organic evolution, variation is supplied primarily by mutations which are random relative to the potential contribution to fitness of the modified trait. That is, being potentially beneficial to the organism does not make a mutation any more likely to occur. New theories, on the other hand, are consciously designed to solve particular problems. That theories do often succeed in solving known problems is thus more than a matter of chance. This is true, but there is still much chance involved.

Problems, of course, logically underdetermine the theories that might be developed to solve them. But more than this, the nature of solutions devised by scientists depend to a great extent on the *resources* available to them, both cognitive and material. Cognitive resources include the kinds of models with which the scientist is already familiar through training and experience. Material resources include institutional support and the availability of knowledgeable colleagues. In the typical case, these resources were not brought together in order to solve the problem at hand. They existed for other purposes which might or might not make them useful in devising a viable solution. And even if the problem itself was selected because it seemed amendable to solution using the available resources, the solution is still not uniquely determined. It remains to some non-negligible degree a matter of chance whether a proposed solution turns out to be viable or not.

It is often remarked that contemporary science is done primarily in small research groups – not by individuals working alone. But this creates no difficulties for an evolutionary model of science. Modern evolutionary theories allow the possibility of *group* as well as individual selection. It is always an empirical matter how much selection is operating on individuals and how much on the groups to which they belong.[15]

Finally, there is the fact that originally inspired Kuhn. Change in science is sometimes quite rapid and dramatic – nothing like the slow, gradual, process described by Darwin. Here again recent work in evolutionary theory provides a ready answer. Rather than being gradual, organic evolution is now widely thought to occur in relatively short sprints which "punctuate" long periods of relative equilibrium.[16] The suggestion that these long stable periods are the evolutionary analogs of Kuhn's periods of normal science is both obvious and irresistible. One can have rapid evolutionary change without "revolutions."

Much more must be done to expand the above remarks into a full-fledged evolutionary model of science. And even more work will be required to show that such a model can genuinely explain major episodes in the history of science. Even in its present embryonic form, however, it seems to me a more promising model of scientific change than either stage theories, political models, or models of rational development.

COGNITIVE MECHANISMS

The fundamental processes in organic evolution are (i) variation, (ii) selection, and (iii) transmission. There is naturally existing *variation* in the traits possessed by individual members of a population. The fitness of individuals to leave offspring varies with variation in traits. The result is *selection* favoring those whose traits make them fitter. The relative prevalence of the favorable trait increases in the population because that trait is disproportionately *transmitted* to the next generation.

One can go far in explaining many facts about the evolution of particular species in terms of these basic processes, as Darwin first showed. Nevertheless, although by the end of the nineteenth century many people were convinced that evolution had taken place, Darwin's own account of the evolutionary process was in low repute. This was because the *biological mechanisms* underlying variation and transmission were poorly understood. The suggestions of Darwin and his followers seemed to many clearly insufficient to the task. What changed in the first half of the twentieth century is that Mendelian genetics provided a promising account of the required mechanisms. The genetical theory of natural selection succeeded where Darwin's own account of natural selection had failed.

The lesson is that for an evolutionary account of scientific progress to succeed, it must be supplemented with an adequate account of the mechanisms underlying the analogous processes of variation, selection, and transmission. There are, I think, two sorts of mechanisms. First are the biological and psychological mechanisms underlying the cognitive capacities of individual scientists. Second are the social mechanisms operative in the social and institutional environment in which scientists work and interact. For a theory of the cognitive capacities of individuals we should look to the newly emerging cognitive sciences. And for a theory of the social environment we should look to the social sciences, particularly the sociology of science. However, just as ecology is inherently more complex than genetics, so the social sciences are inherently more complex than the cognitive sciences. I will restrict my comments to the latter.

The first thing one must realize is that the designation "cognitive science" includes a diverse set of disciplines ranging all the way from cognitive neurobiology to cognitive anthropology. In between are cognitive physiology, cognitive psychology, linguistics, artificial intelligence, and cognitive sociology. For the job at hand, I am most attracted to the "lower" end of this spectrum, neurobiology, physiology, and psychology. Much research in artificial intelligence, in particular, seems to me too committed to a computational picture of thinking that often looks like a computerized version of Logical Empiricism. Humans just do not work like that.[17]

Representation

What distinguishes the contemporary cognitive sciences from earlier work on similar topics is a focus on how humans construct, manipulate, and store *representations*.[18] If we are to use this resource in attempting to understand how scientists construct, use, and learn *theories*, we must recast our thinking in appropriate ways. This reorientation makes much of the philosophical literature on the nature of theories obsolete. Most of that literature assumes that theories are to be analyzed as formal, axiomatic systems, which are a particular type of *linguistic* entity.

A standard view in cognitive psychology suggests associating theories with *families of models*, or "schemata." Particular models can, of course, be described linguistically, but there is no reason to take the models themselves to be linguistic entities. They are, rather, more abstract structures which may, however, be physically encoded in networks of neurons. In fact, most of what Kuhn says about "exemplars" holds for schemata, except that here the notion is applied to any model, not merely to those few that play a particularly central role in shaping the field.

On this approach, one gives up the idea of a "theory" as being an ontologically definite, well defined, entity. The closest one can come to a definitive presentation of a theory is in the standard, advanced textbooks. These contain most of the exemplars of the field together with a range of other models, but none could claim to present "all" the models. That is not a well-defined notion. Nevertheless, this approach does provide the right kind of underpinning for an evolutionary account of scientific change.[19]

Even in relatively specialized fields, different scientists possess somewhat different repertoires of models which they are prepared to deploy in new situations. These differences may be the result of having gone to different schools, having had different teachers, having had different interests, and having worked on different problems since completing their education. So here we have a ready source of *variation* in the acquired cognitive resources of individual scientists. This variation is surely great enough to provide a considerable part of the explanation why different scientists propose different solutions to the same problem.

Scientific Judgment

No one denies that scientists exercise their individual judgment as to which theory or general approach to the field is best. But differences among rival accounts of science are perhaps most strongly revealed when it comes to accounting for how such judgments are made. Logical Empiricists, of course, denied any official interest in how such judgments actually are made. They claimed to be uncovering a *logical* relationship between data and theories which objectively determined which of rival theories is best supported. To make a mistake in judgment on such matters is like committing a logical fallacy, such as affirming the consequent. Later philosophers of science, though officially more concerned with judgments actually made, nevertheless focused their energies on developing an objective criterion for the *rational* choice of a research tradition. Kuhn and recent sociologists of science portray such judgments as essentially *political* in nature, the result of negotiations and struggles among competing interests.

Research by cognitive psychologists into human judgment concentrates on how people actually make judgments about various subjects, including causal relationships and probabilities. But psychologists are also interested the *effectiveness* of peoples' judgmental strategies in leading them to make *correct* judgments. Similarly, they investigate how people could be more effective and why they often are not as effective as they might be given the evidence at their disposal.[20]

This research is sometimes described as an investigation into human *rationality*. But it is rationality in the purely *instrumental* sense of employing effective means (judgmental strategies) to achieve desired goals (correct judgments). The rationality most philosophers have sought has been *categorical* in the sense that whether a judgment is rational is completely independent of whether it turns out to be correct or not. In this sense rationality is like logical validity. Whether an argument form is valid or not does not depend on the truth or falsity of either the premises or conclusion. The nature of the relationship is all that matters.

From the standpoint of an evolutionary model of science, the judgments of scientists about particular models, or whole programs of research, constitute one important mechanism which scientists use to select themselves into or out of sub-populations sharing a particular viewpoint. The apparent "evolution" of the corresponding family of models is an artifact. What really happens is that the relative number of scientists in the field advocating models of a particular type increases as the result of the decisions of individual scientists to join this group. Another source of change, of course, is the training of new recruits.

Experimentation

Everyone agrees that experimentation has played a major role in the evolution of science since the seventeenth century. Once again, however, different accounts of the nature of science offer very different pictures of the nature and importance of that role.

Classical Logical Empiricist writings, for example, contain almost no references to experimentation as such. The reason seems to be that, within this framework, the only point to experimentation could be to produce new *data* which could then be used as premises in one's inductive logic. But whether the data came from experimentation or just observation would be irrelevant. The most one could say, as some more recent authors have pointed out, is that one could use an inductive logic to determine what data, if it were available, would strongly confirm or disconfirm various hypotheses. That could provide a motivation for designing experiments which, assuming the best confirmed of existing hypotheses, are most likely to yield the desirable data.

The writings of Kuhn, and of the major post-positivist philosophers of science, exhibit a similar "neglect of experiment."[21] This can only be ascribed to the general "theory centrism" of philosophers and intellectual historians. The same cannot be said of relativist sociologists of science, whose writings are full of detailed accounts of episodes in the laboratory. Almost without exception, however, the point of these investigations is to exhibit "contingencies" in the design, execution, and interpretation of experiments. The point, of course, is to show that the results of experiments do not provide an "objective" or unequivocal basis for choosing among alternative theories. Rather, like theories, experimental results are also subject to negotiations among scientists with competing interests.

Thinking of individual scientific judgment as a selection mechanism provides another account of experimentation. On this account, experimentation provides the individual scientist with a way of choosing among alternative models – a way that gives the world itself a major role, though of course not totally determining role, in the choice. This way of looking at experimentation focuses attention on the *design* of an experiment as a vital component in the scientists' *decision strategy*. The overall goal of the strategy is not merely to produce data, but to produce data that will sharply *discriminate* among rival models.

Crucial Experiments

To get at least a schematic idea of how individual judgment can act as a selection mechanism, it is helpful to consider a simple model of crucial experiments. Imagine several scientists faced with the task of choosing between two families of models (theories), which provide rival accounts of a recognized domain of inquiry. Call these rival families F_1 and F_2 respectively. We need not assume these scientists to be "disinterested seekers after the truth." Each may have strong professional interests that would be served by choosing one or the other of the two theories. But for many scientists, none of these interests will be so strong as to preclude the *possibility* of coming to choose the initially least preferred theory.

Now imagine a recognized phenomenon for which both sets of models contain some candidate models. Suppose further that this phenomenon has detectable aspects that would be clearly different depending on which model fits best. And suppose that, given existing experimental technology, it is possible to build some apparatus that can reliably detect the aspects in question. Next, we must imagine that all the scientists have sufficient knowledge of the subject matter, the rival models, and the experimental apparatus, to make the following judgments. If one of the F_1 models fits the phenomenon, then it is very likely that the apparatus will produce a measurement in the range R_1. Similarly, if one of the F_2 models fits the phenomenon, it is very likely that the apparatus will produce a measurement in the range R_2. Conversely, if an F_1 model fits, it is very *unlikely* that the measurement would be in the range R_2. And if an F_2 model fits, it is very *unlikely* that the measurement would be in the range R_1.

In these special circumstances, any of the scientists in question could subscribe to the following obvious decision strategy: If the actual result of the experiment is R_1, choose F_1 as the best fitting family of models. If the actual result of the experiment is R_2, choose F_2 as the best fitting family of models.

To see why this is an *effective* decision strategy, one need only review the conditions of the experiments as outlined above. If the F_1 family of models really does provide a good fit to the real phenomena, then it should be very likely that the experiment has result R_1. Following the simple decision strategy, a scientist would choose F_1 as the best fitting family of models – which is the correct choice. Similarly, if the F_2 family provides a good fit, it is very likely that the result of the experiment will be R_2, and a scientist following the strategy will choose F_2 – again the correct choice. In sum, no matter which of the two models provides the better fit, the circumstances described make it very likely that the scientist will make the correct choice, no matter which that happens to be.[22]

Applying this simple model of a crucial experiment requires, of course, that the scientists are correct in their judgments about which outcomes of the experiment are very likely or very unlikely, relative to the rival theories. They may not be. Or they may not all agree on their judgments, in which case only some will take the experiment as a basis for their own decision as to which theory is better. Moreover, it is important that the probabilities in question be quite extreme, that is, close to one or zero. Otherwise scientists' prior commitments will tend to dominate their choices. We may safely assume that scientists want to choose the better fitting model. But if an experiment goes against their prior commitments, and their professional interests, they are not going to switch allegiance unless they are convinced that there is very little chance that their previous choice is correct.

Finally, there is never any guarantee that anything like a crucial experiment can be designed. And, of course, the broader the rival families of models, the more difficult it will be discriminate between them. Nevertheless, good approximations of crucial experiments do sometimes occur, and the resulting judgments by large numbers of scientists may dramatically change a field in a relatively short time.[23]

TOWARD A NATURALISTIC, REPRESENTATIONAL, COGNITIVE, EVOLUTIONARY THEORY OF SCIENCE

By way of summary, I offer the following review of the major features I think future theories of science should exhibit.

Naturalistic

A theory of science must itself be a *scientific* theory. Attempts to construct accounts of science *a priori* by appeal to logic or someone's intuitions cannot succeed. There just is no place to ground a philosophically autonomous theory of science. This does not mean that one is reduced merely to describing the behavior of scientists. The goal is to be able to *explain* how science is done and why it is as successful as it is. Nor is one prohibited from making *normative* claims about how effectively to pursue scientific goals. But any such normative claims must be based on reliable empirical theories of how science in fact works, theories that connect methodological strategies with clearly articulated goals.

Representational

The models scientists construct are clearly intended by them to *represent* various aspects of the world. And it is very difficult to deny that at least sometimes they are moderately successful at this task. An adequate theory of science must make sense of this fundamental aspect of science. That Kuhn and later historically inclined philosophers of science have been satisfied with a *non*-representational account of science seems in large measure an overreaction to the obvious shortcomings of Logical Empiricism. But it is possible to have a representational account without embracing anything like Logical Empiricism.

Cognitive

The ability to construct models of complex and often remote aspects of the world is a deliberate and self-conscious extension of the evolved cognitive capacities for "mapping" their environment which humans share with many animals, particularly other mammals. In constructing their own models of science, philosophers of science should exploit this fundamental insight of contemporary cognitive science. The cognitive sciences also provide resources for constructing a naturalistic account of scientific judgment which replaces the inductive logics of Logical Empiricism.

Evolutionary

Science does not develop in a series of "stages" in any but a trivial sense. Nor does it exhibit "rational progress" in the sense of Lakatos or Laudan. The growth of science is more *evolutionary*, with individual scientists, together with their ideas, being selected from a population exhibiting considerable variation. The focus should be on the evolution of scientific *communities*, at the level of research specialities, rather than on the evolution of "ideas." The ideas follow the scientists, not the other way around. So the family of models associated with a research speciality evolves as the associated community evolves.

Conclusion

If the philosophy of science is itself a theoretical part of a science of science, then, on the above picture, it too should evolve both as a research community and, consequently, as a family of models. I have sketched its evolution since World War II and indicated the ways I would like to see it evolve in the future. Whether it does or not depends, of course, on the individual judgments of people in the field and, perhaps even more importantly, on the people who will be coming into the field in the future.

University of Minnesota

NOTES

[1] For an elaboration of this line of thinking, together with quotations and historical references, see Rorty (1979, Ch. 3). While admiring Rorty's historical analysis, I disagree with his rejection of representation as essential to knowing and, consequently, with his extreme pragmatism.

[2] One might point to "creation science" and other forms of religious fundamentalism as showing that a philosophical justification of science is still much needed. In truth, the threat from religious fundamentalism is not so much intellectual as social and political. Nevertheless, for social and political reasons it is important that the debate be joined, and prominent philosophers of science, such as Kitcher (1983) and Ruse (1982), have done so. What these philosophers have *not* attempted is to find a place completely outside of science from which to demonstrate that "creation science" is lousy science. That, I think, is impossible.

[3] I owe this lovely term to Richard Burian. The characterization of Logical Empiricism appearing in the text applies at best to the version that developed in the United States following World War II. As recent scholarship shows, the European origins of Logical Empiricism were much more complex. For some indication of how complex they were, see Friedman (1991).

[4] The circumstance was that Kuhn had agreed to publish a monograph as part of the University of Chicago Press's *International Encyclopedia of Unified Science*. These were intended to be shorter even than Kuhn's book. Kuhn's book turned out to be the second last in a series that began as a principal organ of Logical Empiricism. Ironically, Kuhn's work was one of the major contributors to the demise of Logical Empiricism.

[5] Kuhn seems to have been influenced in Wittgensteinian ways of thinking by the philosophers Stanley Cavell and Paul Feyerabend, who were his colleagues at Berkeley in the late 1950s.

[6] In this I follow Gerald Doppelt (1978) who, significantly for what follows, has also written on political philosophy. It must be admitted, however, that Kuhn himself has always been fascinated by differences in linguistic practices among different research traditions. This comes out in his accounts of his earliest encounters with the history of science (Kuhn 1977). As someone trained in contemporary physics, he found he could not really understand the writings of much earlier scientists. But this evidence for linguistic incommensurability might be an artifact of the perspective of a historian looking back at the written record of scientific changes that took place a hundred or more years ago. It does not follow that the participants in those changes experienced linguistic incommensurability.

[7] Among the more prominent members of the "historical school" in the philosophy of science are Lakatos (1978), Laudan (1977, 1984), Shapere (1984), and Toulmin (1961, 1972).

[8] A noteworthy exception is Henry Frankel (1982, 1987).

[9] These include Shapin's (1975, 1979) study of phrenology in eighteenth century Edinburgh and Don Mackenzie's (1981) study of the development of statistics in Britain in the late nineteenth and early twentieth century. For further discussion, and many further references, see Shapin (1982). The "foundational" writings of the Edinburgh school are Barnes (1974, 1982), Bloor (1976) and the papers collected in Barnes and Shapin (1979), Barnes and Edge (1982), and Collins (1982).

[10] References to this more radical sociology of science include Latour and Woolgar (1979), Knorr-Cetina (1981), the papers in Knorr-Cetina and Mulkay (1983). Among more recent works are Gilbert and Mulkay (1984) and Latour (1987).

[11] A number of philosophers of science have been exploring accounts of science that, in one form or another, seem to embrace this combination. These include Boyd (1981), Churchland (1979), myself (Giere 1988), Hooker (1987), and Ruse (1986).

[12] The political use of the word "revolution" in its modern sense seems not to have gained widespread currency until the later part of the seventeenth century (Cohen 1985, Ch. 4). The word was at that time also applied to "the scientific revolution." Before that it had long been used to refer to any cyclical phenomenon, particularly astronomical phenomena, but also political phenomena. In this older sense of the term, "the structure of scientific revolutions" could refer to Kuhn's whole stage theory, in which the stages repeat, and not merely to the "revolutionary" stage.

[13] A number of thinkers, including Donald Campbell (1960, 1974) and Stephen Toulmin (1972), have long urged this way of approaching the problem. For the most recent sustained attempt, see (Hull 1988). I have elaborated my own account somewhat in (Giere 1989) and, more extensively, in (Giere 1990).
[14] For a comprehensive and critical review of the recent literature on evolutionary epistemology in general see Bradie (1986). In addition to a wide ranging set of papers, (Callebaut and Pinxten 1987) contains an extensive bibliography.
[15] For a brief historical survey of the literature on group selection see Sober (1984), pp. 215–26.
[16] The classic reference on punctuated equilibria is Gould and Eldredge (1977).
[17] More congenial is the work of Herbert Simon and his associates on scientific discovery (Langley *et al.* 1987; Kulkarni and Simon 1988). Also promising is some very recent work based on systems with parallel distributed processing (Churchland 1989). For an up to date survey of recent work in the cognitive approach to understanding scientific revolutions see (Giere 1991) and the papers in *Cognitive Models of Science* (*Minnesota Studies in the Philosophy of Science*, Volume 15) (Giere 1992).
[18] Such things were, of course, banned from behaviorist psychology as unscientifically "mental." But the advent of the computer made that stance impossible.
[19] This view of theories is developed in chapters three through five of *Explaining Science* (Giere 1988).
[20] Standard references to the recent literature on human judgment include (Kahneman, Slovic, and Tversky 1982) and (Nisbett and Ross 1980).
[21] The phrase is borrowed from the title of Allan Franklin's recent book (1986). Others who have recently emphasized the role of experimentation in science include the philosophers of science Nancy Cartwright (1983) and Ian Hacking (1983), and the historian of science Peter Galison (1987).
[22] The above account of crucial experiments is developed at greater length in Chapter 6 of *Explaining Science* (Giere 1988).
[23] In the final chapter of *Explaining Science* (Giere 1988) I argue that the 1960s "revolution" in geology provides just such an example.

REFERENCES

Barnes, B. (1974), *Scientific Knowledge and Sociological Theory* (Routledge & Kegan Paul, London).
———. (1982), *T. S. Kuhn and Social Science* (Columbia University Press, New York).
Barnes B. and S. Shapin (eds.) (1979), *Natural Order: Historical Studies of Scientific Culture* (Sage, Beverly Hills, CA).
Barnes, B. and D. Edge (eds.) (1982), *Science in Context* (MIT Press, Cambridge).
Bloor, D. (1976), *Knowledge and Social Imagery* (Routledge & Kegan Paul, London).
Boyd, R. (1981), 'Scientific Realism and Naturalistic Epistemology', in *PSA 1980*, Volume 2, P. D. Asquith and R. N. Giere (eds.) (The Philosophy of Science Association, East Lansing), pp. 613–52.
Bradie, M. (1986), 'Assessing Evolutionary Epistemologies', *Biology and Philosophy* **1**, pp. 404–59.
Callebaut, W. and R. Pinxten (eds.) (1987), *Evolutionary Epistemology* (Reidel, Dordrecht).
Campbell, D. T. (1960), 'Blind Variation and Selective Retention in Creative Thought as in Other Knowledge Processes,' *Psychological Review* **67**, pp. 380–400.
———. (1974), 'Evolutionary Epistemology', in *The Philosophy of Karl Popper*, P. A. Schilpp (ed.) (Open Court, La Salle), pp. 413–63.
Carnap, R. (1950), *Logical Foundations of Probability* (University of Chicago Press, Chicago [2nd ed. 1962]).
Cartwright, N. (1983), *How the Laws of Physics Lie* (Clarendon Press, Oxford).

Churchland, P. M. (1979), *Scientific Realism and the Plasticity of Mind* (Cambridge University Press, Cambridge).
———. (1989), 'On the Nature of Theories: A Neurocomputational Perspective', in *Scientific Theories*, C. Wade Savage, (ed.), *Minnesota Studies in the Philosophy of Science*, Volume 14 (University of Minnesota Press, Minneapolis).
Cohen, I. B. (1985), *Revolution in Science* (Harvard University Press, Cambridge).
Collins, H. M. (ed.) (1982), *Sociology of Scientific Knowledge – A Source Book* (Bath University Press, Bath).
Doppelt, G. (1978), 'Kuhn's Epistemological Relativism: An Interpretation and Defense', *Inquiry* **21**, pp. 33–86.
Frankel, H. (1982), 'The Development, Reception, and Acceptance of the Vine-Matthews-Morley Hypothesis', *Historical Studies in the Physical Sciences* **13**, pp. 1–39.
———. (1987), 'The Continental Drift Debate', in *Scientific Controversies*, H. T. Engelhardt, Jr. and A. L. Caplan (eds.) (Cambridge University Press, Cambridge), pp. 203–48.
Franklin, A. (1986), *The Neglect of Experiment* (Cambridge University Press, Cambridge).
Friedman, M. (1983), *Foundations of Space-Time Theories* (Princeton University Press, Princeton).
———. (1991), 'The Re-evaluation of Logical Positivism', *Journal of Philosophy* **88**, pp. 505–19.
Galison, P. (1987), *How Experiments End* (University of Chicago Press, Chicago).
Giere, R. N. (1988), *Explaining Science* (University of Chicago Press, Chicago).
———. (1989), 'The Units of Analysis in Science Studies', in *The Cognitive Turn: Sociological and Psychological Perspectives on Science*, S. Fuller, M. DeMey, T. Shinn and S. Woolgar (eds.), *Sociology of the Sciences Yearbook*, Volume XIII. (Reidel, Dordrecht, 1989), pp. 3–11.
———. (1991), 'Implications of the Cognitive Sciences for the Philosophy of Science', in *PSA 90*, Volume 2, A. Fine, M. Forbes, and L. Wessels (eds.) (The Philosophy of Science Association, East Lansing, MI), pp. 419–30.
———. (1992), *Cognitive Models of Science, Minnesota Studies in the Philosophy of Science*, Volume XV (University of Minnesota Press, Minneapolis).
Gilbert, G. N. and M. Mulkay (1984), *Opening Pandora's Box* (Cambridge University Press, Cambridge).
Gould, S. J. and N. Eldredge (1977), 'Punctuated Equilibria: The Tempo and Mode of Evolution Reconsidered', *Paleobiology* **3**, pp. 115–51.
Hacking, I. (1983), *Representing and Intervening* (Cambridge University Press, Cambridge).
Hooker, C. A. (1987), *A Realistic Theory of Science* (SUNY Press, Albany).
Hull, D. (1988), *Science as Process: An Evolutionary Account of the Social and Conceptual Development of Science* (University of Chicago Press, Chicago).
Jeffrey, R. C. (1985), 'Probability and the Art of Judgment', in *Observation, Experiment, and Hypothesis in Modern Physical Science*, P. Achinstein and O. Hannaway (eds.) (MIT Press, Cambridge), pp. 95–126.
Kahneman, D., P. Slovic and A. Tversky (eds.) (1982), *Judgment Under Uncertainty: Heuristics and Biases* (Cambridge University Press, Cambridge).
Kitcher, P. (1983), *Abusing Science* (MIT Press, Cambridge).
Knorr-Cetina, K. D. (1981), *The Manufacture of Knowledge* (Pergamon Press, Oxford).
Knorr-Cetina, K. D. and M. Mulkay (eds.) (1983), *Science Observed* (Sage, Hollywood, CA).
Kohlberg, L. (1973), *Collected Papers on Moral Development and Moral Education* (Harvard University Press, Cambridge).
Kripke, S. (1972), 'Naming and Necessity', in *The Semantics of Natural Language*, G. Harmon and D. Davidson (eds.) (Reidel, Dordrecht).
Kuhn, T. S. (1962), *The Structure of Scientific Revolutions* (University of Chicago Press, Chicago) (2nd ed. 1970).
———. (1977), *The Essential Tension* (University of Chicago Press, Chicago).
Kulkarni, D. and H. Simon (1988), 'The Process of Scientific Discovery: The Strategy of Experimentation,' *Cognitive Science* **12**, pp. 139–75.

Lakatos, I. (1978), *Philosophical Papers*, J. Worrall and G. Currie (eds.), 2 Vols. (Cambridge University Press, Cambridge).
Langley, P., H. A. Simon, G. L. Bradshaw and J. M. Zytkow (1987), *Scientific Discovery* (MIT Press, Cambridge).
Latour, B. (1987), *Science in Action* (Harvard University Press, Cambridge).
Latour, B. and S. Woolgar (1979), *Laboratory Life* (Sage, Beverly Hills).
Laudan, L. (1977), *Progress and Its Problems* (University of California Press, Berkeley).
———. (1984), *Science and Values* (University of California Press, Berkeley).
Levi, I. (1980), *The Enterprise of Knowledge* (MIT Press, Cambridge).
MacKenzie, D. A. (1981), *Statistics in Britain: 1865–1930* (Edinburgh University Press, Edinburgh).
Merton, R. K. (1973), in *The Sociology of Science*, N. Storer (ed.) (Free Press, New York).
Piaget, J. (1954), *The Construction of Reality in the Child* (Basic Books, Free Press).
Nisbett, R. and L. Ross (1980), *Human Inference: Strategies and Shortcomings of Social Judgment* (Prentice-Hall, Englewood Cliffs).
Putnam, H. (1975), 'The Meaning of Meaning', in *Minnesota Studies in the Philosophy of Science*, Volume 7, K. Gunderson (ed.) (University of Minnesota Press, Minneapolis).
Rorty, R. (1979), *Philosophy and the Mirror of Nature* (Princeton University Press, Princeton).
Ruse, M. (1982), *Darwinism Defended* (Addison-Wesley, Reading Mass.).
———. (1986), *Taking Darwin Seriously* (Reidel, Dordrecht).
Russell, B. (1914), *Our Knowledge of the External World* (Allen and Unwin, London).
Salmon, W. C. (1984), *Scientific Explanation and the Causal Structure of the World* (Princeton University Press, Princeton).
Scheffler, I. (1967), *Science and Subjectivity* (Bobbs-Merrill, New York).
Shapere, D. (1964), 'The Structure of Scientific Revolutions', *Philosophical Review* **73**, pp. 383–94.
———. (1984), *Reason and the Search for Knowledge* (Reidel, Dordrecht).
Shapin, S. (1975), 'Phrenological Knowlege and the Social Structure of Early Nineteenth-Century Edinburgh', *Annals of Science* **32**, pp. 219–43.
———. (1979), 'The Politics of Observation: Cerebral Anatomy and Social Interests in the Edinburgh Phrenology Disputes', in *On the Margins of Science: The Social Construction of Rejected Knowledge*, R. Wallis (ed.) (Sociological Review Monograph, No. 27, Keele).
———. (1982), 'History of Science and its Sociological Reconstructions', *History of Science* **20**, pp. 157–211.
Sober, E. (1984), *The Nature of Selection* (MIT Press, Cambridge).
Toulmin, S. (1961), *Foresight and Understanding* (Harper & Row, New York).
———. (1972), *Human Knowledge* (Princeton University Press, Princeton).
van Fraassen, B. C. (1980), *The Scientific Image* (Oxford University Press, Oxford).
Whitehead, A. N. and B. Russell (1910–13), *Principia Mathematica*, 3 Vols. (Cambridge University Press, Cambridge) (2nd ed. 1925–27).

DIMITRI GINEV

MICRO- AND MACRO-HERMENEUTICS OF SCIENCE

The Main Aspects of a Research Program

Many recent debates on the tasks and purposes of hermeneutic philosophy of science suggest a distinction between micro- and macro-hermeneutics of science. What I refer to here is a distinction between a hermeneutic theory of effective-historical consciousness on the level of scientific community and a hermeneutic-ontological analysis of the nature of modern science. Certainly, the distinction I have in mind has nothing to do with the distinction between micro- and macro-foundations in contemporary philosophy of social sciences. Micro-hermeneutics is not to be considered as a sort of methodological individualism. Micro-hermeneutics does not strive for interpretative explanations of individual behavior. Its aim is not to interpret scientific enterprise by referring only to the "intentional properties" of individuals who are involved in this enterprise. Micro-hermeneutics is rather a kind of social holism because it starts out from the ontological primordiality of the scientific-community's modes of constructing a world of research objects.

Micro-hermeneutics of science might be defined with reference to (the study of) three basic aspects of this ontological primordiality. The *first one* is the "ontological narrativity" that makes possible the historical totality of a scientific-community's research life. The view I am going to suggest here contradicts the conception that immanent narrative coherence of a given communal life is only an outcome of the acts of interpretative imagination (such as inventing, distorting, imposing, constructing meaningful units, etc.). There is a "real narrative totality" that encompasses the research life of a scientific community. In other words, community's research life is inherently storied. From the viewpoint of hermeneutic phenomenology, the narrative of a research life is a manifestation of the original scientific community's horizon which determines the boundaries of possible theoretical thematization. The narrative lies behind all modes of a community's existence within-a-world-of-theoretical-objects. The narrative is presupposed in each interpretative (theoretical or empirical) practice of scientific community. In other words, the "story" that makes from the collection of researchers a homogeneous community is not a story-telling externally imposed upon the members' activities. Rather, the narrative is inherent in all research activities of the scientific community. David Carr (1986, p. 177) makes a similar point with respect to communal life in general when he writes: "To tell the story of community and of the events and actions that make up its history is simply to continue, at a somewhat more reflective and usually more retrospective level, the story-telling process through which the community constitutes itself and its actions."

D. Ginev and R. S. Cohen (eds.), Issues and Images in the Philosophy of Science, 87–93.

But what is this immanent narrative of scientific community about? *It is the story-telling process through which the community constitutes its region of scientific research.* As I pointed out, the narrative is pre-given to the concrete research work of a scientific community as a horizon of possible theoretical thematization that become gradually actualized in the process of this work. The transformation of this horizon into a narrative is a result of projection through which the constitution of the region of scientific research begins. Thus, the narrative is the actualization (in the sense of becoming evident) of the original community's fore-having, fore-sight, and fore-conception by means of which the "region" under study receives its structure and articulation. Furthermore, within the narrative, a community's cognitive aims and methodological orientations become explicit. At the same time, despite the possible diversification of the community into several subgroups following different methodological programs, the narrative preserves an ontological identity of the communal we-subject. The task of micro-hermeneutics is to reveal the scientific community's narrative both as a pre-given project of the actual research work and as a totality of all research possibilities within-the-region that is thematically constituted.

The *second* aspect of the ontological primordiality of a scientific community's modes of doing research work is the openness of a scientific-community's theoretical horizon. I would like to enter into the specificity of this aspect by stressing that from the hermeneutic point of view there is no closed cognitive structure. Each theory or research program is embedded into a horizon of implicit meanings and orientations. During the (potentially unending) process of new interpretations of the cognitive structure the original horizon is open to changes. The openness of the theoretical constructions of science is to be conceived as a consequence of the *radical historical finitude of the scientific consciousness* that produces these theoretical constructions. Speaking in Gadamerian terms, the embededness of each particular scientific consciousness into an open horizon has an ontological priority over the concrete epistemic acts of that consciousness by means of which the cognitive structures of scientific knowledge become articulated. Consequently, the particular scientific consciousness always brings in its epistemic acts its historically finite horizon. Thus, if one takes the whole structure – the cognitive structure as it is articulated within a historically finite horizon – one is to state that each particular scientific consciousness is characterized by a "situated transcendence": Since the whole structure (or, the cognitive-structure-being-embedded-into-an-open-horizon) is a "fore-structure of completion" (Gadamer), the cognitive structure produced by a particular scientific consciousness appeals to new interpretations of it that may close the horizon into which it is embedded. Through these new interpretations the so-called "fusion of horizons" takes place. The particular scientific consciousness transcends its horizontal situatedness and becomes an ingredient of the historical process of the "fusion of horizons." In other words, through its "situated transcendence" the particular scientific consciousness transforms itself into a "effective-historical consciousness" (Gadamer). The subject of this consciousness is the (potentially open) community of interpretations brought into being by the process of the "fusion of horizons."[1] Now, I am in a position to introduce a hermeneutic notion of scientific tradition. In contrast to the epistemological notion of tradition (e.g., Laudan's notion from *Progress and Its Problems*), the hermeneutic one does not refer to the rationally reconstructed history of science, but to that historical dynamics of science which is governed by the fusibility of horizons in which the cognitive structures of scientific knowledge become articulated in accordance with rational norms and standards. From hermeneutic point of view, the notion of scientific tradition is defined by the series of interpretations that aim at a closure of a given "fore-structure of completion." (The series of interpretations of Thomson and Tait's abstract dynamics provides an illustration of the hermeneutic notion of scientific tradition. Until the early '90s of the last century Thomson and Tait's theoretical construction received many interesting interpretations whose aim was not a revision of its mathematical formalism and conceptual apparatus but rather a completion of the horizon of possibilities left after the famous *Treatise on Natural Philosophy* (1867). The interpretations led to a new hydrodynamics (Horace Lamb's mathematical theory of the motion of fluids), a new acoustics (Lord Rayleigh's mathematical theory of sound), and a productive development of the theory of electricity and magnetism whose final outcome is Maxwell's version of classical electrodynamics.) In each new interpretation the fore-structure of completion becomes partially closed because the initial horizon is fused with a new one. Yet this partial closedness signals at the same time the rise of a new (potentially open) synthetic horizon. Scientific traditions are to be characterized by different degrees of openness with respect to the theoretical constructions that are built up within them. (Thus, the tradition of interpretations of Thomson and Tait's abstract dynamics is characterized with a much higher degree of openness as compared with the tradition of the late-nineteenth century theoretical physics that followed structural principles and hypotheses about "hidden machinery of matter." The latter tradition proved to be not open for interesting generalizations or specializations.)

(The claim of the principal openness of the cognitive structures of science does not preclude the possibility of defining a criterion for a "closed theory" from a semantic point of view. However, in so doing one should isolate the theory from its embeddedness into a horizon and treat it as a pure formal structure whose semantic interpretations constitute a closed set of models. This approach, precisely, is demonstrated by the so-called "structuralist view" of empirical theories, where the claims of empirical theories are rendered as set theoretical versions of Ramsey sentences. Against the background of Heidegger's "ontological difference" (i.e. the difference between ontical and ontological levels of reflection) there is no logical contradiction between the hermeneutic claim of the openness of science's cognitive structures and the structuralist thesis that each empirical theory is characterized by a closed set of semantic models. The hermeneutic claim is raised with reference to the hermeneutic ontology of existence, whereas the structuralist thesis is formulated in ontological terms.)

The *third* aspect of the ontological primordiality of a scientific community's modes of doing research work is the way of articulating a contextual network of methodological rules and cognitive aims. My claim is that the particular rules, norms and aims of scientific knowledge are to be specified not in the form of isolated hypothetical imperatives (as this view is developed in Laudan's normative naturalism) but rather as undetachable moments of a contextual network that is to be understood by studying the "research everydayness" of a scientific community (Ginev 1997). As a milieu of creating the interconnections among the different epistemic acts of a scientific community's research work, this everydayness is distinguished by a kind of holistic rationality that is not reducible to the instrumental rationality of the individual research work.

Reflection on these three aspects brings to light the "hermeneutical-ontological foundations" of the "cognitive order" as they are constructed by scientific communities. In opposing the normative rational reconstruction of "finished scientific knowledge," the task of micro-hermeneutics of science is to disclose the dynamics of scientific life in *status nascendi*. Vis-à-vis this task, micro-hermeneutics of science is essentially akin to the programs of "ethnography of science in the making". Yet unlike these (purely empirical) programs, the hermeneutics of science on the level of scientific communities is trying to unfold the "ontological pre-structuring" of the cognitive orders that are produced by employing theoretical and practical activities.

* * *

The beginning of the investigations into macro-hermeneutics of science can be traced back to Heidegger's theory of the "mathematical projection" as the essence of modern science, formulated in his 1936 lecture course "Basic Questions of Metaphysics." What the theorist of the "ontological difference" offers here is not a new conception about the "foundations of mathematics" but an understanding of the "metaphysical meaning of the mathematical." The starting point is that each kind of modern theoretical discourse is only "the execution and consequence" of the modern position taken toward Being and toward the way in which "beings" become constituted as "things" of experimental study and technological manipulation. Taken in such a broad perspective, the mathematical proves to be the foundation not only of modern science, but of modern metaphysics as well.

Heidegger's ideas are to be applied to the hermeneutic study of the main epistemic types of modern science conceived as types of "mathematical projection." This extension of Heideggger's conception is what I call "macro-hermeneutics of science." In my view, there is not only one modern "position taken toward Being" but a set of such positions distinguished by different sorts of "mathematical projection."

The task of macro-hermeneutics is twofold. On the one hand, it is to provide an interpretative account of modern science as a continuum of theoretical modes of Being-in-the-world by studying the existential genesis of the epistemic types of science. On the other hand, macro-hermeneutics should study the epistemic types of modern science as a diversity of general thematizing projects.

The task of a macro-hermeneutic study of science is to be understood in the context of the debates on the so-called "double hermeneutics." The central issue of these debates is the question, whether the unavoidability of using theoretical interpretations of objects constituted by self-interpretations of human agents may legitimate a special epistemological status of the human sciences. Authors like Hubert Dreyfus and Charles Taylor, who support new versions of Diltheyan dichotomism, claim that the "double hermeneutics" (theoretical interpretations of self-interpretative objects) defines the epistemological *differentia specifica* of the human sciences. On the other hand, authors like Richard Rorty and Joseph Rouse reject any significant epistemological difference between natural and human sciences by raising the radical hermeneutic argument that the interpretative constitution of research objects is a universal feature of all types of science, and, as a consequence, there is not stronger sense in which hermeneutic interpretation is specific to the human sciences. (In addition, these authors claim that the self-interpretation is not a fact that can distinguish human beings as research objects.) Beyond these two extreme positions of neo-Diltheyan dichotomism and radical hermeneuticism, two authors – Mary Hesse and Karl-Otto Apel – who were involved in the debates suggest the idea of the continuum of scientific attitudes. According to Hesse (1980), regarding the role of hermeneutic interpretation, all the scientific disciplines lie on a continuum between the poles of interpretative constructivism and mathematical objectivism. Although from a quite different philosophical standpoint, Apel (1992) defends the same idea by claiming that there is a continuum of scientific attitudes between the poles of nomological theorizing and critical hermeneutics. The idea of a continuum avoids both the revival of old epistemological dichotomies and the post-epistemological unification of science by means of radical-hermeneutic arguments. Yet from my point of view, this idea should be articulated in the framework of hermeneutic phenomenology. In outline, the reasoning is this: One can characterize the pre-scientific mode of concernful everydayness as a zero-level of theorizing. Against the background of this zero-level, a theoretical mode of Being-in-the-world is to be defined as a general scientific attitude, which objectifies the world through developing a specific thematizing project. With respect to the degree of this objectification, the theoretical modes of Being-in-the-world lie on a continuum between the pre-scientific concernful everydayness and the total formalization of the world. Thus, one can reconstruct the continuum by analyzing the ways of objectifying the pre-scientific world in accordance with the different thematizing projects (Ginev 1995).

It is a basic aim of this program to argue that the general theoretical modes of Being-in-the-world can be reconstructed not only in terms of hermeneutic phenomenology, but in terms of a holistic epistemology as well. In this case, one is looking for a continuum of general epistemological types of science. It is the correspondence between the hermeneutic continuum of thematizing projects and the continuum of epistemological types that invites the search for a complementarity between hermeneutics and epistemology.

To a certain extent, my distinction between micro and macro-hermeneutics echoes the recent disputes over the "hermeneutics of scienceworld." According to authors who are closely related to approaches in analytical philosophy of science, the aim of the hermeneutical study of science should be limited to an internal interpretation of the theoretical constructions of "scienceworld" (Eger 1993). The hermeneutic investigations of these authors lead to a specific (non-referential) semantics of "texts of the book of science." Bevilacqua and Giannetto (1995) are right when stressing that the hermeneutical study of science being so limited touches only the "fore-conceptions" in constructing theories and the linguistic "fore-sights" of science in the making. However, it does not deal with the "fore-having" of research work, which belongs not to the "scienceworld" but to the pre-scientific lifeworld. Against the background of the disputes over the "hermeneutics of scienceworld" one can distinguish between hermeneutic semantics of the theoretical constructions in science and cultural hermeneutics of illuminating the existential meaning of science by analyzing lifeworld contexts. In a similar vein, I draw elsewhere the distinction between moderate and radical hermeneutics of science (Ginev 1995). The representatives of the former do not go beyond the "epistemologically-centered paradigm." Their aim is a hermeneutic extension of this paradigm by shifting the boundary between subject and object of scientific knowledge. By contrast, the representatives of radical hermeneutics are trying to elaborate a "hermeneutically-centered paradigm" in philosophy of science by surmounting epistemology. Yet the hermeneutical issues in studying the scientific-community's lifeworlds are neglected by both camps. I completely agree with Martin Eger's claim that the issues raised in the investigation of the lifeworlds of scientific communities can bridge the gap between hermeneutic semantics of scientific languages and cultural hermeneutics of the roots of science in the pre-scientific lifeworld (Eger 1995).

In my view, micro-hermeneutics of science bears an essential resemblance to what Bevilacqua and Giannetto qualify as semantic hermeneutics of the "texts in scienceworld." But it includes also Eger's hermeneutics of scientific lifeworlds. On the other hand, macro-hermeneutics of science is closely related to what several authors call cultural hermeneutics of modern science (or, hermeneutics that is relevant to the whole corpus of modern science). Yet, despite these resemblances there is an important difference between the two distinctions. The distinction between semantic hermeneutics of theoretical constructions and cultural hermeneutics of the "book of science" simply parallels the distinction between "context of justification" and "context of discovery." By contrast, the distinction between micro- and macro-hermeneutics refers to problematique that is independent of these contexts. In other words, both micro and macro-hermeneutics are defined in the context of a hermeneutico-phenomenological problematique which is irreducible to both the semantics of theoretical constructions and the cultural (and socio-psychological) aspects of research work. Furthermore, only within this context does the dialogue between hermeneutics and epistemology (on a macro-level) become possible.

This paper has moved rapidly over a variety of ideas that are not familiar in analytical philosophy of science. Its purpose is to serve as a portrayal of the dominant paradigms of hermeneutic philosophy of science.

University of Sofia, Bulgaria

NOTE

[1] In my view, Bohr's initial formulation of the principle of correspondence can be construed as an attempt to actualize the role of effective history of scientific tradition in constructing new theories. It was Bohr's intention to show that though no ordinary spatio-temporal description of quantum processes in a classical theoretical framework is possible, the interpretative transition from quantum theory to classical electrodynamics still exists. The claim that every description of atomic processes in terms of space and time must be interpreted in a framework of ideas which have been introduced and defined by the classical theory, requires not a "rational reconstruction" of scientific development but a historical hermeneutics of science.

REFERENCES

Apel, K. O. (1992), 'The Hermeneutic Dimension of Social Science and Its Normative Foundation', *Man and World* **25**, pp. 247–70.

Bevilaqcua, F. and E. Giannetto (1995), 'Hermeneutics and Science Education: The Role of History of Science', *Science and Education* **4**, pp. 115–26.

Carr, D. (1986), *Time, Narrative and History* (Indiana University Press, Bloomington).

Eger, M. (1993), 'Hermeneutics as an Approach to Science: Part I and II', *Science and Education* **2**, pp. 1–29, 303–28.

———. (1995), 'Alternative Interpretations, History and Experiment: Reply to Some Commentaries', *Science and Education* **4**, pp. 173–88.

Ginev, D. (1995), 'Between Epistemology and Hermeneutics', *Science and Education* **4**, pp. 147–59.

———. (1997), *A Passage To The Hermeneutic Philosophy of Science* (Rodopi, Amsterdam, Atlanta).

Hesse, M. (1980), *Revolutions And Reconstructions In The Philosophy of Science* (Harvester Press, Brighton).

ROM HARRÉ

THE REDUNDANCY OF SPACETIME: SPECIAL RELATIVITY AS A GRAMMAR AND THE STRANGENESS OF 'c'

The argument that the indexical aspects of spatial and temporal concepts are ineliminable and incompatible with an interpretation of spatial and temporal relations in terms only of the locations and moments of a spatio-temporal manifold that is independent of the material system of the world would be greatly strengthened if it could be shown that the same conclusion can be arrived at from an analysis of the conceptual structure of physics itself. I shall try to show that by adding an argument drawn from the principle of covariance to the argument from indexicality, an overwhelming case for the redundancy of any form of independent spatio-temporal manifold can be made *that is conceptually tied to the level of covariance at which the proof is obtained.* For example, it can be shown that flat Newtonian space and time are redundant at the level of covariance expressed in Galilean relativity, that is for frames of reference in relative, rectilinear uniform motion to one another. I take the principle of covariance to assert that the laws of nature are of the same form relative to whatever frame of reference they are measured, in a set of such frames that are mutually related by some such relation as relative spatial or temporal displacement, or that differ in relative uniform linear velocity. And this, even though the *measures* of the common, frame-independent process the laws describe differ from the measures obtained by investigators in other frames of refence related in the relevant way. If we want to know the results obtained by investigators in another frame of reference we can translate them into the terms of our frame by the use of a coordinate transformation, serving as translation manual. Of course these are measures of the same process only if the laws of nature that describe the process are covariant under the very same coordinate transformation that we used to translate the exotic measures. Only the process whose measure is 'c' yields the same result in every frame and so is in need of translation. The concept of covariance is applicable only on the assumption of an independent spatio-temporal manifold relative to which the laws of nature are of the same form. But if the laws of nature are of the same form regardless of their relations to some alleged independent manifold then that manifold plays no role in physics and can be discarded as redundant.

To refute all absolutist interpretations of the spatial and temporal concepts of physics, the Leibnizean and other conceptual arguments against the possible existence of an independent spatial, temporal or spatio-temporal manifold are not enough. Even if there were no independent manifold there might be a privileged system of spatial and temporal relations embedded in some material substrate, which could serve the same absolutist role. We shall need

D. Ginev and R. S. Cohen (eds.), Issues and Images in the Philosophy of Science, 95–105.

Einstein's elegant argument to show the redundancy of the electromagnetic aether to complete the task.

A frame of reference will be taken to be a multidimensional set of material entities, the relations between which do not change during the relevant time. How do we know they do not change? Perhaps because there are no signs of a causal process which would alter their mutual relations, or perhaps because there is a larger frame to which their relative positions can be referred, and these relations can be seen to be conserved. But the larger frame will need a yet larger frame to guarantee its integrity. And so on. So the notion of a frame of reference is a pragmatic or working notion; defined *ad hoc* for some purpose. An inertial frame is a frame of reference in which a body subject to no impressed forces is at rest or moves in a straight line with respect to the fixed bodies of the frame. How do we know that a body is not being acted on by an impressed force? Because it is at rest or moving rectilinearly with respect to the fixed bodies of the frame. The circularity is obvious, but need not trouble us in practice. Again in the concept of an inertial frame we have another pragmatic *ad hoc* concept.

COVARIANCE IN PERSPECTIVE

Let us suppose that there is a manifold of locations and moments independent of the material system of observed things and events. Let us call this the 'absolute background'. It might be immaterial (Newton) or material (the luminiferous aether). One important aspect of the history of physics is the progressive realisation that the processes of nature are indifferent to the ways that physical processes might be related to that absolute background. Instances of a type of physical process might be occurring at different places, at different times or both. They might be moving with different uniform velocities relative to that background, and so to each other. They might be rotating at different angular velocities relative to it, and so on.

The history of physics discloses a progressive realisation that physical processes are indifferent to differences in location, moment or uniform linear velocity with respect to any such imagined absolute background, and so to a difference of the same type from each other. An elegant explanation of the ubiquity of these principles of indifference is that there is no such background, material or immaterial. This is the thesis of the redundancy of the hypothesis that there exists an absolute space, time and space-time for the conduct of physics.

From Cusa we have a principle of the indifference of the unfolding of material processes to the place or moment at which they occur. From Galileo we have a principle of the indifference of the unfolding of mechanical processes to their velocity relative to some absolute background. Finally from Einstein we have a principle of the indifference of all physical processes, mechanical and electromagnetic, to their relative velocity to some absolute background.

Each level of indifference can be expressed as a principle of covariance, in that the laws of nature describing physical processes take the same form under progressively more restrictive coordinate transformations. Displaying the development of relativistic physics as a progressively more stringent application of the covariance requirement involves, in a queer way, the retention of the absolutist *concepts* for the exposition of relativity. We set up the progressible tighter relativity 'theories' by affirming, step by step, 'No change in the laws of nature with respect to changes in certain absolute positions, moments or motions, in an imagined basic manifold or set of manifolds.' Expressing the laws of nature in such forms as meet the covariance conditions for each level of stringency, we can then safely ignore such absolutes in physics. We can throw away the ladder once we have ascended it! Making the totality of the laws of physics, mechanics, electromagnetism, quantum mechanics, and so on, covariant under the Lorentz transformation, is just to make them independent of the most general candidate for absolute spatio-temporal manifold, the Minkowski space-time. Which is now redundant.

THE PROGRESSIVE ELIMINATION OF ABSOLUTE LOCATION, ABSOLUTE MOMENT, AND ABSOLUTE LINEAR UNIFORM VELOCITY FROM PHYSICS: CUSA TO EINSTEIN

The elimination of these concepts from physics begins with the *De docta ignorantia* of Nicholas of Cusa (1440). One condition for the possibility of establishing an absolute location would be that it should be able to identify the 'centre of the world' and locate oneself in relation to it. But Cusa shows that any material thing will serve as a marker for the centre of a system of spatial reference.

We can explain the cosmology of Nicholas of Cusa with the help of the idea of 'perspective'. Cusa's analysis of the foundations of knowledge in *De docta ignorantia* leads to the conclusion that all knowledge is perspectival, that is, infected with the point of view from which it is obtained. For philosophy of physics the central Cusan insight is that all points of view from which to learn about the physical universe are equivalent. The view of the universe from our point of view is infected with our perspective. The views from each of the indefinitely many other points of view are similarly infected. None of these points of view is privileged as a location from which a true view is to be obtained. Our ignorance, which is profound, is irremediable. This conclusion is based on two arguments.

The first argument was repeated by both Copernicus and Galileo. The situation of the cosmic observer is likened to that of someone on a ship. '. . . if someone did not know that a body of water was flowing and did not see the shore while he was on a ship in the middle of the water, how would he recognize that the ship was being moved?" (Cusa 1440: II, 12:117). It would be natural for such a person to take his position on the ship as the physical and geometrical centre of the world. Cusa continues, 'Because of the fact that it would always seem to each person (whether he were on the earth, the sun, on another star) that he was the 'immovable' centre, so to speak, and that all other things were moved: it would assuredly be the cause that if he were on the sun, he would fix a set of poles in relation to himself; if on the earth, another set; on the moon another; on Mars, another, and so on' (Cusa 1440 II, 12:117).

This argument leaves the concept of 'the centre of the universe' still intact. It shows only that no human observer can say with certainty where that centre is and whether he is standing at it. Cusa's second argument goes to deny the intelligibility of the very concept of the centre of the universe. This is the famous argument that as far as cosmology is concerned the centre and circumference of the universe are identical. The reasoning turns on the question of the boundedness or unboundedness of the cosmos. 'For if it [the cosmos] had a centre, it would also have a circumference . . . hence it would be bounded in relation to something else, and beyond the world would be both something else and space' (Cusa 1440 II, 11:114). But if the universe encompasses everything it is unbounded and so 'cannot be enclosed between a physical centre and a circumstance'. For a circle the 'maximum' is the circumference and the 'minimum' is the centre: 'For with reference to motion we do not come to an unqualified minimum, that is a fixed centre. For the minimum must coincide with the maximum, therefore the centre of the world coincides with the circumference. Hence the world does not have a circumference.' Without a fixed circumference there can be no fixed centre, so any point can be chosen as a centre. To suppose there is a fixed centre is to suppose something contradictory.

From this brief sketch of Cusa's seemingly mystical but actually hard-headed geometrical cosmology a quite modern sounding argument can be drawn, without, I believe, much distortion. It follows from the cosmology that (1) there is no physical centre. Wherever an observer is stationed will appear to be such a centre but the argument shows that appearance is not a registration of a true physical centre. A privileged geometrical centre could only be identified if it were possible to identify a physical centre and to colocate the two. But there is no physical centre. So (2) there is no physically identifiable location for a privileged geometrical centre. But (3) all locative acts must be referred to some geometrical centre. Taking (2) and (3) together we reach the physically significant conclusion that (4) there are no privileged locative acts.

The next step involves the explicit formulation of Galilean relativity, as a principle of indifference, or covariance. It is not until relatively recently that the Galilean transformation was algebraically formulated, so that data obtained by the measurement of some process can be transmitted to an observer on another frame of reference, moving with uniform velocity relative to the former as if they had been obtained by the dweller in the second frame. Here is Galileo's presentation of covariance.

Shut yourself up with some friend in the main cabin below decks on some large ship, and have with you there some flies, butterflies, and other small flying animals. Have a large bowl of water with some fish in it; hang up a bottle that empties drop by drop into a wide vessel beneath it. With the ship standing still, observe carefully how the little animals fly with equal speeds to all sides of the cabin. The fish swim indifferently in all directions; the drops fall into the vessel underneath; and in throwing anything to your friend, you need throw no more strongly in one direction than in another, the distances being equal; jumping up with your feet together, you pass equal spaces in every direction. When you have observed all these things carefully (though there is no doubt that when the ship is standing still everything must happen in this way), have the ship proceed with any speed you like, so long as the motion is uniform and not fluctuating this way and that. *You will discover not the least change in the effects named, nor could you tell from any of them whether the ship is moving or standing still* (Galileo 1632, pp. 186–7).

The formulation of the Galilean transformation (the rules for transforming the results of measurements made on a physical system in one frame of reference into the results that would have been obtained by measuring the same properties of the same physical system in another frame of reference moving with uniform linear velocity relative to the first) had to wait for their formulation until the late nineteenth century.

Before the innovations suggested by Einstein, it was known that the laws of mechanics are Galileo indifferent but it was thought the absolute motion could be detected relative to an electromagnetic aether. This idea can be expressed in terms of the boat and stone analogy. If one is boating on the surface of a lake and the shores are hidden in the mist, it is possible to determine one's velocity relative to the stationary water in the lake by dropping a stone overboard. The stone will produce concentric rings of ripples on the surface of the lake whose velocity is independent of the both the vertical and horizontal components of the velocity of the stone. So the velocity of the ripples will be independent of the velocity of the boat, since it is this which endows the stone with a horizontal component of velocity. The 'absolute' velocity of the boat with respect to the lake can be determined simply by observing the relative velocity of the boat with respect to the motion of any one of the ripples. We could see this as an analogy behind the setting up of the Michelson-Morley experiment. To take the next step, both Minkowski space-time as the abstract post-Newtonian absolute background and the luminiferous aether as the best candidate for a material absolute background must be disposed of.

An Interpretation of Einstein's Way of Unifying the Laws of Nature as Disposing of Minkowksi Space-time as a Candidate for the Absolute

If we can find a coordinate transformation under which the electromagnetic laws are covariant for frames in relative motion to one another, we should be able to work out a corresponding metaphysical principle of indifference. There is just such a transformation, namely the Lorentz transformation. To express this in an indifference principle we shall need to imagine a possible absolute space and time with respect to relocation in which, and motion relative to which, the electromagnetic processes are indifferent. This imagined background is the space-time invented by Minkowski.

The troubling situation for Einstein was that the laws of mechanics were known to be covariant with respect to the Galilean transformation but those of electromagnetism were not (Voigt's theorem of 1891). On the other hand, under the Lorentz transformation the laws of electromagnetism are covariant but the mechanical are not.

Which set of laws should we favour? To which space and time should we suppose the processes of nature to be indifferent – Galilean transformation privileged Newtonian Space and Time. The Lorentz transformation we can see in hindsight privileged the Minkowskian space-time. Which absolutist background was to be declared redundant? The only material candidate was the luminiferous aether. Einstein thought that there was reason to deny that the electromagnetic aether served any useful purpose, but his argument was not an indifference argument.

If a version of the laws of mechanics, covariant under the Lorentz transformation could be created by revising such leading mechanical parameters as proved necessary, the whole of physics would be unified at the level of Galilean relativity, that is to the level of the processes described by laws which would be indifferent to motion with respect to the corresponding absolute manifold, Minkowski space-time. And so that candidate for the absolute must be declared redundant.

There still might be some relation to some absolute material background, for example relative acceleration of processes, one to another and so to the absolute frame, were there to be one, to which the laws of nature and hence the processes they describe are not indifferent.

Einstein's Argument Against the Idea That There is a Absolute Material Background Manifold

If there were to be a candidate for an absolute material background it would surely be the electromagnetic aether, the alleged medium by which light is propagated. Einstein's reason for declaring this 'stuff' redundant had nothing to do the Michelson-Morley experiment, but was based on an analysis of the seemingly paradoxical role it played in the explanation of electromagnetic induction. I quote the first paragraph of this famous paper in full.

It is known that Maxwell's electrodynamics – as usually understood at the present time – when applied to moving bodies, leads to asymmetries which do not appear to be inherent in the phenomena. Take, for example, the reciprocal electrodynamic action of a magnet and a conductor. The observable phenomenon here depends only on the relative motion of the conductor and the magnet, whereas the customary view draws a sharp distinction between the two cases in which either the one or the other of these bodies is in motion. For it the magnet is in motion and the conductor at rest, there arises in the neighbourhood of the magnet an electric field with a certain definite energy, producing a current at the places where parts of the conductor are situated. But if the magnet is stationary and the conductor in motion, no electric field arises in the neighbourhood of the magnet. In the conductor, however, we find an electromotive force, to which in itself there is no corresponding energy, but which gives rise – assuming equality of relative motion in the two cases discussed – to electric currents of the same path and intensity as those produced by the electric forces in the former case.

Examples of this sort, together with the unsuccessful attempts to discover any motion of the earth relatively to the 'light medium', suggest that the phenomena of electrodynamics as well as of mechanics possess no properties corresponding to the idea of absolute rest. They suggest rather that, as has already been shown to the first order of small quantities, the same laws of electrodynamics and optics will be valid for all frames of reference for which the equations of . . . .

The analysis is directed essentially to throwing into doubt the importance or even the role, of the aether as a universal and absolute material foundation for electromagnetism. The structure of the argument is perhaps best appreciated diagrammatically.

A (moving magnet)		**B** (moving coil)	
Phenomenon	: Current	Phenomenon	: Current
Explanation	: Electric Field	Explanation	: Electromotive Force
Model	: Aether Tension	None	
Common ontology	: Aether		

Einstein's argument: The phenomena of A and B reveal the same electromagnetic effect and involve the same relative motion – but the explanations built upon the assumption of the electromagnetic aether as a common ontology and an absolute manifold of places are different. Should not symmetrical phenomena have symmetrical explanations? The source of the asymmetry in explanations is the aether. Abandon it.

We have now shown that absolute space and time, however conceived, are redundant for physics up to and including the Galilean degree of indifference.

RELATIVITY AS THE GRAMMAR OF SCIENTIFIC COMMUNICATION

Grammar of What?

What then is one learning if one learns the Special Theory of Relativity? Is it a physical theory in the sense that a theory of blackbody radiation is a physical theory? With the absolute background in both its abstract and its material manifestations declared redundant up to the Galilean degree of indifference, a possible role as the physics of Minkowski space-time is ruled out. Yet it is of immense importance to physics. I shall try to show that it is best interpreted as a grammar for a certain kind of communication.[1]

A set of coordinate axes, 'bolted' to a frame of reference, can be used to describe locations in that frame unambiguously. A coordinate transformation is a set of rules for rewriting the results of making measurements on some physical process in one frame of reference as if they had been made in another, related to the former in some determinate and physically relevant way. For example, the Galilean transformation permits an observer in one frame of reference to translate the results of measuring the motion of a body relative to that frame into the results that would have been obtained by another observer making measurements of the motion of the very same body in another frame of reference moving with uniform linear velocity relative to the first frame. Or to put the point slightly differently, the Galilean and Lorentz transformations allow observers in different frames of reference to communicate their results one to another in such a way that their results are mutually intelligible and do not contradict one another. The reason why we use the Lorentz rather than the Galilean transformation is just that that transformation guarantees the widest possible range of physical processes to be accommodated in a unified physics.

In examining Cusa's, Galileo's and Einstein's (Special) relativity theories we have compared two frames of reference which differ from one another in at least the following physically relevant ways:

- The origin of the spatial axes of the second frame does not coincide with the origin of the axes of the first frame.
- The origin of the temporal axis of the second frame does not coincide with the origin of the temporal axis of the first frame.
- The second frame is moving with uniform linear velocity with respect to the first frame.

We have not considered cases in which frames are mutually accelerating, for example the case in which the second frame is rotating with uniform angular velocity with respect to the first frame.

The Thesis of Reciprocity

The measurements made by an observer in the second frame have as much validity for physics as those made by an observer in the first frame. Since the postulate applies to the measurements obtained in any two frames of reference, it is equivalent to the principle of relativity, that there is no privileged frame. Expressed mathematically the thesis of reciprocity amounts to the requirement that coordinate transformations always form a group, that is by repeated transformation through a sequence of coordinate systems suitably fixed to frames of reference related in the way required for the application of the transformation, the original measurement's results are always recovered when we reach the frame from which we started the sequence of transformations.

So far these principles represent the grammar of exchanges of information between observers. Built in to this is the assumption that all observers measure the properties of one and the same process. This assumption enters into the physics of the universe through its expression as the principle of covariance, that the forms of the laws describing physical processes are independent of the frames of reference within which they are studied.

But where is the physics in all this? So far we seem to be recording only the rules of grammar for successively communicating the results of observations and measurements from one observer to *any* other, whose situations differ either by location, by epoch or by uniform relative rectilinear velocity (and perhaps by the scale of the measuring devices they are using). Why this grammar and not some other?

The answer highlights the first fragment of physics reflected in coordinate transformations. It amounts to the thesis that there is no causal relation between a frame of reference and a physical process that occurs in it. Or to put the matter more clearly, only for such frames is a coordinate transformation under the constraint of covariance possible.

The second fragment of physics that has to be accommodated in whatever grammar we devise, and is responsible for the Lorentz transformation taking the form that it does, is the fact that the velocity of light, *as measured*, is the same in all frames of reference. The one measurement result that can be sent from observer to observer without being subject to the relevant transformation is the velocity of light. It follows also that we cannot use the physics of light propagation to identify a privileged frame of reference. But are not Maxwell's Laws the laws of just that process? How can they possibly be covariant if the measure of 'c' is the same in all frames of reference?

By showing that the best reading of the relativities is as grammars for exchanging information obtained by the making of measurements in different frames of reference, we undercut any argument for a substantival interpretation of space and time. By Ockham's Razor there is no need to introduce a manifold of locations and a manifold of moments, or a joint manifold of locations at moments, in addition to the things and events of the material world. Spatial and temporal concepts refer to nothing but relations between things and between events, relations that reduce to rules of mutual exclusion and coexistence.

THE EXCEPTION: THE TROUBLE WITH LIGHT

The result of measuring the velocity of light propagation is the same in every frame. So there is no need to use the rules of the relevant transformation grammar to enable dwellers in different frames to incorporate the measurements of this velocity made by every observer in the histories written by each. But for those processes which are indifferent in their relations to the absolute, and so were the same for all those who wished to measure their properties in whatever frame, the results of their measurements could only be coordinated by the use of the transformations. The results of the measurements of the velocity of the propagation of light would be wildly paradoxical if the process of propagation was unaffected by its position, epoch or velocity relative to the absolute background. (Is it that it is affected in just such a way as to compensate for the effect on measurement by different frame dwellers? This has never seemed to me remotely plausible.)

If, as Einstein's symmetry argument suggests, there is no aether, then there is no material mechanism for the propagation of light. So the fact that its measure is the same in all frames and that it is in no need of a translation grammar to be incorporated into all histories is not paradoxical. However, it is utterly mysterious. Thus, having shown:

- that independent manifolds of spatial places, temporal moments and spatio-temporal 'location' are redundant, and that there is no good argument for an immaterial absolute background;
- that measurements conducted on the same process in different reference frames can be coordinated and mutually translated;
- that there are no privileged frames of reference, since there is no good argument for supposing there is an absolute material background;
- by the covariance test being able to identify which processes are unaffected by their relations either to an absolute immaterial or to a material background, and found[2] that all physical processes are of this sort;

it is surely astonishing that *c* is a measure of a process and that it can be exchanged between inhabitants of different frames without modification. The same processes of mechanical propagation have different measures, why does the same process of luminiferous propagation have the same measure? The process, the properties of which are being measured by different frame dwellers in a certain well defined set of frames, is *the same process*. What guarantees to all frame dwellers that it is the same process? The covariance of the laws which describe that process under coordinate transformation. So if there is covariance, that is the process measured is the same for each frame dweller, then there must be a grammar for coordinating the measures each obtains for the same property of the common process. Sameness of process is expressed in the coordinate transformation under which the laws are covariant. It must be that very same transformation which serves to coordinate the results of measurements, made, in principle, by any frame dweller whose frame is in that relation to all others in the set. It seems quite astonishing that this simple and elegant relationship does not hold for the propagation of light.

There remain a goodly number of possible frame relations that need to be subjected to the same kind of treatment, until the spectre of absolutism is finally exorcised. For example:

- The 'screw sense' or 'handedness' of the second frame is the opposite of that of the first frame.
- The units of measurement in the second frame bear a constant relation to the units of measurement in the first frame.
- The second frame is rotating with respect to the first frame.
- The second frame is moving with uniform linear acceleration with respect to the first frame.
- The second frame is rotating with uniform angular acceleration with respect to the first frame.

Oxford University

NOTES

[1] Of course this is not a novel or original suggestion. It is worked out in detail by Lucas (1973: Section 44) and by Lucas and Hodgson (1990: Ch. 2).

[2] 'Found' meaning it turned out that Maxwells Laws are covariant under LT, Newton's Laws can be modified to be so; and that all subsequent laws of nature are so formulated that they are, and still fit the phenomena.

REFERENCES

Einstein, A. (1904), 'On the Electrodynamics of Moving Bodies', in *The Principle of Relativity* (ed.) (Dover, New York), p. 37.

Galileo, G. 1632 (1953), *Dialogues Concerning the Two Chief World Systems*, trans. S. Drake (California University Press, Berkeley).

Lucas, J. R. (1973), *A Treatise of Time and Space* (Methuen, London).

Lucas, J. R. and P. E. Hodgson (1990), *Spacetime and Electromagnetism* (Clarendon Press, Oxford).

Nicholas of Cusa 1440 (1954), *De Docta Ignoratia*, trans. G. Heron (Routledge and Kegan Paul, London).

PATRICK A. HEELAN

CONTEXT, HERMENEUTICS, AND ONTOLOGY IN THE EXPERIMENTAL SCIENCES

I want to challenge the metaphysics of modern philosophy that is so deeply embedded in our language and culture, particularly when it is presupposed by modern culture's most successful enterprise, modern science. I will show that everyday culture, artistic presentations, and experimental science give evidence that in practice the metaphysics of modern philosophy is often shunted aside and overcome in favor of a more hermeneutic approach. Among philosophies of a hermeneutic kind one thinks of the works[1] of W. Dilthey, M. Heidegger, H-G. Gadamer, P. Ricoeur, and those influenced by them,[2] as well as, perhaps, the later E. Husserl. I propose to take Heidegger as the key figure of this group, interpreted where necessary, since natural science is not well represented within this tradition, by works drawn from related sources. It is difficult, and probably rash, to challenge the entrenched establishment of modern philosophy rooted in the grand rational tradition of Greece with the sling-shot of Heidegger & his friends. A futile job? A difficult one. Goliath stands on the high ground of well-established facts, objective theory, logical argument, and awe-inspiring technological control, while David is perceived to be a native of secret and hidden places protected by a shield of myth and nihilism that border on the irrational. Ironically, whatever you may think about myth and nihilism, they are not far from what the general educated public believe they spot in today's culture of scientism. Going to the heart of the matter, the problem is: Whose rationality? Both the fundamental uneasiness of analytic philosophers with continental philosophy and the general public's suspicion of institutional scientism turn on criteria of rationality and, of course, its contrary, irrationality. Questions are raised: What are the general characters of rationality? Are there many kinds of rationality? Who can pronounce on their legitimacy or illegitimacy – scientists? philosophers? (Who among them?) The citizen rulers of a republic of reason? Such questions are fundamentally philosophical questions and they have no agreed answer today. The message of this paper is: how much better would the contemporary need for universal rational guidance in science, art, and in everyday affairs be served by a hermeneutical philosophy than by the contemporary progeny of the rationalist tradition of Greek philosophy and modern scientism!

HERMENEUTICAL PHILOSOPHY

I want to test the limits or edge of modern and contemporary philosophy and a series of philosophical experiments. In the words of Gary Larson's well-known cartoon series, I am concerned with the "far side" of modern philosophy. The experiments will attempt to use hermeneutical philosophy stemming

D. Ginev and R. S. Cohen (eds.), Issues and Images in the Philosophy of Science, 107–126.

from the work of Dilthey, Husserl, Heidegger, Gadamer, Ricoeur, and others – to address some critical questions in an imagined laboratory of mixed everyday, artistic, and scientific experience. Some of the questions I will address are also being addressed by American philosophers inspired, say, by the works of J. Dewey, C. S. Peirce, L. Wittgenstein, and R. Rorty, or by the works of various sociologists and decision theorists concerned with "bounded rationality" or the "social construction of reality." I am at odds in some fundamental, infra-structural ways with all of these approaches, while appreciating the beauty and value of some of their superstructure.

You will recall that Heidegger pronounced modern philosophy, and with it the grand Western philosophical tradition of metaphysics to be dead. Let us look at the metaphysics in question. It is the project originating with the ancient Greek philosophers that sought for objective knowledge of the universal, unchanging, principles or arché of Nature, and in particular, to the form, scientific (or better, scientistic) in inspiration, that modern philosophy has followed since Descartes. I take the "death of philosophy"[3] to make the claim that the Enlightenment story of Being – indeed the entire Western preoccupation with rational Greek values – needs to be replaced by a new and, of course, more contemporary, story of Being about which more below. As for the announced death of metaphysics, we recall Mark Twain's comment on the report of his death, that it was greatly exaggerated. Modern philosophy still lives on as the life and soul of institutional Western culture, and is the largely unquestioned background of the Received View in the philosophy of science. What my philosophical experiments do is to begin to test the claims of modern philosophy by analyzing and commenting on why we have come to accept or reject some of the knowledge claims that we do accept or reject, in everyday experience, in the arts, and in experimental science.

To talk about philosophy is already to presuppose that one is "beyond the philosophy in question," that the "philosophy in question" is an object to be questioned in the light of a new grasp of what philosophy should be. Since philosophy treats only of the most general questions, of Being, Knowledge, Truth, Causality, etc., it is something of an oxymoron to speak about the "edge of philosophy" or the "far side of philosophy"; for what topic is beyond philosophy? It will be a meaningful project only insofar as we have a philosophical platform that extends beyond traditional philosophy and gives a vantage point from which traditional philosophy can be brought into question. This vantage point I take to be hermeneutic philosophy, also known as hermeneutic phenomenology, as grounded in the phenomenology of facticity, historicity, and contextuality. This is the core of what is usually today called "continental philosophy." I will return to this in a moment.

The "metaphysics" or "philosophy" whose death Heidegger proclaimed certainly included modern philosophy, though his claim went beyond this. Modern philosophy rests on the discovery and elucidation of objective, universal, and unchanging concepts that possess upon reflection a privileged character of generating subjective certainty because they are clearly and distinctly laid out, like mathematical proofs. Such concepts, it is supposed, are purged of all hidden, confused, or implicitly contradictory elements. A philosophy based upon them will comprise only normative, interior (i.e., mental) representations of fact that are transparent to the subject's gaze and that can be manipulable via mathematics and logic to assure and deliver to the agent/subject control over external (i.e., physical) Nature. Such is the character of modern philosophy: it privileges theories that are capable of subjective grounding through the transparency of their objective representations, and that are effective in granting predictive control.

Modern philosophy of this kind is often called today by its critics a "philosophy of presence," for its goal is to ensure that nothing is concealed from the philosopher's gaze. Critics challenge the possibility of such a goal, on the grounds that things never show themselves fully to human subjects but merely reveal what historical facticity, contextuality, and existential hermeneutics permit to be revealed of the object through the resources of language and external representations. Despite its claim to the contrary, modern philosophy then has a far side, a complementary "philosophy of absence," and this too is a part of the story of Being. Thus, there is always a play between presence and absence, between an uncovering and a concealing in any attempt to express philosophically Truth about the world. If, to use a traditional image, "Truth is a woman . . . ," then that image is no longer the undraped Sophia of classical culture but Nietzsche's Woman or *haute couture*'s Woman of many runways. This brings us to Heidegger and his word for Truth, *alethia*, an "uncovering" and at the same time a "concealing" with a philosophical grounding in what phenomenological and hermeneutical philosophy call, "*die Sache selbst*," the way things are given in experience, that is, within the context of facticity, historicity, contextuality, and hermeneutics. About this, more below.

Hermeneutical philosophy is, in the first instance, a *philosophy* and should not be confused with the methodological use of hermeneutics in history, literature, art and the humanities in general to infer meaning from texts and other human artifacts, though it does ground it. Hermeneutical philosophy, in its philosophical role, makes universalistic and ontological claims, for that is what philosophy is as distinct from psychology, sociology, and other specific subject matter studies. Hermeneutic philosophy starts from the position that all explicit knowledge in the form of statements depends on a prior understanding or fore-understanding present in some form of praxis that unites subject and object ontologically before they are separated and distinguished by the uttering of objective statements. This fore-understanding is then an ontological condition of active involvement with the world prior to predication; it is pre-predicative, therefore, relative to statement formation, though not necessarily prior to the formation of language or languages, and it is ordered to the formation or correction of language – not, say, to some mystical intuition of Being, nor to the formation of clear and distinct ideas in the Cartesian sense. Gadamer called this state, after Augustine, the *verbum interius*[4] – the pre-condition for and interior judge of outer expression.

For Heidegger, this state was exemplified in everyday *praxes* within which we use equipment. Consider, he says, the praxis of hammering a nail with a hammer. It is a simple ordinary action with a common tool that does not require expertise to perform. Because the tool is so simple and easy to use, a person does not need to understand what a tool is or what a hammer is as a tool; it suffices to know how to use the tool. The hammering is performed successfully – the nail is driven in – as long as the hammer functions as it should. Attention is focused on the nail which is present to focal or explicit awareness.[5] The hammer as such is absent from such awareness, since using a hammer does not imply knowing its concept – what a hammer is as a tool – which defines how it is constituted by design, materials, construction, and naming as a tool. If, however, the hammer breaks and you can no longer continue hammering, you begin to fuss about this thing in order to find or make a replacement, and the fuss may include an analysis of what precisely a hammer is, because you may need to explain to somebody what you need or find something to replace it temporarily to complete the job on hand. At this point, what was absent from your focal awareness during the hammering becomes present through the need to theorize about it and to formulate a concept of the hammer as such. Let me suppose this is done and a replacement is made or bought; this thing now has a name, "a hammer," to which is attached a precise concept. But this physical thing, say, of iron and wood, and labelled "hammer," is not uniquely, essentially, and universally a hammer such as is represented by the content of a theoretical concept or universal; this physical thing can serve different uses, for example, as a paper weight, a door stopper, a nutcracker, and so on. Only within the praxis of hammering can it appear in experience as a veritable hammer, "*zuhanden*," or "ready-to-hand," as Heidegger called it. When not in use, it is still a "hammer," or more correctly, it carries the label "hammer," but the name does not denote something essential about the thing, Heidegger refers to it as merely a resource, "Gestell," or "present-at-hand," available for a plurality of human uses, hammering being just one.

One of the characteristics of modern philosophy is the wish to compile encyclopedias. What now of an encyclopedist who wants to compile a catalogue of everything existing in the world, including tools and other equipment? For the reasons given above, our encyclopedist would face serious frustrations in trying to enumerate the cases that fit into any set of *a priori* equipment categories. Would things be counted by their labels, by something essential to their being or functions, or by how they are actually used? If by labels, the traffic is in words, not things. If by essential character, equipment has no essential character, only a multiple resource relationship. If by actual use, then we must note that equipment actually in use tends to be absent from the agent's focal or explicit awareness and easily passed over and not counted. The encyclopedist will be helped by language that has names for the categories of equipment, but language does not distinguish between equipment used ("ready-to-hand") and equipment available for use ("present-at-hand") and between names used for concepts and names used to refer to things like equipment that in use tend to be absent from thought.

In place of the hammer, consider any one of the multitude of physical links or environmental conditions that accompany hammering in a concrete situation. In general, it is the business of science to study and record the endless plethora of linkages and conditions involved in human praxes, especially those that are concealed in fact or in principle from the human agent when acting. Some of these physical links, such as tools, are connected with the use of equipment, and others, such as bones, muscles, and nerves, etc., are functional parts (parts in a scientific sense) of the human body used in performing the required actions. Some of the environmental conditions are physical but passive, like walls and beams, others are social and intentional, like building codes and sensitivity to the comfort of one's neighbors, and so on. The encyclopedic cataloguer, aware of Heidegger's analysis, must be suspicious of all such lists on several grounds. The play between presence and absence in the pre-predicative praxes of life challenge the view basic to modern philosophy, that philosophical consciousness is objective and can, and should be transparent to itself. In addition, the hermeneutic philosopher will look with suspicion on lists drawn from both language and science because they are immersed in a metaphysics that edits their contents, and has its own way of sorting the real from the unreal, the necessary and essential from the contingent, the genuine from the superstitious or illusory, imposing its historical will on both language and science. The Heideggerian philosopher, seeing Being as characterized by historicity, facticity, contextuality, and hermeneutics, will be suspicious of all lists and data-bases and the texts that use them, questioning critically the metaphysics that lurks behind them all.

Phenomenology is the study of the pre-predicative praxis as shadowing forth phenomena, before expression but with a view to expression. Husserl and Heidegger spoke of a focus in this way on "*die Sache selbst*" as the starting point for a new kind of philosophy. Return to "*die Sache selbst*" is, as it were, a return to the *Cogito* with a new and critical look. Husserl's return brought forth the *Cogito cogitatum* as a correction of the Cartesian *Cogito*, and thereby introduced contextuality into the analysis of experience. Heidegger probed deeper into the *Cogito* to discover the role of fore-understanding which in turn uncovered the hermeneutical pre-predicative dimension of the *Cogito cogitatum* as "*die Sache selbst.*"

Briefly, let me comment on four of the characteristics of hermeneutical philosophy that were mentioned above: existential hermeneutics, facticity, historicity, and contextuality.

Existential hermeneutics is the activity of translating pre-predicative experience into language, using the existing resources of language, correcting these resources, or developing appropriate new linguistic resources. Such activity is never comprehensive, complete, or historically unrevisable. Moreover, while admitting a variety of expert but open communities of knowers, such activity is universalizable at every socio-historical epoch; it is not relativistic in the corrupting sense.

Facticity refers to the philosophical focus on what is being uncovered here and now within the experience of the socio-historical praxis; this is the focus on "*die Sache selbst*," on what is trying to appear in experience and be expressed in language. It is always, as Husserl never tired of pointing out, the experience of invariance amid variations.

Historicity means the recognition that phenomenological and hermeneutical analysis is performed by somebody living in a certain culture, at a certain time, and with a certain education, all of which are the product of historical traditions. Philosophical hermeneutics accepts the historical story of Being which I can summarize: The Being of knowing is the knowing of Being, where Being and knowing are both socio-historical.

Contextuality (which overlaps all of the above) refers to the "work space" of the praxis, the environmental conditions that make possible the praxis and are necessary for it to unfold in time; it is an ontological invariant of the praxis' temporal dimension. Among them are physical resources, moral interests, and ensconced traditions and skills, and an unknown indeterminate continuum of implicit pre-predicables including possible future scientific predicables.

In my philosophical experiments I will inquire as to whether one can find in the practices of everyday life, art, and modern science a need for, even a justification of the hermeneutic critics of modern philosophy. Particularly with respect to modern science, this inquiry breaks new ground. Firstly, no one familiar with the literature of continental philosophy would expect that modern science would be invoked in its support. In fact, one would expect the contrary, a downright criticism of modern science for exercising a certain hegemony over all modern intellectual, social, political, economic, and artistic concerns and even over philosophy.[6] Secondly, few modern philosophers, continental or otherwise, know much about science or its culture. Few are equipped to handle scientific research with an insider's familiarity. Although I am no longer a research scientist, I received an insider's formation in physics research and I have reflected deeply on it. I can also boast that some of the fruits of my own research even after 40 years are referred to perennially in the Science Citation Index and some of its results are visible from my office on the skyline across the Potomac. However, this should not impress the reader overly; philosophy is something else. But I am as a result also able to assure you that I am not over-impressed by science – whether as a harbinger of salvation or global ruin. But something like a critique of science is necessary to do philosophy of science, that is, for the philosophical sake of the philosophy of science, not just for social, political, economic, or ethical reasons, but to bring science back into contact with the history of Being, for there was a time when science was natural philosophy.

In the following philosophical experiments I will try to look critically at the fundamental implicit grounding of certain pieces of knowledge drawn from science, performance art (theatre), and everyday perception, first, from the point of view of modern philosophy to point out how it is implicitly sifting our current perceptions, and then by way of contrast, from the point of view of the continental criteria mentioned above to study how our contemporaries have become accustomed to choosing these novel and (to many of our colleagues) invalid criteria over the old ones. In this way I hope to make plausible the claim that the hermeneutical criteria of facticity, historicity, and contextuality are universal and apply in principle to all knowledge.

EXPERIMENT 1. A HERMENEUTICAL VIEW OF VISUAL PERCEPTION AND THE MÜLLER-LYER ILLUSION

The display in Figure 1 is a familiar one, it is associated with the Müller-Lyer Illusion. What can we see in the display? We can see presented in Figure 1 a visual object that violates the structure of Euclidean geometry and Euclidean space. A, though physically equal to B, appears – or can be made to appear – to be both shorter than B and in depth more distant than B. Let me focus on this mode of appearance of the M-L Illusion. Incidentally, I believe I was the first to point out the depth dimension in the M-L Illusion. Taking geometrical optics, which assumes light travels in straight lines in Euclidean space, to embody the physics and metaphysics of natural philosophy, Figure 2 shows why, according to geometrical optics, the apparent depth and size characters of the three-dimensional M-L Illusion are incompatible with a Euclidean interpretation. You may well conclude that such an object cannot be constructed in the real world and must then be an illusion caused by a breakdown of the visual system. The case is not so easily dismissed. Shapes such as that of the M-L Illusion are not as anomalous in visual experience as we may believe.[7] Everyday language about shape and space was in historical times profoundly non-Euclidean. Euclidean criteria began to permeate the scientifically motivated elite only in the 14th century. Compare, for example, the architectural spaces pictured in the Arena Chapel

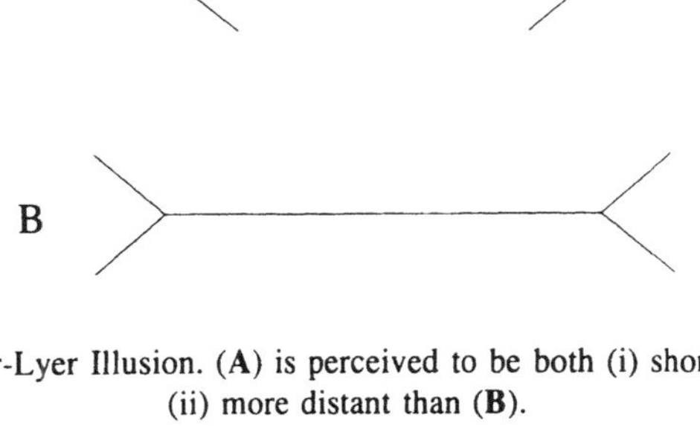

Figure 1. Müller-Lyer Illusion. (**A**) is perceived to be both (i) shorter than (**B**) and (ii) more distant than (**B**).

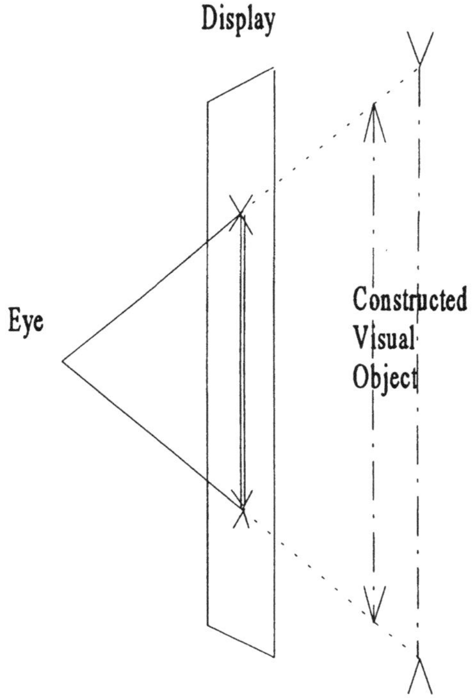

Figure 2. Euclidean Analysis of M-L Illusion. This analysis shows that (i) and (ii) are not compatible with Euclidean space.

frescoes in Padua by Giotto which were praised at the time for their life-like realism with those in Brunelleschi's painting of the Baptistery of the Duomo in Florence which ushers in the modern world of scientific picturing. There is an ancient historical tradition in the West that took metaphysical and even visual space to be non-Euclidean, while at the same time knowing that for the purposes of carpentry and engineering, the carpenter and the engineer must use Euclidean geometry. Aristotle and Plato said as much.[8] What ontological value does such a tradition bear? Is it just an error to suppose that there is no sense in connecting non-Euclidean visual structure of things to ontology? Returning to the display of the M-L Illusion, we ask: Should we judge that what we see in the display of Figure 1 is the representation of a non-realizable thing, a non-reality, and the output of the breakdown of vision?

The answer given by science and modern culture is Yes! Vision has the capacity of falling into systematic errors or illusions, and the M-L Illusion is just one of them. The reasons for this conclusion are fundamentally metaphysical and relate to the way modern criteria of objectivity are entrenched in our culture; these demand the possibility of total presence to an arbitrary (universal) subject via a clear, distinct, and unique description of the thing and its inertial motions in a visualizable Euclidean space/time. I want to comment on these criteria as they apply to what we can see and have seen in Figure 1.

- What is seen in Figure 1 lacks a precise quantitative description based on reproducible measures. The external cues on which measures are based are there in the display, but the measurement itself is a product of the *verbum interius*, by which I mean that an existential hermeneutic of these cues that is followed by a public statement of the outcome; and these actions are performed by the viewer whose individual intellectual skills, integrity, and linguistic ability are on the front line. *Comment*: how serious an obstacle is all this to knowing? Have we not just agreed about the character of what we have seen? On what grounds do we rule out the possibility that Figure 1 is a representation of some thing that is "realistic" in a different way, let us say, that it relates the object to the viewer as a thing in the world laden with social, historical, and biological values? Is this not knowledge too?
- You may have noticed that other visual interpretations of Figure 1 are possible, depth can be reversed and even the relative disparity of lengths can be changed by subjective means, using the narrative imagination to contextualize the visual object. *Comment*: don't human-centered perspectives also constitute knowledge, and shouldn't a philosophy of knowledge take account of ALL knowledge?
- Since the visual object does not have precise quantitative measures, it does not father a progressive cultural technology. It is then a poor challenge to Newtonian criteria. Had it been normative, one argues, our kind of developed society would not have been possible. *Comment:* why must ALL knowledge be tied to progressive technological control?
- The praxis of such anomalous vision creates a challenge to the traditional assumption of objectivity, that every thing has a unique in-itself essential being – natural or God-given – that admits of only one correct metaphysical description. *Comment:* are things knowable in isolation from other things, from human subjects and historical societies, from global and cosmic connections? If not, will it not be the case that knowability necessarily involves a diversity of perspective?

Let me try a more hermeneutical approach and state the question more generally. Could it be that there are more than one philosophically valid descriptions of a thing? And what are the conditions that would give this multiplicity legitimacy within hermeneutical philosophy?

I made a preliminray study of this problem in my 1983 book, *Space-Perception and the Philosophy of Science*. I have continued to think about the problem and I will summarize some of my current thinking about it. What creates the issue in question is that in interpreting Figure 1 there are apparently two measurement praxes invoked, an explicit scientific Newtonian praxis based upon comparisons with a movable physically rigid standard or its equivalent and an implicit visual praxis that does not rely on such standards but is more hermeneutic, that is, it seems implicitly to ask other than scientific questions and rely on visual constructions cued most likely to angles subtended at the eyes by the visual object. For socio-historical reasons, the culturally privileged praxis is the Newtonian one. But why is the alternative praxis banished to the realm of non-reality and illusion, if it is not to confer exclusive legitimacy on a scientific picture that is useful for converting things into equipment but deprives the viewer of a legitimate everyday picture that presents things in a way relevant to the socio-historical context of everyday life?

Let us examine today's socio-historical context of everyday life. Particularly in Western societies, people are born into a world that has been "carpentered" by human activity. It contains not just natural things, such as trees and flowers, but human artifacts that can be used for measurement. These include laboratory as well as ordinary equipment, and among other things, an abundance of architectural constructions that can be used to impose scientific measures on ordinary life. We learn (from childhood) to see things "stereoscopically" – to borrow a phrase from Wilfred Sellars[9] – both as value-laden everyday objects or "Manifest Images," and as value-neutral scientific objects or "Scientific Images." Consider the following perceptual experience described by Rudolf Arnheim.[10] It is one that we can all repeat. He enters a large Gothic Church. The interior is architecturally composed of a nave bounded by two parallel rows of arched bays separated by columns of equal height, completed by an apse in which the bays are continued until they meet the end wall behind the altar. Arnheim notes that his immediate unreflective (pre-predicative?) view of the interior of the church is double; he sees simultaneously two rows of equal arched bays bounded by columns of equal height marching in parallel straight lines down the nave of the church, and two rows of unequal bays and columns curving inwards toward an imaginary meeting point behind the altar with the bays and columns reducing in size with distance from the viewer. We apparently have the capacity of seeing the church in two ways, Euclideanly and non-Euclideanly. Before asking the obvious question, which is correct? we need to grasp the fact that there are two possibilities latent in the pre-predicative situation, and that to describe these two possibilities we would have to enlarge the linguistic capacities of any possible audience so that the two possibilities could be described, contrasted, and discussed. It should not be too hard to do this since there was a time when the spatial vocabulary we now use did not connote the privilege given to Newtonian/Euclidean space in modern metaphysics. Let us suppose that we have enlarged our linguistic capacities appropriately. In the hermeneutic situation just described there is the pre-predicative potentiality of using a Newtonian descriptive language, L_A, and a non-Newtonian visual language, L_B. (There are, of course, several scientific spaces, Newtonian, Relativistic, etc. and a plethora of visual spaces, but let us suppose for argument's sake that we are dealing with just one of each.) L_A is governed by the assumption of Euclidean geometry and a part of a Scientific Image, and L_B is governed by the assumption of a non-Euclidean geometry and a part of a Manifest Image. We have to assume that the two measurement praxes, scientific and visual, interfere with one another pre-predicatively in the sense that vision while responding to its angular cues (supposing these are the relative latent variables to which it is responding) could be distracted by the Newtonian (architectural) cues and fail to be able to develop a totally coherent non-Euclidean perception of the church space. Let us suppose, however, that the viewer can develop two distinct and different church spaces and can respond coherently, if prompted, to two sets of questions, those that aim to explore the Newtonian vision (spoken by L_A), and those that explore some alternative non-Newtonian vision (spoken by L_B). There is no reason for thinking that there is a one-for-one translation of one description into the other. Thus, scientific and visual spaces show a kind of "complementarity," analogous with the roles of position and momentum in quantum physics.

I propose the following *compatibility thesis*:[11] The logical conditions for philosophical compatibility are that L_A and L_B must be partially ordered within a complemented non-distributive lattice of descriptive languages (each contextualized by its experimental praxis) ordered by statement inclusion. By statement inclusion, I mean, for example (see Figure 3 below where arrows

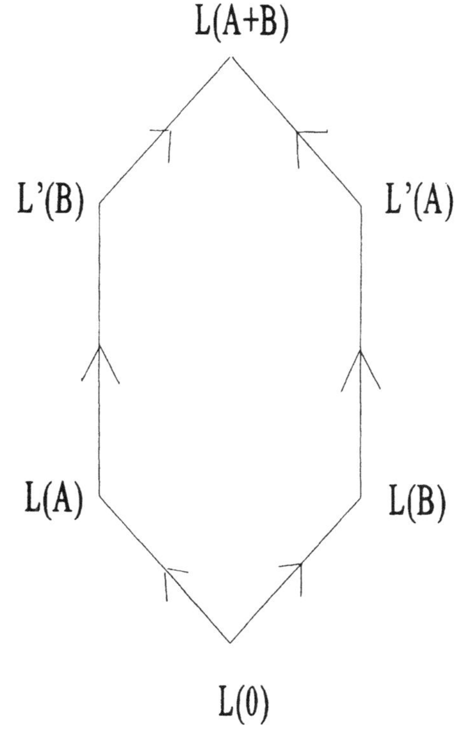

Figure 3. The complemented Lattice of L_A (Scientific) and L_B (Visual). L_0 is the GLB that defines the identity of the subject of predication throughout. L_{AB} is the LUB that is the most general descriptive language that is metaphysically legitimate.

"——>—" indicate inclusion), that if p and q are statements, then, (p) ("p in L_O" implies "p in L_A, L_B, $L_{A'}$, $L_{B'}$, and L_{AB}"), while (q) ("q in L_A" implies "q in $L_{B'}$ and L_{AB}").

The lattice will have a greatest lower bound (GLB) praxis with a minimum corresponding language L_o, that is capable of defining by ostension or description the identity of what is taken to be the subject of predication throughout; it is then a language common to L_A, L_B, and all the languages of the lattice, within which a thing or event can be contextually identified and can then become the subject of further predications in the languages higher in the lattice (L_A, L_B, etc.) provided that the appropriate praxis and context for each predication is fulfilled.

The lattice will have a least upper bound (LUB) praxis with a corresponding descriptive language L_{AB}, that is capable of dealing with situations of mixed cues and mixed measures, leading to mixed spaces, like that described by Plato and Aristotle when they treated earthly and heavenly spaces as structurally, i.e., geometrically, different, or like Arnheim's double visual response to the church.

Finally, in order to ensure that there is real novelty in the addition of L_A and L_B, the lattice must be (what is called) a "complemented" and "non-distributed lattice"; that is, there must be complements $L_{B'}$ to L_A, and $L_{A'}$ to L_B, which, to ensure novelty, do not reduce respectively to L_A and L_B. This last condition is equivalent to non-distributivity of the lattice under sums and products of its elements. The shape of the complemented lattice (for just two complementary languages) is given below in Figure 3.

The lattice should be read from the top down since L_{AB} expresses the contemporary story of Being as available hermeneutically to someone immersed in modern culture. L_{AB} includes two basic descriptive languages, L_A and L_B, and the elements, L_o, $L_{B'}$ and $L_{A'}$, that link them together as descriptive of the same thing. The common referent in the everyday world is given through some basic common praxis which is reflected explicitly in L_O. The referent has both a purely visual description in L_B and a scientific description in L_A. These two descriptions "dress" the same thing with different "clothes" for different purposes. Playing the game of presence and absence, the first description reveals the thing as measured by human life interests, while concealing all that science can teach about it; the latter reveals the thing as measured by physical standards, most likely for the purposes of control, while concealing its capacity to bring fulfillment to the human senses. L_{AB} relates the twofold "presences" and "absences" reflectively to different contexts of appearances relating to Manifest and Scientific Images.

A formal lattice of the kind of Figure 3 can be generalized. I give it the name, *context logic*. As such it could be proposed as a logical test for the philosophical coherence of hermeneutical frameworks of knowledge. The test has some contemporary virtues: it satisfies the taste for non-linear connections, with strong affinities to, and maybe helpful lessons for, quantum and high-energy physics, chaos phenomena, the social sciences, the arts, multi-cultural perspectives, and even, with an added historical dimension, philosophical perspectives.

EXPERIMENT 2. A HERMENEUTICAL VIEW OF THEATRE

We begin with the element of facticity and historicity in the theatre. Theatre is a part of our cultural lives. Most of us have some, perhaps limited, involvement with the theatre; all of us know something about the culture of the theatre. Robert Crease in his book, *The Play of Nature*, has produced a study of Dasein-as-theatre-maker through an analysis of the experience of people today who are involved with the theatre to a significant degree. Theatre-making is organized around the contemporary historical praxes of producing plays, presenting or performing them, and participating as a theatre-audience. Associated with these primary theatre praxes are a host of support activities involving technicians, musicians, critics, promoters, salespersons, and many more. All of these are mere resources and contribute as needed when called upon to bring to realization a particular play, say, Hamlet. *Hamlet* as a work of theatre art is a single thing, one and undivided, but what is the character of its unity? We note that no matter where or when *Hamlet* is presented, it has to cope with local conditions, and using limited material and artistic resources, it must be presented in a way that "speaks" to its local audience. These conditions are addressed in the design and execution of a production that tailors the performance to the resources, and shapes the interpretation of the script and story for a local audience. To make the work "speak" to its local audience is the aim of every production. To adapt a work, such as *Hamlet*, to a particular audience while leaving it "the same" and "unchanged" as a work of art is the challenge of a theatre production. The challenge has to be met, says Crease, by what Aristotle called *phronesis*, that is, by intellectual and practical skills, here, of the producer and director. However, phronesis or practical knowledge is not a theory and does not work by rules; it is learned only by experience. It neither adds nor subtracts anything from the work of art, but decides how a work is to be presented to a local audience. Different productions are "profiles" – Husserlian "profiles" – of the work of art; some are past profiles and contribute to the tradition that shapes this generation's productions of *Hamlet*, others are hidden in the future and will be revealed in their historical contingency only when their character has been specified by practical decisions made well, badly, or indifferently in accordance with the uniqueness of *Hamlet* as a work of art and the historicity of its being. This historicity is revealed through the variety and succession of *Hamlet* productions.

Let us call the philosophical language that is competent to speak of *Hamlet* as a work of art, *Hamlet* language. The *Hamlet* language can fairly be described – and this is my inference, not Crease's – as a least upper bound of a lattice like that represented in Figure 3 within which past productions ("complementary" to one another in the QM sense) are related by a partial ordering.

In the game of presence and absence, *Hamlet* permits itself to be revealed only historically and only one local production at a time. The lattice of Figure 3, which in Experiment 1 was interpreted synchronically, has now been shown to have in addition a possible historical or diachronic dimension. All such structures of presences and absences, whether synchronic or diachromic, fail to fulfill the criteria of modern philosophy. For everyday perception, then, and for works of art, they do point away from the transparency of Being claimed by modern philosophy toward the historical Story of Being and a hermeneutic philosophy. I now turn to question experimental science as to whether it too is in accord with this observation.

EXPERIMENT 3. A HERMENEUTICAL VIEW OF LABORATORY SCIENCE

Crease, in the book just cited, also develops a parallel between theatre and experimentation, applying the categories of his hermeneutic analysis of theatre-making to scientific experimentation. This he sees as writing/interpreting a script – in science, it is generally a theory – capable of serving the "theatre" of the laboratory in which experiments are like "productions," profoundly shaped by practical reason or phronesis, in the course of which scientific entities are "presented" by measurement processes in a sequence of experimental "runs," for the critical appraisal of an "audience" of the expert scientific community. Crease calls this an argument for a hermeneutical philosophy from the analogy between the theatre and the laboratory.

While he does indeed establish a parallel between experimentation and theatre, the argument in favor of a hermeneutic over an analytic philosophy of science needs more than is presented in the book. The mentality created by modern philosophy has to be attacked in its home in our everyday culture, and the scientific paradigms that underlie this cultural persuasion need to be debunked. I have tried to do this by my study of visual perception and the brief account of theatre as one of the arts. Let me now address science.

The scientific paradigms that underlie modern philosophy and its hold on public culture are mostly in physics. I will show that in many areas of physics, old and new, the criteria of modern philosophy are violated, and violated without regret or regard for the consequences that would befall philosophy or our culture at large, if it became generally known that such philosophical principles were to be abandoned. Physicists are acutely aware of the moment when on the basis of laboratory experience they can say: a phenomenon (such as has been suggested by theory), whether new, such as the top-quark or a black hole, or old, such as an electron beam, does in fact exist at a specific location and time. At that moment they are closest to being natural philosophers, and their instincts should be respected and studied. Those instincts have in many cases moved beyond the edge of the metaphysics of modern philosophy.[12]

Consider, for example, such a classic piece of physics as Newtonian Mechanics and the Law of Gravitation. It is now known that when many bodies are involved in gravitational interaction, such as the solar system, the outcome can be unpredictable and chaotic. One of the major favors that heaven has given us is to have established the orbits of the solar system so that they are not presently in danger of degenerating into chaos. However, we know now that the metaphysical expectation of predictable natural order that the mechanical model seemed at one time to offer and that therefore was thought to be the very epitome of rational transparency to the human mind fails this test. This indicates both a theoretical and an experimental failure in Classical Physics. The chaotic outcome was revealed rather by the use of computational techniques than, heaven forbid! by experimentation with the solar system.

What is chaos? It is the experimental emergence of unpredictable movement when the boundary conditions that represent the initial starting point of a dynamic system are close to those where the system is no longer governed by the usual assumption, the "principle of the differential equation," that small changes produce small changes. Each domain of physics is delimited by critical conditions that circumscribe and define the "work space" or "context" for the successful applicability and the physically possible solutions of the equations of motion of that domain. Consider, for example, what happens when a ping-pong ball is dropped vertically on a knife edge. The outcomes are predictable except when the ball is directly over the knife edge. When this is so, small changes in the way the ball is dropped can effect great changes in the outcome, causing the ball to bounce to one side or the other of the knife edge irregularly. This chaotic outcome occurs either when it is no longer possible to ascertain or control the dynamic parameters that yield solutions of the equations of motion, or when non-linear or higher-order terms in the equations of motion which are generally ignored contribute significantly to the relevant solutions of the equation. Other examples of physical chaos can be brought to mind: the smooth flow of liquids rendered turbulent when critical dynamic parameters are reached; oscillating chemical reactions, population dynamics, etc. All mathematical accounts of nature fail or "become critical" at some point because the real ultimately is not the mathematical; the real behaves mathematically and predictably only up to a point. The problem is that while initial conditions of any degree of magnitude no matter how small can be imagined, the real world is different, and chaotic behavior begins at some indeterminate point that depends on its previous history. Recall the imaginary butterfly on a beach in Japan that by beating its wings produces a hurricane in the Caribbean! There is neither in real life nor in the laboratory pure transparency to the *n*th degree; there is always a point beyond which our theoretical representation of the system goes over the metaphysical edge of the predictable and controllable, and chaos ensues.

Not that chaos lacks a system. It has its own ordering principles but these do not govern the individual but only large populations of systems. If L_A is the proper experimental language of the condition where small changes produce small changes and L_B the experimental proper language of the condition where small changes produce chaotic movements, then to the extent that these experimental languages can describe the same set of systems through their common resource L_O, these systems could be involved either in the classical experimental praxis A or the chaos experimental praxis B; in each case, there is an appropriate experimental language (L_A and L_B respectively) and these languages are complementary to one another in L_{AB} in something like the quantum mechanical sense.

Moving to non-classical physics, the General Theory of Relativity (GTR) and Cosmology have affinities with the Müller-Lyer Illusion, since general relativistic space-time is non-Euclidean and probably finite in size. It also has a beginning and maybe an end in time. Gravitational force disappears in GTR and is replaced by an equivalent curvature of space-time, so defined that it produces locally the inertial acceleration of an equivalent gravitational field. So gravity, the baneful irritant of the Cartesian mind because its non-perspicuous and theological character is removed from science and replaced by equivalent inertial forces in curved space-time. The GTR is, Oh! So beautiful! And simple! However, leaving aside its mathematical complexity which, of course, is non-linear and horrendously so, GTR creates special problems for the experimental physicist with a metaphysical inclination – and even for one with no such inclination. Let me mention just a few: clocks run slow or fast depending on the local gravity or acceleration. This affects airplanes, and it has been verified that the on-board clocks of planes that make many international trips no longer keep universal time. More troublesome still, is that since our own bodies have gravity, they produce of themselves changes in the local measures of space and time, involving our bodies in the experimental determination of the ever-changing local dynamic parameters of cosmic nature. We are not ourselves outside of cosmic nature and we cannot even imagine under the purview of GTR what that kind of objectivity would be like. Think, in addition, of the possible ubiquity of black holes, cosmic "worm holes," dark matter, cosmic organization balanced on an improbable knife-edge with the threat of unpredictable chaos. Our attempt to imagine in a normative way the universal "metaphysical" space and time of GTR is impossible, because it is dynamic and historical, and never the same. We must conclude finally that one would be hard put indeed to defend either the thesis that GTR as an experimental praxis is transparent, orderly, and predictable and should be embraced on that account or its contrary, that it is not transparent, etc. and should consequently be rejected. Physicists have for long embraced parts of both theses, the absence of experimental transparency and the presence of an experimental GTR universe.

Moving to high energy physics, you have probably heard of symmetry-breaking and renormalization as great break-throughs in physics that have created important syntheses. There has been a great push to show that all physics at the highest level is, on the one hand, theoretically one and universal, while, on the other, it is transparently simple and reductive in terms of its smallest building blocks. Behind this demand are the conflicting impulses of spiritual holism, like Stephen Hawking's, reaching upward in response to the distant religious call of the Neo-platonic One and materialistic reductionalism, like Francis Crick's, reaching downward for a world without spirit explained by its smallest material parts. In a recent paper, however, in *Physics Today*,[13] Sylvan Schweber of Brandeis University, commenting on recent trends in physics, pointed out that the two engines of theoretical unity, symmetry-breaking and the renormalizing-group have created a "hierarchical layering of quasiautonomous domains," each with its own ontology of particles and "fundamental laws (or effective Lagrangian)" but from which "it is next to impossible to *deduce* the complexity and novelty that emerge through composition" (emphasis in the text).[14] The hierarchy of stabilized domains is strongly correlated with the level of energy exchanges within a domain and is increasingly interdisciplinary. Such a picture tends to reject both reductionism and holism in favor of emergent domains populated with new "particles" and "fundamental laws" derivable neither from the bottom up nor from the top down. Each would have its own unifying theoretical characters in symmetries, conservation laws, the dimensionality of space, and range of interactions. Increasingly, these levels are united in their use of common interdisciplinary praxes, for example, imaging based on the recording of finely tuned photon and particle resonances and computer modelling using visualizations to explore the solutions of nonlinear equations that define the emergence of chaos or describes its dynamic states.

Schweber sees physics today as in the service of creativity and novelty. He writes, "The scientific enterprise is now largely involved in the creation of novelty – in the design of objects that never existed before and in the creation of conceptual frameworks to understand the complexity and novelty that emerge from the *known* foundations and ontologies. And precisely because we create these objects and representations we must assume moral responsibility for them."[15]

Returning to the theme of a hermeneutic philosophy for physics, theoretical discontinuities are often covered by a unity of praxes. And such would be the case in physics, if Schweber's account is correct. The logical (synchronous) condition that seems to be required is that the set of experimental languages – usually more than two differentiated by their quasi-autonomous domains form a non-distributive complemented lattice with one branch for each of the domains. The lower bound, L_O, would treat each "particle" ostensively in a world of standardized technologies and laboratory equipment; such ostension prescinds from concepts and depends exclusively on established praxes of identification. The upper bound, L_{AB} includes the theoretical models and equations of each domain as well as the technique (and principles) that isolate the "context" of one domain from those of the other domains. There will also be theoretical and practical linkages among the "particles" of different domains. To the extent that this condition is fulfilled, the contextual criterion I stated above for philosophical coherence of hermeneutic framework of contextual knowledge would be fulfilled. In practice, the demand would be logically normative for the coherent development of a science like physics.

By far the most insightful application of context logic and the hermeneutical lattice is in quantum physics. But I have written at length on this subject and I will not then comment on it here.[16]

CONCLUDING REMARKS

Returning to the topic of this paper, I have shown that an alternate metaphysics to that of modern philosophy, a hermeneutical metaphysics, is already long in possession of the everyday community, the art community, and more recently even in possession of the scientific community. This has been supported by some arguments of my own and some supplied by other writers on this topic. In the domain of science, I have shown how much experimental science has in common with the culture of ordinary life, how much it shares criteria which the general public hold in common, and which are commonly found in sophisticated branches of human experience. Theatre is just one such example. Like everyday perception or music for which a like argument can be made, theatre today is an example of rational knowledge grounded in implicit rational and hermeneutical principles of historical Being, and these aim at a universality that is the special character of philosophical principles. The criterion is the *compatibility thesis*. Life, common practice, and science constitute one of the traditional routes into philosophy. It was the way shown by Plato and Aristotle and the route taken by many pivotal philosophical writers since. This is not to deny that the philosophical tradition was also nourished on texts, mathematical or linguistic. But at critical intellectual moments, the greatest of philosophers have turned for further enrichment and orientation from texts to what is implicit in the historical pre-predicative experience of their epoch – Descartes to the praxis of the new geometrical science of nature, Kant to the implicit antinomies in the praxis of this new science, Hegel to the new science of history, Husserl to the intentionality latent in logic and psychology, Heidegger to the hermeneutical character of experience and especially of technology. This is what I have attempted to lay out programatically in this paper.

Georgetown University, Washington, D.C.

NOTES

1 See references below for some relevant works of these authors.

2 See, for example, the works of Babich, Crease, Heelan, Kisiel, Kockelmans, Rouse. On the question of *rationality*, McIntyre, though not a continental philosopher, has written much that needs to be seriously pondered within the context of this paper.

3 See Heidegger (1973) and the other works referenced below.

4 Gadamer. *Truth and Method.*

5 Some of you will recognize a similarity between parts of this analysis and Polanyi's discus-

sion of "focal" and "tacit" awareness in his book, *Personal Knowledge*, which was published in 1958. By that time, Heidegger's work was well known in Europe, the first edition of *Sein und Zeit* having been published in 1927. This book, however, did not become available to a general English-speaking audience in an English translation until 1962, four years after *Personal Knowledge* was published.

[6] Parenthetically, I ask, why criticize science for its cultural hegemony? Once upon a time science was natural philosophy and sat on the board of trustees of philosophy. But since science became an independent profession in the 19th century, should not the criticism fall on those trustees of philosophy who became complicit with the hegemony of science? Ironically the kind of science with which it became complicit no longer rules many of the most significant thrusts in scientific research.

[7] See Heelan, *Space-Perception and the Philosophy of Science*, especially Chapter 5.

[8] Consider that Plato and Aristotle approved of the use of (what we would call) "Euclidean" measures for carpenters. But for the heavens and beyond, there is a different story. "It is therefore evident," as Aristotle wrote, "that there is also no place or void or time outside heaven," *De Coelo*, 279a. Plato seems in the *Timaeus* to hold the same view, since for both writers perceptible things and the space itself – whether conceived in terms of place or the void – ended at the finite heaven, and so space cannot be Euclidean because Euclidean is necessarily infinite in extent.

[9] Sellars, pp. 35–6.

[10] Arnheim, p. 266.

[11] See Heelan 1983 (1988), chapters 10 and 13.

[12] There is a cadre of distinguished physicists who have attempted to wrestle with the metaphysical issues underlying contemporary physics. Among the more thoughtful are Silvan Schweber, Steven Weinberg, Paul Davies, Freeman Dyson, Eugene Wigner, Max Dresden, just to mention a few.

[13] Schweber, 'Physics, Community, and the Crisis of Physical Theory,' pp. 34–40. See also the authors cited there.

[14] Schweber, p. 36.

[15] Schweber, p. 39.

[16] See Heelan 1983 (1988), 1989, 1995, and the references therein.

REFERENCES

Arnheim, Rudolf (1974), *Art and Visual Perception* (University of California Press, Berkeley).

Babich, Babette E. (1994), *Nietzsche's Philosophy of Science* (SUNY Press, Albany).

Crease, Robert (1995), *The Play of Nature* (University of Indiana Press, Bloomington).

Dilthey, Wilhelm (1989), *Introduction to the Human Sciences*. Trans. by Rudolf Makkreel (Princeton University Press, Princeton).

Dresden, Max (1985), *Physical Reality and Mathematical Description* (Reidel, Dordrecht).

Gadamer, Hans-George (1981), *Truth and Method* (Crossroads Press, New York).

———. (1981), *Reason in the Age of Science*. Trans. by Fred Lawrence (MIT Press, Cambridge, MA).

Gadamer, Hans-George, Specht, E. and W. Stegmuller (1988), *Hermeneutics vs Science? Three German Views*. Trans. by John M. Connolly and Thomas Keutner (University of Notre Dame Press, Notre Dame, Ind.).

Heelan, Patrick A. (1983), *Space-Perception and the Philosophy of Science* (Univesity of California Press, Berkeley and Los Angeles, paper 1988).

———. (1987), 'Husserl's Later Philosophy of Science', *Philosophy of Science* **54**, pp. 368–90.

———. (1989), 'After Experiment: Research and Reality', *Amer. Philos. Qrtly* **26**, pp. 297–308.

———. (1991), 'Hermeneutical Phenomenology and the History of Science', in *Nature and Scientific Method: William A. Wallace Festschrift*, Daniel Dahlstrom (ed.) (Catholic University of America Press), pp. 23–36.

———. (1994) 'Galileo, Luther, and the Hermeneutics of Natural Science,' in *The Question of Hermeneutics: Festschrift for Joseph Kockelmans*, Timothy Stapleton (ed.) (Kluwer, Dordrecht), pp. 363–75.

———. (1995), 'An Anti-epistemological or Ontological Interpretation of the Quantum Theory and Theories Like It,' in *Continental and Postmodern Perspectives in the Philosophy of Science*, B. Babich, D. Bergoffen and S. Glynn (eds.) (Avebury Press, Aldershot, Brookfield (USA)), pp. 55–68.

Heidegger, Martin (1962), *Being and Time*. Trans. from the German by John Macquarrie (Blackwell, Oxford), 1962. [Originally published as *Sein und Zeit*, 1927].

———. (1966), *Discourse on Thinking: Translation of Gelassenheit*. Trans. by John Anderson and E. Hans Freund (Harper and Row, New York).

———. (1967), *What is a Thing?* Trans. by W. B. Barton, Jr. and Vera Deutsch (Regnery, Chicago).

———. (1972), *On Time and Being*. Trans. by Joan Stambaugh (Harper and Row, New York).

———. (1993), *The End of Philosophy*. Trans. by Joan Stambaugh (Harper and Row, New York).

———. (1977), *The Question Concerning Technology and Other Essays*. Trans. by William Levitt (Harper and Row, New York).

Husserl, Edmund (1970), *The Crisis of European Science and Transcendental Philosophy*. Trans. by David Carr (Northwestern University Press, Evanston).

Kisiel, Theodore (1993), *The Genesis of Heidegger's Being and Time* (University of California Press, Berkeley).

Kockelmans, Joseph J. (1993), *Idea for a Hermeneutic of the Natural Sciences* (Kluwer, Dordrecht/Boston).

———. (1985), *Heidegger and Science* (CARP and University Press of America, Washington, D.C.).

Kockelmans, Joseph J. and Kisiel, Theodore (ed.) (1970), *Phenomenology and the Natural Sciences: Essays and Translations* (Northwestern University Press, Evanston).

McIntyre, Alasdair C. (1988), *Whose Justice? Whose Rationality?* (University of Notre Dame Press, Notre Dame).

Polanyi, Michael (1958), *Personal Knowledge: Toward a Post-Critical Philosophy* (University of Chicago Press, Chicago).

Ricoeur, Paul (1981), *Hermeneutics and the Human Sciences: Essays on Language, Action and Interpretation*. Ed. and trans. by John B. Thompson (Cambridge University Press, Cambridge).

Rouse, Joseph (1987), *Knowledge and Power: Toward a Political Philosophy of Science* (Cornell University Press, Ithaca).

Schweber, Silvan S. (1993) 'Physics, Community, and the Crisis of Physical Theory', *Physics Today* (November), pp. 34–40.

Sellars, Wilfred (1963), *Science, Perception and Reality* (Humanities Press, New York).

Weinberg, Steven (1992), *Dreams of a Final Theory* (Pantheon, New York).

WILLIAM. E. HERFEL AND CLIFFORD. A. HOOKER

COGNITIVE DYNAMICS AND THE DEVELOPMENT OF SCIENCE

I. IN SEARCH OF DYNAMICS

Professor Polikarov has spent considerable time surveying various dimensions of the science-technology process, and scientific method in particular (Polikarov 1973; 1983). Therein he has expressed a view which has radical implications *vis-a-vis* traditional philosophy, namely that philosophy of science should itself be considered an extension of the science-technology process, and in this precise sense: Philosophy of science is a more general theory of human cognitive processes than is any particular science and *methodologically* it stands to the sciences as its domain of phenomena in just the way that a particular scientific theory stands to its domain of phenomena. This view issues in the general naturalist program to understand the world as a natural unity and hence to see knowledge as a natural phenomenon within it. This is a radical enough program philosophically, but the naturalist program further requires seeing the science-technology process and our epistemological understanding of it as a single dynamic system, mutually interactive and with a distinctive historical development.

This naturalist, systems program contrasts sharply with the traditional conception found on both the rationalist and empiricist sides. According to both these views, e.g., method comprises an eternal set of formal rules, focused on logic. Whence only the logical content of science-technology is relevant, and objective content is specified in a manner wholly logically distinct from the actual historical realisation of the science-technology process.

It is easy enough to argue that this latter, formalist view does not do justice to the facts. Scientific method, along with scientific theory and technology, have all changed dramatically over the last 500 years and continue to evolve. Professor Polikarov's studies in effect document part of this development (see particularly Polikarov 1983, Ch. 5). Even our conceptions of what it is to be intelligent and rational, and our conception of what logic itself is, have changed and, more importantly, have changed through exposure to developing scientific theories themselves. Science as an institutionalised activity has emerged, on a planetary scale, from an earlier scholastic tradition and the discursive practices of Renaissance gentlemen. There has been a massive transformation in the role of science and technology in the historical process over the past 200 years, bringing them into intimate interaction with governmental and commercial processes and resourcing.

Nonetheless, Western philosophy of science has remained largely formalist. This century has seen a steady progression in the dominant model of rational science, from austere empiricism to Popperian falsificationism, to Lakatosian

D. Ginev and R. S. Cohen (eds.), Issues and Images in the Philosophy of Science, 127–172.

research programs and Kuhnian normal and revolutionary science, to an increasingly critical and diverse set of responses ranging from anti-rationalist sociology of science and Feyerabendian epistemological anarchism through evolutionary epistemology, computationalism and various realisms to revamped empiricism.[1] Briefly, our analysis of this developmental path is that it reflects a search for an adequate formal model of rational science where reason would be represented by a set of formal rules of method satisfied by the model, a search that has failed across successive models, and the later diversity reflects reactions to that failure which range from sociological and political nihilism at one extreme to reworked empiricism at the other.

The alternative theory we are exploring is that procedural rationality can only be fundamentally characterised non-formally – that formal rules of method are special cases that need to be explained – and that therefore the proper model within which to characterise it is an 'embodied' dynamical one, not an abstract static one. The dynamical models we consider most promising are the new irreversible non-linear models of complex adaptive self-organising systems, which characterise all living systems.[2] A dynamical conception of reason, we claim, is then able to properly encompass the fundamentally social and historical nature of science while grounding it in the psychological capacities of individuals, all of which are sundered from it in static formal models.

The Marxist tradition has always had an integrated, embodied, dynamical conception of history and Polikarov's diverse analyses reflect that orientation, giving his work some natural connections to our own work. However, while we regard the insistence on integrating scientific work into its economic and socio-political contexts as sound, we believe that this should not be at the expense of understanding what is distinctive about the scientific enterprise, and not at the expense of abandoning any attempt to characterise it normatively. We would seek to súbstitute instead a normative theory of social design for science, embedded in a dynamical theory of society (see section V below). And we regard dialectical material dynamics, insofar as it can be made scientifically precise, as we do the normal/revolutionary dynamics of Kuhn, namely as a suggestive early model pointing to the far more complex dynamics characteristic of complex adaptive self-organising systems (see section III below). Although Polikarov's work is too diverse to characterise narrowly, and it sometimes shows a tendency to abstract formal characterisation which sits ill with a dynamical grounding (cf. Piaget below), we understand him as also searching for an appropriate dynamical conception of science and we approach his work within the framework indicated here.

Methodological Problems of Science is particularly prescient on this count: Polikarov (1983) was advocating a dynamical modelling approach to philosophy of science while thinkers in the West were often busy constructing logical models of scientific method (e.g. Newton-Smith 1981) or arguing about whether realism or empiricism was the best way to interpret scientific results (Leplin 1984; Churchland and Hooker 1985). Nevertheless, a similar commitment to the actual dynamics of science is seen contemporaneously with

Polikarov's (1983) work (e.g. Campbell 1974; Hooker 1982; Shapere 1984). Unfortunately, in the two books widely available in English (Polikarov 1973; 1983), the dynamical approach remains schematic; we get only a glimpse of how the dynamics of science can be systematically captured. Throughout this essay we will explore some of Polikarov's inchoate yet fertile observations, showing parallels with our own work. We begin with a brief analysis of some important models of philosophy of science in order to emphasise the systems-theoretic, historically dynamic character of the process we are considering.

II. PHILOSOPHY OF SCIENCE IN SEARCH OF DYNAMICS

Our aim in this section is to summarise the dynamics of science tacit in conventional formalist philosophers of science (Herfel and Hooker 1995; Hooker 1991; 1995a). These all turn out to be at best quasi-dynamical, logical sequences used to model historical sequences (analogous to the way the dynamical naturalist Piaget nonetheless tried to model cognitive development with logical sequences of formal 'stages' when dynamical models were not available, Hooker 1994b). Nonetheless, they provide an important clue to the kind of model that is needed to understand science. Meanwhile, capitalising on their formal character, we can model the twentieth century sequence of dominant philosophical theories of science in this way:

For empiricism, rational scientific knowledge is generated solely from observation using a logic of evidential support. Here observation is both the component supplying substantive content and itself an object of scientific study. Including observation within science for these reasons, we have a simple empiricist model where logic constrains the development of the observation-theory structure of science S. We express this as $L \rightarrow S\ (T \rightleftarrows O)(t)$, with t time, $T \rightleftarrows O$ representing the conjunction of the inductive evidential support relation (or deductive falsification relation) $O \rightarrow T$ and the deductive explanation relation $T \rightarrow O$, and $L \rightarrow S$ indicates that these relations are logical constraints on S imported from L. In the inductivist case the logic inferentially generates T from O, $O(t) \rightarrow_L T(t)$. The essential critiques of this position are of two kinds:

- Observation O does not offer the requisite epistemic foundation from which a logic of support may work.
- A logic L sufficiently powerful for this task does not exist.

Moreover, weaker, more plausible evidential support structures require excluding fundamental scientific decisions from the scope of rationality – including all those in the 'context of discovery', both observations and theoretical hypotheses.

For Popper the model of science is essentially the same, though now with logic restricted to deduction and hence $O \rightarrow T$ signifying falsification only, and observational foundationalism eschewed. In this case, the essential critique is again that it requires excluding fundamental scientific decisions from the scope of rationality – including all those in the 'context of discovery', both observations and theoretical hypotheses, and many others as well, for example termination decisions about when to stop or continue testing, whose importance is reinforced by the absence of any epistemic foundations.

In the case of Lakatos we have a Popperian model with an added layer of scientific research programs P: $L \rightarrow S\ (P \rightleftarrows [T \rightleftarrows O])(t)$, where $[T \rightleftarrows O]$ signifies the conjunction of the deductive explanation relation $T \rightarrow O$ and the falsification relation $O \rightarrow T$, $P \rightleftarrows (T, O)$ indicates the core/belt structuring of T together with a relevance structure on O by P, $P \rightarrow [T, O]$ at each time t, and falsification $[T, O] \rightarrow P$ over some time sequence τ of $[T \rightleftarrows O](t)$, $[T \rightleftarrows O]\tau$, and $L \rightarrow S$ indicates that logic constrains the development of the structure of science. Here the essential critique is again that it requires excluding yet further fundamental scientific decisions from the scope of rationality, both by now including P in the scope of discovery and including decisions to retain or drop P in the scope of termination decisions.

Kuhn's conception of normal science is essentially represented by the same formal model, but with P standing for a paradigm and where $[T \rightleftarrows O]$ signifies the conjunction of the deductive explanation relation $T \rightarrow O$ and the falsification relation $O \rightarrow T$, $P \rightleftarrows [T, O]$ indicates the paradigm structuring of T together with a relevance structure on O by P, $P \rightarrow [T, O]$ at each time t, and $[T, O] \rightarrow P$ indicates the evidential support and falsification relations which operate in normal science over some time sequence τ of $[T \rightleftarrows O](t)$, $[T \rightleftarrows O]\tau$, and $L \rightarrow S$ indicates that logic constrains (and in the inductive case inferentially generates) the development of the structure of science. In this way Kuhn's conception of normal science is intended to provide a model $(L.P) \rightarrow S(T \rightleftarrows O)$, which would rectify the formal deficiencies of empiricism – but Kuhn's rejection of observational foundationalism would remain. The essential critique is analogous to that for Lakatos, it requires excluding yet further fundamental scientific decisions from the scope of rationality, both by now including P in the scope of discovery and including decisions to retain or drop P in the scope of termination decisions. Discussion of Kuhn has focused on the non-rationality of scientific revolutions, that is, of P changes.[3]

From this perspective an obvious move to consider is adding a 'super paradigm/program' SP such that $L \rightarrow S\ (SP \rightleftarrows \{P \rightleftarrows [T \rightleftarrows O]\})(t)$, with P standing for a paradigm or research program, the $P \rightleftarrows [T \rightleftarrows O]$ relations identical to those of either Lakatos or Kuhn and $SP \rightleftarrows \{\ \}$ indicating the same relations as $P \rightleftarrows [\ ]$. The aim would be to bring revolutionary change under formal rational procedure, to encompass even revolutionary change within normal science. Here we might consider a Kantian categorical framework, for example, as providing SP. Kantian version: $(L.SP) \rightarrow S(P \rightleftarrows [T \rightleftarrows O])$. And here we can understand Feyerabend's critiques as aimed at providing a pre-emptive argument against any such attempt to rescue a formal rational model for science. Feyerabend argues that revolutionary change requires counter-induction (and propaganda and the like), so that no logical structure will suffice to generate it. He argues that rival theories are incommensurable, precluding their logical comparison as required by such a model. And he argues that observation is so theory-laden as to render O incommensurable between rival theories, again precluding their logical comparison. Although we are not persuaded by Feyerabend's criticisms, we are equally unpersuaded that a move to introduce SP will alleviate the problems. For unless SP could be given some non-empirical foundation it does not constitute an advance in principle beyond the previous position. And while we have no objection to recognising deeper layers of contingent principle as empirically required, we do not accept that there are non-empirically determined substantive principles.[4]

The upshot is that the succession of philosophical models of rational science shows increasing logical complexity but does not offer any escape from the central difficulty, that of dichotomising scientific decisions into the logical and the non-rational, with many of the most important decisions consigned to the latter category. But in our view, the fact is that whatever scientists are doing, they often have good reasons for their actions and are typically constrained by at least some considered reasons for their actions, and this applies even in situations of fundamental change. Once we recognise all the decisions involved we can begin to see that science must be understood instead in terms of a dynamics of decisions resulting from multiple interactions among scientists and with their environment (Hooker 1991; 1995a). And once we adopt this perspective we can see that each of the foregoing philosophical theories of science tacitly specifies a characteristic scientific dynamics.

Within this section we confine ourselves to brief remarks on the (quasi-)dynamics entailed by formalist verificationism (logical empiricism) and falsificationism (Popper). (For further analysis, from which these remarks are drawn, see Herfel and Hooker 1995.) In section III we return to Kuhn and extract a more complex dynamical account.

Though twentieth century logical empiricist philosophy of science is the least dynamical account, where the analogy with physics is least obvious, it also best illustrates the degenerate features of the standard formalist logical approach to method and rationality. Empiricism aims to base scientific knowledge solely on observation and logic. For logical empiricism, science is modelled as a collection of propositions describing observations and various levels of generalisations together with the set of logical relations among them. Whether a proposition is justified, whether it explains and predicts, and all else concerning its rational role in science, depends solely on its logical relations to other propositions, especially to observational descriptions. Such analyses assume a fixed, structured set of propositions, ignoring how they were arrived at. Thus we reach Reichenbach's notorious distinction between the contexts of discovery and justification (Reichenbach 1961, p. 7). These logical structures are context-independent because free of other considerations than their content, and hence timelessly, a-historically, a-socially, specified. Philosophically, science amounts to no more than a collection of these 'logical snapshots' about which only formal questions can be asked.

From our point of view these models of science stand to the complex processes of real science roughly as the theory of rigid body statics stands to mechanics. In statics, a physical system is modelled by a static structure of forces subject to constraints (e.g. derived from lengths of rigid bodies). One is restricted to asking structural (cf. logical) questions about the character of the force structure under the constraints, in particular whether the structure is stable, that is, whether the forces are in equilibrium. Like the logical snapshot models of science, in this theoretical representation the force structures have effectively been extracted from their dynamical contexts, which do not appear. Statics is a very limited, degenerately a-dynamical fragment of mechanics.

A stability/equilibrium analysis also provides quasi-dynamical information, even if it is very restricted and abstracted from its dynamical context. We may consider sequences of force structures formed by successively introducing new displacements and/or new forces; if the changes are small we can approximate a dynamical process by a suitable sequence of static structures. So too with the corresponding analysis of logical snapshots. We can consider sequences of these snapshots in 'logical snapshot space', each 'virtually generated' from its neighbour by introduction of new data and/or new laws through discovery. We can then try to approximate the process of science by a suitable sequence of related snapshots, thinking of the succeeding snapshot as 'virtually generated' from its predecessor by adjoining to its predecessor the additional displacements its successor possesses. (In the case of logical models of science, but not statics, we can extend this quasi-dynamical information a little by introducing an ideal end-state, fully axiomatised and complete in data and laws, the final state of a completed science, and order sequences of logical snapshots according to their proximity to that ultimate end-state.)

Quasi-dynamics for science provides a weak notion of generation for empiricists if only deductive logic is involved, as it is for the severest form of logical empiricism, since there are essentially only the explanation and falsificaiton relations to constrain allowed theory-data relations (cf. Polikarov 1983, p. 88). Reichenbach, for example, argued that a more complex inductive logical probability structure was required. Adding induction generates a simpler set of pathways in snapshot space because of the additional logical constraints it imposes. Now consider the inductivist ideal where only data displacements need be considered because they suffice to determine the theoretical superstructure. Accumulating observation statements then logically induce a sequence of higher level structures in which the pattern of laws unfolds towards completeness. In effect, induction defines a collection of pathways connecting snapshots in snapshot sequence space by placing constraints on which snapshots can inductively succeed one another under addition of new data. The foregoing characterisation is not restricted to sufficiently small changes, but empiricists typically also assumed simple accumulativity of change which supports an incremental assumption. In any event, inductive pathways specify logically all the ways that rational science can unfold. There is still a timeless logical structure at the heart of this process, the snapshot logical structure; the sequencing is generated externally, from the context of discovery, through the addition of data; but induction then logically structures the allowed snapshot successions. This is inductivist quasi-dynamics.

The required constraint for quasi-dynamics, that is for a static analysis to apply at each point, is that each of the resulting states, the force structures and snapshots, be sufficiently 'near' equilibrium. If a force structure is not near equilibrium then we need to apply dynamics to find out where it will end up, before statics can then be reapplied to analyse the resulting structure. Similarly, each accepted snapshot is constructed so as to be in logical equilibrium, only then can its confirmation and explanation relations be logically analysed. If initially the structure is out of equilibrium, for example if an observation contradicts a law, then the structure is 'adjusted' somehow, through non-logical processes of discovery, until it is in equilibrium, that is, until the theory is all and only that which observational data inductively support. A consequence of this is that each of these equilibrium states is path-independent. Consider two states S_1 and S_2 such that S_2 is a successor of S_1. In force statics it is irrelevant in what order the displacements are added, the net result is always the same. Similarly, for empiricism all that matters for 'generating' S_2 from S_1 is adding to S_1 the logical conjunction of all the data in which S_1 and S_2 differ. The order in which they are added is irrelevant. All possible paths from S_1 to S_2 are logically equivalent. So at equilibrium S_2 contains no information about the pathway by which it was reached. Equilibrium obliterates history.

Further, for empiricism scientific method consists in imposing the relevant set of logical constraints. If these constraints are respected, empiricism held that science will actually develop historically in a rational manner, since science will be constrained to the allowable snapshot-snapshot pathways. Logical structure is invariant across snapshot space, hence along all snapshot pathways. Method remains eternal and universal. All that is allowed to vary are the data and empirical laws. Thus science is characterised by a global and static rationality structure, a set of global constraints on snapshot space residing beyond historical change. It is as if science were developing within an infinitely undisturbable constraint structure that ultimately made all pathway information irrelevant, that rendered all scientists logically equivalent at each scientific stage and in which we could learn about the world but not about being rational. These are further important reasons why we refer to this model as quasi-dynamics.

Rigid body statics and its quasi-dynamics, though useful within its limitations is only a small fragment of physics; without a full dynamics for a system very little about it can be analysed except at or very near its equilibrium states. In the case of empiricist epistemic statics the limits are very severe because of the restriction to logical structure and typically also to simply accumulating observational data. Feyerabend made this point early on (Feyerabend 1962; cf. Sellars 1965) when he criticised empiricism for placing too narrow constraints on successor states of a science, for example requiring that observational data be strictly accumulative and that the predecessor theory be deducible from it as a special case. In addition, inductivism assumes that logical constraints are sufficient to determine relevant snapshot structure, but this turns out to be false, as the failure of empiricism shows (see note 3 references). Moreover, it was not long before Feyerabend was also attacking the constraint to a fixed method, arguing in *Against Method* (Feyerabend 1978a) that there were no rationally defensible logical rules of method that were static across time or universal across science (cf. Hooker 1972; 1991). In all those cases where quasi-dynamics fails to provide logical determination, appeal has to be made to the non-logical process of 'conventional decision', a piece of the dynamics excluded by, and hidden behind, empiricist statics. Let us turn then to briefly consider empiricism's successor philosophy of science, Popperianism.

Since Popper rejected empiricism's epistemically privileged observational base and its positive inductive theory selection criteria, and with that any direct approach to the truth, he had to focus on the characteristics of the process of science as constitutive of its rationality. This represented an important development in the theory of science. However, while not tempted to assume that logic plus observation could determine science, as empiricist were, he nonetheless adopted the empiricist assumption that the only rational constraints on science were those of logic, so that the promise of a genuine dynamics for science never eventuated. As noted, quasi-dynamics for science under constraint to deductive logic, especially in the absence of a privileged observation base, provides only a weakened notion of generation since there are essentially only the explanation and falsification relations to constrain allowed theory-data relations. Popper's response was to move to a more complex implicit dynamics which he came to see as a generalised evolutionary selection process (Popper 1979; cf. Polikarov 1983, p. 84).

It is controversial whether and, if so, how, the orthodox Darwinian evolutionary process can accommodate increasing phenotypic and ecological complexity, but Popper certainly makes it a central feature of his account, with controversial ontological as well as dynamical results. For human beings the inherited basis from which human knowledge grows includes all our basic capacities,[5] especially the capacities underlying rational thought, including consciousness itself and the capacity for symbolic thought (e.g. Popper 1979, pp. 137–9). It is here that the 'natural selection of theories' takes place.[6] This internal development of variation, *evaluation* and selective retention (VESR) processes will, according to Popper, lead to a new kind of externalisation of the evolutionary process with 'the evolution of language and . . . world 3 of the products of the human and mind' (Popper 1987, p. 152), where world 3 is 'the world of objective contents of thought' (Popper 1979, p. 106). Through language and the development of critical evaluation of abstractions, the evolutionary process is removed from the biological world and transferred to a VESR process in an abstract symbolic world accessible only to consciousness.[7]

Whatever one thinks of this symbolic extension of evolutionary models, it is clear that Popper's conception of science is intended to be explicitly dynamical and of a general evolutionary kind. Yet it transpires on closer examination that his methodology, like that of empiricism, derives its structure entirely from formal logic. Everything turns on the properties of classical deductive logic. Both prediction and falsification are interpreted as species of deduction, and other methodological concepts, such as informative content and severity of test, are also defined in deductive logical terms.[8] Moreover, Popper places the process of theory creation outside the realm of rational processes precisely because he does not believe that it can be modelled as a valid deductive inference, reproducing a sharp discovery (theory creation)/ justification (theory corroboration) distinction. Of course the decision to adopt Popper's methodology itself cannot be one dictated by deductive logic, and Popper also insists that this and like decisions are purely conventional, not rational, decisions (see Popper 1980, section 11). These parallels between the Popperian and empiricist frameworks could be extended. Despite his severe philosophical criticisms of empiricism, Popper largely shared the same metaphilosophical ideals and assumptions about the philosophical enterprise (cf. Hooker 1981a; 1987, chapter 3). For Popper, then, scientific development is ultimately again modelled as a formal sequence of logical structures of propositions in an abstract 'snapshot' space. However, careful examination shows that this works no better for Popper than it did for the empiricists (Hodges/Hooker 1995).

This is important to understand, since for Popper the motivation to introduce evolutionary models is to avoid having to relinquish rational control over so many basic scientific decisions. For Popper scientific progress becomes a major, non-trivial issue, as it was not for empiricism, since Popper denies foundations and much of the force of his position derives from the argument that it was through highly counter-intuitive, counter-inductive theories that science has made its most striking progress historically. In effect, Popper wants the freedom of an evolutionary dynamical system, while continuing to demand the control provided by a formal structure.

However, Popper's attempted marriage with evolutionary dynamics turns out not to eliminate this tension, for it turns out that scientific evolution is not a causally structured process but a purely symbolic one, occurring in an abstract symbolic world 3, whose only rational structure is given by logic.[9] By giving up inductive logical constraints, more components are free to change than they are in the empiricist model, enabling Popper to characterise the sequence of scientific states in snapshot space so that they have a formal, structural similarity to the sequence of states in an evolutionary process. But this does not make Popper's models more than quasi-dynamical. Unlike real evolutionary processes, Popper does not have access to constraints corresponding to genomic structure, for example to the powerful adaptive processes of crossover and inversion. Instead he is forced to rely on a logicist version of random mutation (viz. logically arbitrary generation of new hypotheses). This latter has been shown to be an insufficient genetic method for exploring possibilities, principally because mutation has no historical memory and so cannot learn from past successes and failures (Holland 1992; cf. Hooker 1995a).

Ultimately then, and despite his greater emphasis on understanding the development of scientific knowledge, for Popper (and Lakatos too) the rationality of science lies in its logical structure. Scientific propositions are justified just in case they stand in the appropriate logical relations to other scientific propositions. Scientific method consists in imposing the relevant set of logical constraints. If these constraints are respected science will actually develop historically in a rational manner, since science will be constrained to the allowable snapshot-snapshot pathways. Logical structure is invariant across snapshot space, hence along all snapshot pathways. Method remains eternally fixed and universal, despite the 'evolutionary' accumulation of information and the local core constraints. The rationality structure is again global and static, a set of global constraints on snapshot space, infinitely undisturbable, residing beyond historical change. Popper and Lakatos also provide no more than an epistemic quasi-dynamics.

It is time to look elsewhere for a more dynamic conception of science.

III. DYNAMICS EMERGING: KUHN, PHASE TRANSITIONS AND A RESEARCH PROGRAM FOR MODELLING SCIENCE

To begin to develop the foundations for an alternative, more dynamical, approach to science it is useful to begin with Kuhn's analysis of science. Kuhn's *Structure of Scientific Revolutions* (1962) explicitly shifts focus away from the logical structure of theories towards the full panoply of factors involved in science as a real historical, social entity. Correlatively, it replaces the traditional emphasis on the formal justification of theory with a dynamical model of the development and evolution of science. Kuhn's account is still concentrated too narrowly on theory (see e.g. critiques by Hacking 1983 and Galison 1987). Nevertheless it is an explicit attempt to capture the process of science rather than its logical structure.

Kuhn's account, as is well known, analyses mature science into two distinct phases: normal science and revolutionary science, each phase characterised by its unique dynamic. A gross dynamical account of the Kuhnian theory may well equate 'normal science' with the structure of logical empiricism. This certainly captures an important insight. We could then quickly pass on to a reading of his account of scientific revolution from our dynamical perspective. Nevertheless we should pause here for a moment and put normal science in our framework.

For Kuhn, normal science is a rule-governed activity. The problems of normal science are *puzzles* for two reasons:

- the researcher fully expects that any individual puzzle is solvable within the framework provided by (logic and) the paradigm, and
- there are clear procedures, provided by (logic and) the paradigm, for ascertaining that a solution, once it is found, is indeed a solution.

Normal science is thus rule-governed in this sense: given a paradigm, logic then suffices to determine what is justified. Here we see the affinity with the logical empiricist account. For both Kuhn and the empiricists normal science – and for the logical empiricists there is only normal science – is governed by a body of narrowly formal rules of rational procedure. Another point of affinity is that both see normal science as producing incremental changes in theory as the body of science and its procedures are articulated. This results in a near-equilibrium thermodynamic-like quasi-dynamics for normal science: Each move is a small linear departure from the equilibrium provided by the previous state of articulated, justified theory. Furthermore, each move is reversible. Finding out that a result produced in normal science was mistaken would result in the relaxation back to the previous equilibrium state. Once again, this achieves a certain path-independence for normal science; given a data set the resulting theoretical elaboration of the paradigm is determined independently of the sequence in which the data was discovered.

But though this comparison gets the formal similarities between Kuhnian normal science and logical empiricist science right, it would in our view be wrong to simply identify the two conceptions. First we should note that Kuhn rejects the kind of observational foundationalism empiricism requires, agreeing with Popper that observational data are not theory-neutral.[10] Moreover, the puzzle-solving activity of normal science is the result of real work by real researchers. It takes place in time; a static logical-snapshot does not do proper justice to the difficulty and complexity of *even* normal science. Despite the circumscription by the tight constraints of the paradigm, normal science requires ingenuity (Kuhn 1962, p. 36). Ingenuity is clever and intelligent. Whereas for empiricism, discovery – along with other methodological decisions not dictated by formal rules – is classed simply as non-rational, for Kuhn, normal science is both rational and innovative. The point that Kuhn attempted to articulate (with limited success) was that rationality could be rule-governed without being algorithmic. The view that there can be rational activity not reducible to mere rule following, but not arbitrary or capricious *vis-a-vis* the task either, has been developed and clarified more recently (Brown 1988; Hoyningen-Huene 1993). Galison (1987) has argued that experimentalist's decisions are reasonable, despite the fact that they are not made on the basis of deductive proofs. In a Kuhnian context, even normal science cannot be accounted for in the context of classical statics.

This subtle difference emerges still more clearly in Kuhn's treatment of anomalies in normal science (Hoyningen-Huene 1993, pp. 230ff.). Whether an unexpected result is interpreted as a significant anomaly, or rather as a new puzzle to be addressed later, is a matter of historical detail, and individual judgment (Hoyningen-Huene 1993, p. 224). On the logical empiricists' account it is never rational to 'shelve' a puzzle: the only logically coherent reaction to data that conflicts with a theory is to reject the theory and (non-rationally) discover a modified theory consistent with the data. But in Kuhnian normal science, as in a Lakatosian research program, it is held reasonable to 'bracket' conflicting data while exploring the capacity of the paradigm to generate a theoretical explanation for the anomaly. Of course, the accumulation of anomalies judged significant *vis-a-vis* the structure of the paradigm is one of the causes of the transition to the crisis state which precedes a revolution and the eventual establishment of a new paradigm. This state of affairs within normal science is itself extraordinary in relation to governance by formal rules. We would expect normal science's path-independence to break down at this point and for the sequence of anomalies to be important in the formation of scientific judgment about what to do next.

This discussion of Kuhn demonstrates the centrality of the notion of a paradigm to his account. We will not recount the very long and detailed philosophic analysis of this concept. (One could start with Masterman 1970 and work right through to Hoyningen-Huene 1993.) Our interest here is in the dynamical significance of the notion. On our view the main function of the paradigm is to provide constraints, in addition to those provided by formal logic (both inductive and deductive) and observation, on what counts as a scientific result in normal science. We have noted above that formal logic does not suffice by itself to drive science forward, placing both too few constraints to generate well-defined outcomes and too restrictive constraints to enable innovation. Polikarov (1983, p. 88) also argues that the constraint of formal logic is too weak to determine rational decision.) The only way to generate the fine structure of a mature science is to apply tighter constraints than those given by logic alone, yet loosened in their strategic tolerance of anomaly and the like. (For Polikarov (1983, Ch. 5) these additional constraints are provided by heuristic methods.) The constraints provided by the paradigm serve a dual function:

- they allow normal science to be rule-governed (in the sense expounded above), and
- the rigidity of expectation they generate through training within a paradigm facilitates anomaly generation (see Hoyningen-Huene 1993, pp. 226ff.).

In other words, the constraint of adherence to a paradigm both assures the solvability of solvable puzzles and otherwise forces scientists to articulate the presently dominant theory.

Rather than a weakness, as Popperian emphasis on criticism might suggest, this latter function is one of the chief sources of the success of scientific method. 'By focussing attention upon a small range of relatively esoteric problems,' Kuhn (1962, pp. 24–5) remarks,

> the paradigm forces scientists to investigate some part of nature in a detail and depth that would otherwise be unimaginable. And normal science possesses a built-in mechanism that ensures the relaxation of the restrictions that bound research whenever the paradigm from which they derive ceases to function effectively. . . . In the interim, however, during the period when the paradigm is successful, the profession will have solved problems that its members could scarcely have imagined and would never have undertaken without commitment to the paradigm. And at least part of that achievement always proves to be permanent.

This unique form of tentative commitment to a paradigm, combined with critical appraisal of it against empirical evidence and abandonment when it fails sufficiently compelling tests, sets science apart from other attempts at knowledge pursuit (cf. Hooker 1995a). It is the dynamics of the unfolding content of science generated by the institutionalised pursuit of this complex process that we wish to understand, and whose understanding we take to be the primary business of a theory of science.

When we move to Kuhn's crisis and revolutionary phases, the dynamical situation becomes more complex. The commonest reading of Kuhn is that scientific behaviour in these stages can only be characterised as non-rational, as competing groups of scientists first try various ways to shore up a crumbling paradigm that cannot meet its mounting anomalies, then develop rival alternative theories and preach their merits, seeking converts and allies, until finally one direction of theoretical development triumphs and a recognisable dominant paradigm emerges to which all again adhere. There is substantial textual support for this view of Kuhn, yet it is not the only reading. There is an alternative view that, while the conduct of science in this period, as in any period of normal science, may contain non-rational features, on the whole individual scientists proceed reasonably (e.g. Brown 1988). This position too can cite textual support in Kuhn. As we see it, both readings are compatible, their apparent opposition turning on an ambiguity in the notion of rationality, an ambiguity of importance to progress in our quest for a dynamics of science.

Kuhn's retention of a formal rule-governed analysis of scientific rationality in normal science, only relativising it to a particular paradigm, encourages the following line of reasoning: Only when given a paradigm can logic specify rational procedure; so logic cannot specify rules for a change of paradigm; so paradigm change is not rational *quà* rule-governed. (This conclusion is reinforced by the additional argument that laws themselves are not independently verified, since each theory defines observational terms in its own way; however this does not in itself prevent contrary evidence being provided from theory-laden data, see Hooker 1975.) This line of reasoning, consonant with preceding sections, is however compatible with another approach: While the decisions and actions of scientists in periods of crisis and revolution are not governed by formal logical rules, they are governed by individual reasons particular to the institutional scientific contexts concerned, and are reasonable or rational in this sense. Their rationality, as Galison holds, lies in the details. For example, he examines in close detail the decisions to close three sets of experiments in the history of particle physics. The decision to close is an admission by all parties that an experimental result has indeed been achieved. Galison asserts that the rationality of such decisions remains opaque when stripped of its historical context. With respect to the third episode he examines, Galison states,

> Taken out of time there is no sense to the judgment that Anderson's track 75 is a positive electron; its textbook reproduction has been denuded of the prior experience that made Anderson confident in the cloud chamber, the magnet, the optics, and the photography. . . . Experimental physics cannot be rewritten as a logical fantasy in which *all* theorizing is forbidden until the 'facts' clinch

the argument. Nor can experimentation be parodied as if it were no more grounded in reason than negotiations over the price of a street fair antique. (Galison 1987, p. 277)

We shall want to develop this latter conception of rationality, with its switches from commitments of science as an abstract whole to those of individual scientists, from decision-making using context-independent rules to decision-making in full historical context, and from a formal rule-governed conception of rationality to rationality as an informally judged, reasonable procedure in context. For the moment we note that it represents a movement that has been developing for a long time and, while once largely rejected, is now increasingly accepted.[11]

From our perspective, this alternative treatment of Kuhnian crisis and revolutionary science suggests a striking analogy to that of the physics of phase-transitions. Normal science corresponds to a macroscopic equilibrium state within some fixed phase, in particular a crystalline solid state phase, its macroscopic properties globally determined by equilibrium thermodynamics independently of any reference to its individual microscopic components. The detailed dynamics (micro-states) of individual particles is unnecessary to keep track of since we can specify the macrostates in terms of the averaged behaviours of the individuals. This macroscopic behaviour will in turn be determined by the boundary conditions on the system. In normal science, logic plus paradigm rules are the constraints playing the role of the boundary conditions and these constrain the behaviour of scientists to macroscopic uniformity (a few 'renegades', representing unlikely fluctuations, aside). These rules are like the constraints of the crystalline state, constraining the individual molecules to macroscopic uniformity.

Extending the analogy beyond normal science we see that scientific revolutions play the role of nonequilibrium phase transitions, the melting of the crystal. For Kuhn a scientific revolution, viewed from within a particular paradigm, is like a non-equilibrium phase transition viewed from within a thermodynamic framework. The dynamics of a phase transition cannot be described within the equilibrium theory because the phase constraints have been sufficiently weakened that they no longer serve to determine a unique macroscopic state. Similarly, there is no logic to the revolutionary scientific transition because the constraints have been sufficiently weakened that they no longer serve to determine a unique macroscopic state of science. Consider the theory of ideal gases. Even at equilibrium, the ideal gas law is limited. It describes, not the details of the dynamics, but only those of quantities averaged over all molecules. Pressure, for example, is average molecular momentum change per unit area and temperature is average kinetic energy. The assumption of equilibrium assures us that these averages are well-defined and dominate macroscopic behaviour, with the Boltzmann distribution (or equilibrium) specifying the magnitude of the small deviations one can expect about the average. But as one might expect, and *even for describing the average behaviour of the molecules*, the ideal gas law breaks down at the phase transition to liquid. From the point of view of equilibrium thermodynamics, a phase transition represents a singularity whose treatment lies beyond that theory's capacity to model or explain.

Nevertheless, we can utilise nonequilibrium thermodynamics to describe the transition. On this view, we can no longer rely on boundary conditions to specify even macroscopic behaviour near the point of transition. For instance, the reversal of the relationship between density and temperature in near freezing water can not be explained on the equilibrium theory. Such events as the freezing of water must be explained in terms of a fluctuation triggering a non-linear transition to a new phase. Boundary conditions are not sufficient to determine behaviour; we must instead examine (at least some of) the details in terms of molecular dynamics to explain the phenomenon. This is literally so because, although there has been some progress in developing a non-equilibrium thermodynamics employing only or basically macroscopic thermodynamic quantities, the only rigorous theory presently available is a quasi-equilibrium theory for which departures from equilibrium are infinitesimally small and reversible (the Onsager relations hold) – see (e.g. Caplan and Essig 1983). For treatment of far-from-equilibrium transitions, which mark the dynamics of all major physical phase transitions and all significant changes in living systems, we need explicit consideration of individual molecular dynamics. (This is well illustrated, e.g., in Chandrasekhar's elegant analysis of the formation of Bernard cells and other hydrodynamic phase transition phenomena; Chandrasekhar 1961.)

By analogy with the theory of phase transitions, we obtain the result that we will also need to model paradigm shifts as far-from-rule-governed phenomena. Sufficiently 'small' anomalies might be removed through small modifications of theory or even paradigm (though Kuhn does not discuss this latter possibility) that might be modelled in terms of small, rule-governed modifications, in analogy with near-equilibrium thermodynamics. But revolutions are not minute, simply reversible departures from an existing paradigm; they represent large, irreversible changes driven by an accumulation of significant anomalies too powerful to encompass with minor adjustments. This means that here also we will not be able to model these transitions by appeal only to rule-governed features, for such features 'fracture and flow' under paradigm shift and are typically not well defined during the transition. (Temperature, e.g., is only well defined for a Maxwell distribution of molecular velocities, which breaks down during a non-equilibrium phase transition.) Instead, the transition between paradigms will only be able to be understood by considering the dynamics of the individuals involved, the scientists.

To understand molecular dynamics we have to introduce considerations of the properties of individual molecules and the dynamical interactions they support. Similarly, to understand the dynamics of scientists we have to introduce considerations of the cognitive properties of individual scientists and the dynamical interactions they support. The relevant properties of those 'microscopic' components of science will be their individual commitments to experimental techniques, data runs, particular theoretical principles, methodological procedures, risk/pay-off trade-off structures and the like (Hooker 1987, 1995a). Analogously with molecules, these will vary significantly across the relevant scientific community and fluctuate according to the local context in which they are operating.

Nevertheless, there will also be partially macroscopic, i.e. institutionally extended, structures, each covering some of the various kinds of interacting groups of scientists (and relevant others). These provide constraints on local behaviour and so partial coherences among groups of scientists. Galison (1988, p. 526), for example, argues that traditions of theory building, experiment and instrumentation are not rigidly locked together and, while each of these traditions exhibits periodic discontinuities, they do not break simultaneously. Instead they are interleaved in such a way as to provide coherence to scientific enterprises. Alternatively, Brown (1988) has argued that in the transition from Aristotelian to Galilean science there were not only discontinuities in fundamental principle at various orders of science, but equally there were sufficient points of agreement to make a rational debate possible.

At the 'crisis points' in the gas-to-liquid and liquid-to-solid phase transitions too there are quasi-macroscopic structures present of great complexity and highly dynamic, continually fracturing/dissolving and reforming, and it is fluctuations in individual molecular behaviour in interaction with these structures that ultimately explains the passage through the phase. So too in scientific revolutions, there will be 'regions' of the scientific institutions, not necessarily spatially localised (especially in this age of electronic communication), of great complexity and highly dynamic, continually fracturing/dissolving and reforming, in which particular methods, or theoretical principles, or experimental traditions, and so on, persist. And it will be fluctuations in individual scientists' commitments and behaviour in interaction with these quasi-structures that ultimately explains the passage through the phase.

Consider a collection of elements initially taking on randomly assigned values of some property but interacting with each other in such a way that each element tends to move toward the values of its neighbours. (This arrangement might be realised by electron spins in a magnetisable medium (Anderson 1985; Stein 1989) or mutually persuadable voters facing an election.) If the interactions are appropriately non-linear, these systems will show macroscopic shifts in alignment in which all the elements adopt the same value, or shift from one common value to another. Such shifts also model the way a collection of 'persuadable' interacting scientists, willing to alter their own paradigm allegiance as a function of the allegiances of those 'near' them, can show macroscopic shifts in alignment in which they all adopt the same allegiance or all switch allegiances. These shifts are dynamically analogous to those characterising the rejection of one paradigm in favour of another – although we expect real paradigm shift processes to be much more dynamically complex (e.g. because of the complex patterns of institutionalised influences playing upon working scientists).

Polikarov (1983, Ch. 12) fundamentally agrees with this view. He explicitly interprets contemporary accounts within the philosophy of science as *models* of science's development (Polikarov 1983, p. 197). Several of his (Polikarov 1983, p. 199) categories of analysing the results of the theories of Kuhn, Lakatos and Holton characterise the dynamics they describe: 'continuous/discontinuous,' 'differentiation/integration' (these refer to the evolutionary dynamics of the scientific process) and 'extrapolation/complex approach' (these refer to the way the dynamics are to be analysed). Throughout this essay we have articulated, refined and extended these categories (with particular attention to what Polikarov terms the 'complex approach').

One example of where attention to details of the development of science provides a more accurate account is Polkarov's (1983, pp. 207ff) discussion of the bifurcation of scientific disciplines into subdisciplines. As the volume of scientific literature in a specific field increases new subdisciplines self-organise around specific topics. This is analogous to the symmetry-breaking bifurcations observed in self-organising far-from-equilibrium systems. Physical systems differentiate along these lines with increasing energy throughput (presumably a necessary, but not sufficient, condition). The observable (number of papers published) that Polikarov borrows from the scientometric literature to explain scientific differentiation is actually a crude parameter that could be cashed out in terms of money (or person-hours) invested, information accrued, or perhaps even energy throughput (see Adams 1988). Further investigation can shed light on the relevant variables involved.

The bifurcation into subdisciplines is in contradistinction to the progress through the revolutionary bifurcations that Kuhn is famous for (Polikarov 1983, p. 209). We would add that the former type of bifurcation can eventually lead to divergence in method. Although successive subdisciplines may well emerge within a particular paradigm, new methodologies – new modelling techniques – emerge which, while they are compatible with the dominant paradigm (here we agree with Polikarov that incommensurability is largely a myth), are not methodologically reducible to it. An example is recent work on superconductivity. While initially studied within the framework of classical electrodynamics, research into superconductivity eventually became an autonomous pursuit. The progress in superconductivity has come about by employing models that are only locally valid within the realm of superconducting materials (Cartwright *et al.* 1995). Although the story is sketchy, the lesson is clear. This elaboration of Polikarov's account can be seen as both a critique and extension of Kuhnian revolutionary dynamics. The line between discontinuous revolution and divergent evolution can be thin indeed (see also Polikarov 1983, pp. 312ff). As science differentiates through a series of theoretical 'cascades' (Polikarov 1983, p. 209), the network of results upon which empirical work depends becomes increasingly complex.

Polikarov identifies a set of constraints, wider than paradigms, that shape (govern?) this dynamic. He terms these constraints '*dominants of scientific research* (DSR).' They are characterised as follows: 'By DSR we understand an actual and prospective trend of investigation which leads to significant (fundamental or applied) results, develops successfully and exerts growing influence in a given (narrow or broader) domain' (Polikarov 1983, pp. 213–4).

We see 'DSRs' as playing the role in Polikarov's account that 'long-term', 'middle-term' and 'short-term constraints' play in that of Galison (1987, Ch. 5). Like Polikarov (1983, p. 216), Galison (inspired by Braudel) typifies these constraints in terms of their duration. 'Braudel found it useful to parse events into what he called a "geographical time," a "social time," and an "individual time."' (Galison 1987, p. 246) An example Galison (1987, p. 246) cites as a 'long-term theoretical constraint' is the conservation of energy which 'by the late nineteenth century . . . any experimentalist would assume in the design or interpretation of an experiment; if a result contradicted it, the experimentalist would look again at his instruments and procedures.' (Hooker 1976 makes a similar observation, suggesting that the more general the level of theoretical frame the longer the time scale for change.) Polikarov (1983, p. 214) appropriates the scientometric results (as discussed above) to measure the effects of these 'dominants'.

Similar to Galison's view on scientific decisions, Polikarov (1983, p. 214) sees the dominants as embedded within a micro-sociological context. In his analysis, he draws on the sociological literature to measure the effects of the DSR. Nevertheless, we believe that Polikarov would agree with Galison (1987, p. 258) when he argues that scientific results can not be reduced to the interests of the researchers involved.

However, Polikarov (1983, p. 219) does acknowledge that micro-social dynamics significantly enter into scientific development. For instance, it is crucial that there be open channels of communication so that a DSR may be extended: 'The existing conditions of communication are also to be included in the factors accounting for the social dependence of the dominants . . .' Here again we see fundamental agreement with Galison; although sociology is crucial (Galison 1987, p. 258), science can not be reduced to the interests of its individuals.

Polikarov (1983, p. 218) sums up his characterisation of the dynamics of the DSRs by comparing it with both punctuated equilibrium and phyletic gradualism (for a recent review see Gould and Eldredge 1993 and the bibliography therein). Both dynamics are possible in the evolution of DSR's. Bifurcation can occur when a dominant branches to form the basis of a new subdisciplines. On the other hand, new dominants can spontaneously emerge, forming the rugged landscape of punctuated equilibrium (Polikarov 1983, p. 219).

The traditional models of science are plagued by the appearance of scientific decisions undetermined by logical rules (and held non-rational on that count). Polikarov's discussion (1983, p. 221) of the inability of formal logic alone to explain the evolution of DSRs is in accord with this assessment. In many cases, however, we might expect to find, as Galison and Brown have found, that scientific dynamics, even revolutionary dynamics, can be characterised in terms of informal, context-dependent reasons for each individual's changes. This need not be uniformly or completely so; Solomon 1992, e.g., has argued that the record of recent science is mixed on this score, with other institutional factors influencing scientists' decisions, and she may well be correct. But this is not to render our model invalid, it is only to say that in many contexts the dynamics of individual scientists is more complex and diverse than can be appropriately labelled reasonable. And this is to say that the institutions characterising science do not always operate so as to constrain individual scientists to scientifically relevant reasons for their decisions and behaviour. And this is to call for their structural improvement in this respect, which can only be done when individual dynamics has been understood. In the details, where we find the dynamics, we will also find the rationality that eluded Kuhn. (In the terms of rhetoric, we seek a re-examination of what Kuhn would call 'persuasion' or 'conversion', but in a dynamical context.)

IV. SCIENCE-TECHNOLOGY AS A DYNAMIC, HISTORICAL REGULATORY SYSTEM: AN INITIAL CHARACTERISATION

Historical Dynamics

A general conception of science as a partially hierarchical, distributed, partially self-organising regulatory system is explored in Hooker 1995a (see also Hooker 1987, Chapters 7, 8 and Hahlweg/Hooker 1989), so here we shall be brief.

Self-maintaining, self-organising living systems grow and differentiate, and behaviourally adapt, under an increasing flow of energy and information through them, partly organised by themselves. Consider, from this perspective, the story of civilisation. To each major transition in human civilisation there has corresponded a major increase in energy throughput, driving in turn the accumulation of population, urban agglomeration and economic and social role specialisation, and so the production and use of information. The agricultural revolution removed most plant species from river bed ecologies so as to increase the flow of energy (food) to human settlements, driving population surplus, and so city-states, inter-city trade and wars, and so driving social organisation, empire religions, and the early sciences (geometry for land measurement, astronomy for flood prediction). Then came increasing external energy subsidies (wood, wind, water) increasing the scale of city states and their activities through to medieval times. Later came the fossil subsidies (coal, oil and gas), making the industrial revolution possible and so to the nuclear and large scale fossil subsidies today associated with economic and political globalisation, and with that the globalisation of organisation, culture and science. In short, our historical species dynamics looks like that of a growing, spatially differentiating, functionally complexifying, self-organising system, showing similar features both to ecosystem evolution and to embryogenesis.

Science itself shows similar dynamics, growing from a few Renaissance 'gentleman' exchanging letters to the global billion dollar institutions of today employing millions in exquisitely specialised roles. The approach is nicely illustrated by seeing how technology interrelates with it. Cognition evolved because adaptive control of behaviour was reinforced. Central to that control is anticipative intervention to disturb the environment in order to learn by its response and, subsequently, to modify the environment so as to reduce risk. From this flows the central role of technology in the scientific process. Our earliest technologies were provided by our bodies. Subsequently they evolved in delicate and increasingly intimate interactions with both practice and theory. Technologies are essentially amplifiers. With respect to science they

- extend its information base through instrumentation, e.g. electron microscopes,
- extend accessible methodologies, e.g. through numerical approximation techniques and automated data processing,
- generate new concepts, e.g. computer models of cognition,
- extend epistemic institutions, e.g. through supporting global communications, and
- provide the basic resource base for scientific activity, from economic surplus generally to rare earth metals and other specific resources.

Conversely, the development of technology is directly impacted by science through

- new theoretical conceptions,
- new theory-driven designs, and
- performance evaluation and redesign learning (fluid mechanics gained more from the development of reliable aeroplanes than vice versa).

This dynamical co-evolution of method, theory, and technology is of the essence of science and a vivid demonstration of its open-endedness. Its dynamics form a set of generic positive feedback/feedforward loops which can be represented as in Diagram 1. Theory provides the framework for developing new experiments, whose outcomes in turn provide information to develop better

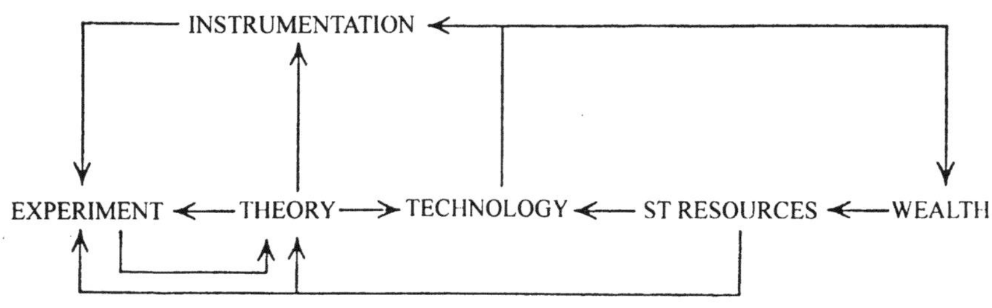

Diagram 1. The Science-Technology Machine. Arrows indicate positive feedback connections to produce mutually reinforcing loops that stimulate scientific, technological, and economic development.

theory. At the same time, applied theory provides the basis for better technologies, in turn providing both new technologies for new experiments (so to new theories) and providing new bases for the creation of wealth. Increasing wealth in turn provides increasing resources both for technology and for the conduct of experimental science itself, thus reinforcing the whole process. Like any irreversible biological system with an increasing flow of energy and resources through it, these positive feedback driven processes have produced an exponential expansion of science-technology across the 450 years since its modern inception, from its initial handful of individual practitioners to the planetary institutions with billion dollar budgets today. Correlatively, it shows increasing internal differentiation (specialisation) into myriad interconnecting fields linked together by complex technological (instrumental), mathematical, experimental and theoretical practices as its stored information increases.

The mutual interaction of method, theory and technology is nicely illustrated by the evolving role of the senses in science. On the one side, there has been an increasingly refined critique of natural sensory perception for its limitations (e.g. limited discrimination), biases (e.g. tracking delays) and imperfections (e.g. illusions). Technology and theory were essential here, e.g. the camera and optics for detecting perspectival bias, infra-red and x-ray technologies and backing theory for the use of the non-visible electromagnetic spectrum. On the other side, there is the development of extensions to, and substitutions for, the senses, e.g. telescopes, microscopes, micrometers, x-ray photography. These allow us to confine use of our senses to those narrow circumstances where they work best (e.g. identifying and counting appropriately shaped and coloured human-sized objects).

So we need to think of science in these respects as a dynamic system, transforming its own instrumental 'body' as it evolves/develops. Through technology science also transforms its social and natural environment (e.g. respectively motor vehicles, world agricultural gene trade) and the natural and policy processes which yield the dynamics of that environment (e.g. respectively greenhouse effect, economic modelling for policy determination). And through the development of various science studies and science policy studies, which together form the so-called science of science, it is also transforming its own institutional design and social relations (e.g. respectively, the trend to larger specialised, research groups, governmental science advice processes) including its own social evaluation. (This latter is marked by increasing ambivalence as felt powerlessness increases, e.g. reliance on, but distrust for, doctors or nuclear engineers.) This sophisticated and thorough transformative capacity is a crucial part of understanding science as an increasingly autonomous, dynamic cognitive system. (But this perspective is suppressed in the logic-centred conception of method and epistemology.)

This instrumental transformation of science is itself part of a wider transformation of western culture and organisation in which science-technology is the 'Change Machine', generating conceptual and technological changes which are fed into a wider economic and socio-political ecology of institutions designed to take advantage of them and which in turn feed back to reinforce the development of science-technology. The economic market is the central institution in Western society and it rewards and amplifies (saleable) change at the momentary technological production frontier: the better mousetrap, the new mouse gene technology. Science-technology is our most effective generator of such changes. Like evolution, the market has no global direction to it, it exploits change opportunistically. Parliament is a mini-market in votes, it rewards and amplifies (saleable) change at the momentary political production frontier: the better export stimulus, the new deal for mentally ill. Like the market, there is no intrinsic global directedness constraining the process, nothing to damp down the impact of momentarily advantageous change. Even Western religion has collaborated in getting out of the way of change. The Protestant's connection to God is individual and private, not through a public social form (as it was for the feudal Catholic), the church is wherever Protestants gather, so Protestant Christianity is compatible with change in a way that communal faiths are not. The result is a highly dynamic, currently relatively unstable, system which has been increasing in scale, intensity and pace since the Renaissance and is now transforming the planet.

An account of social dynamics of this kind forms the wider proper setting for a theory of epistemic institutions. These will not be separate self-contained institutional fragments in a collection of such disparate fragments, but partially autonomous sub-systems, together forming a cognitive institutional dimension or aspect of the whole institutional ecology of Western society. From the point of view of epistemic theory, relativist sociologies of knowledge and the like present useful information but fail to make a principled design distinction between epistemic and other social institutions, or epistemic aspects of social institutions. They thereby fail to do justice to the distinctive cognitive features of either science or technology (cf. Hooker 1987, sections 7.3, 8.4.2). But understanding these matters has an urgent practical dimension as well.

For this contemporary crystallisation of human organisation around cognition-driven institutions is the major contemporary transformation in planetary historical, ecological and evolutionary conditions. If we are not to undermine our own survival (and that of many other species as well) with a resulting chaos, these developments must come to constitute the emergence of a new order of stable planetary regulatory structure. Designing such an integrated system is now our most urgent task. A small but important part of this task is to develop an adequate dynamic systems conception of science. It is to this that the present work is designed to make an initial contribution.

Content Dynamics

To provide some initial detail to this idea, we begin with an unconventional reading of the conventional information representation of science as abstract propositional contents, namely to represent it as specifying a regulatory system. Theories regulate the development of practices (technologies) and data structures (facts); methods regulate the development of theories. The case of mechanics provides an illustration, see Diagram 2.

THEORY	METHOD
Order Structures	Pure Mathematics, e.g., Analysis
Logic/Algebra	Geometric Quantization
Geometry: Topology	Topological Dynamics
Projective	
Conformal	Global Stability
Affine	Analysis, Chaos
Metric	Methods
General Dynamics	Deterministic and Statistical Methods
State Specific Dynamics	Instrumental Design and Error Analysis

Diagram 2. A Regulatory Structure for Physics. Each level on the left-hand side regulates the constructions at levels below it using the methods at or below its level on the right-hand side.

Here, complemented by method, each order provides a recipe for constructing the orders below it. Given the logic/algebra to be employed in dynamics, e.g., together with a characterisation of time (say, modelled by the real numbers), then the general form of the dynamics is fixed (e.g. as between classical and quantum dynamics). Generalising across science, theories are used to provide both the foundation for methods of data generation in their domain and a normative critique of the nature and status of that data. So we may think of theories themselves, at their more modest order, as providing a set of instructions in a program for the construction of lower order methods, technologies and data. Even data have a residual regulatory character in virtue of their theoretical embeddedness.

The regulatory power of science derives from the depth of the regulatory orders it provides. As the theory and method orders are ascended, the range of data structures which can be incorporated increases. At first only the specific empirical structures corresponding to specific empirical generalisations can be encompassed (explained), e.g. how this ball falls, this pendulum swings. Then classes of generalisations can be derived through alteration of parameter values in a theory, and thereby their classes of empirical generalisations encompassed, e.g. the law of free fall, the pendulum law. Next, classes of structurally different theories can be derived at successively higher orders, e.g. from Newton's laws; and so on, e.g. to Lagrangian and Hamiltonian theory and thence to symplectic structure on differentiable manifolds and so on 'up'. Each ascent allows regulated response to be conditioned one nesting deeper: 'If the pattern is of kind K, then if the conditions are of kind C then . . .' The widening range of data structures encompassed means a widening range of situations to which regulated adaptation becomes possible. This gives science an increasing adaptability.

But no simple linear hierarchy is involved. All of the orders of science potentially interact, modifications in any one of them may induce modifications in any (or even all) others. This is quite clearly going 'down' the structure. On the other hand, new data can lead to the modification or even abandonment of theory, even fundamental theory, and/or of the methods based on a given theory. Theory and method are also intimately interlocked: methods are used to confirm and disconfirm components of theory, while each method is built on and requires a set of theoretical presuppositions. One does not use, e.g., a statistical analysis appropriate for independent particles if the theory specifies interacting fields. Furthermore, theories interact across domains with other theories and methods, e.g. gas and fluid mechanics interact with thermodynamics in heat transfer theory and even in Einstein's quantum models of radiation. (See also Hooker 1975.) Thus we may very appropriately think of all the content components of science as together forming a highly interactive complex regulatory system.[12]

Methods, theories and technologies may all be refined and extended to new cases. Many refinements to lens-making and optical theory and practice over the previous two centuries, e.g., gradually led to the improvement in performance of microscopes and telescopes thereby providing important information, e.g. for a germ theory of disease and for cellular biology generally, and engendering further extensions of optical experimental methods (cf. Brown 1987; Hacking 1983). This is the 'normal' situation. Even so, the consequences of refinements can reverberate throughout science, as those of optical instrumentation did. Or consider that our refinement of methods for measuring solar system constants, combined with our refinement of computer modelling techniques, now lead us to believe that several planetary orbits can exhibit chaos.

But science may also change in more radical or revolutionary ways. Consider first Einstein's generalisation of the Newtonian mechanical concept *of reference-frame invariance* so as to include electromagnetic as well as mechanical phenomena. This led to a profound transformation of the structure of physics which illustrates a fundamental regulatory process, the forcing of development up regulatory orders by forcing a retreat to less committed (cf. more neotenous) assumptions. Einstein was not the first to explore the idea of modifying classical mechanics in the light of electromagnetic theory, nor even the first to discuss the idea of some kind of relativity principle. But he was the one who clearly grasped that the modifications required reached as deep in the regulatory structure of physics as kinematics and were not to be confined to dynamics alone. This proved to be the crucial regulatory insight. The Einsteinian shift stimulated the development of generalised theories of space-time structures and the abstract representations of dynamics on them as configuration spaces (Lagrangians) and on their Hamiltonian co-spaces, which ultimately led to still further generalisations of dynamics, e.g. to symplectic flows on abstract differentiable manifolds, as ascent of the regulatory structure through adding new 'top' layers. The more dramatic case of quantum theory presents similar lessons (see Hooker 1994c for discussion).

So for both the relativistic and quantum revolutions we find partial shifts at several different regulatory orders plus the stimulation of higher order theory and method development. This complexity to revolution is not confined to modern science; a similar treatment can be given of the Copernican-Galilean revolution (see Brown 1988; Hooker 1995a). The same kind of regulatory representation can be given even of logic and mathematics and general theory of scientific method, where there has been historical change as knowledge improves.[13] The power of revolutionary change lies in the degree to which it disturbs the existing regulatory order in such a way as to stimulate development of a richer regulatory structure to replace it. In Galileo's and Newton's hands the Copernican Revolution stimulated the development of new method and high order theoretical structures (see note 18), as did the relativistic and quantum revolutions. And from these latter jointly flowed eventually the various quantum cosmologies and cosmological dynamics (black holes, big bang, baby universes, etc.) that currently captivate imagination. However, these revolutions would not have been possible without the plethora of small regulatory changes produced by the 'normal' science that preceded them, e.g. Tycho Brahe's refined stellar observation instruments or the measurement of atomic spectra with refined spectroscopic methods.

Regulatory systems may show two importantly different dynamical processes which, following Hooker 1995a, we shall call refinement ('horizontal' extension of regulation) and ascent ('vertical' extension of regulation). Regulatory refinement is the process of increasing regulation within the existing regulatory architecture available to a system. Its correlates in science include increasing precision, increasing scope of generalisations about stably characterised classes of entities, increasing technological applicability and diversity of laboratory procedure and so on. In short, it corresponds to increasing adaptation and is roughly what Kuhn refers to as normal science. Regulatory ascent is the process of adding new regulatory layers to the system that override and conditionalise those already in place. This results in a more explanatorily powerful and adaptable regulatory system. Einstein's distinctive contribution to the development of mechanics was offered earlier as a simple example. In short, it corresponds to increasing adaptability and is roughly what Kuhn refers to as revolutionary science.

These two processes interact in complex ways, especially for creatures and institutions operating under resource constraints. The combined process is called superfoliation. The situation is summarised in Diagram 3, which provides an information characterisation of both processes.

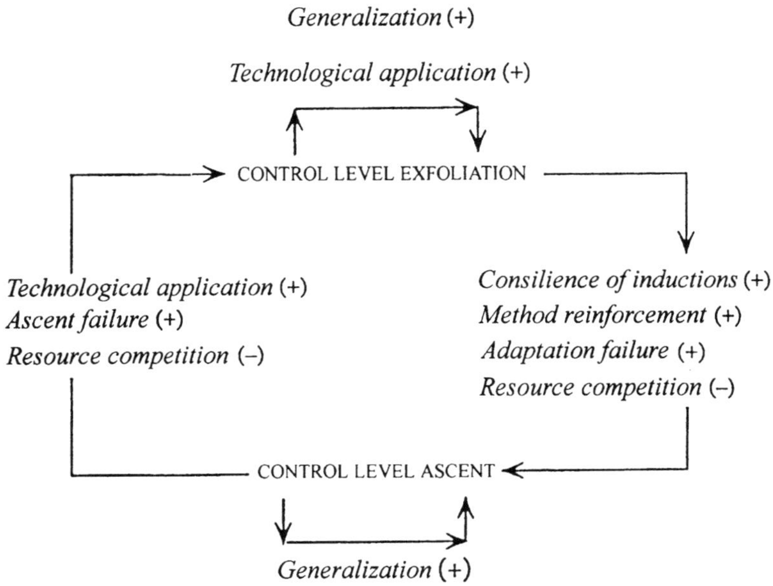

Diagram 3. Regulatory System Development for Science. The arrows indicate positive feedback (+) and negative (–) feedback relations between the processes of exfoliation of cognitive commitments (i.e., refinement and extension of existing commitments) and regulatory ascent in cognitive commitments (i.e., shifting categories of cognitive commitment so as to achieve increasingly higher order regulatory control).

Positive and negative feedback relations between the two processes, now to be discussed, are represented schematically on Diagram 3.

The conditions for regulatory ascent are not well understood in any domain of complex regulatory systems. One process is that of response to regulatory failure. As the attempt to achieve regulation fails, commitment to specific regulatory processes may be suspended in an effort to free up resources to develop alternative, more adequate regulatory structures. The example of relativity theory just discussed provides an example of suspending kinematic assumptions (about simultaneity), a 'backing up the regulatory hierarchy', in the cognitive domain in order to open up further possibilities for theory development. This is a case of positive feedback between *failure* of refinement and exploration of ascent. A related process is evidently Whewellian consilience of inductions: several instances of a higher order regulatory structure (e.g. of generalised Hamiltonian dynamics) which have emerged during the course of horizontal exfoliation (i.e. from specific Hamiltonians) may then be compared in order to extract explicitly the higher order structure instantiated in them. Here is a case where *success* in refinement leads to success in ascent.

There is also negative feedback between the two processes. Ascent may lead to theoretical representations which make the development of laboratory practices and social technologies extremely laborious and expensive (e.g. high energy accelerators), or dangerous (e.g. genetic engineering) etc., which may detract from the extension of refinement (or even from both processes). Again, refinement may so alter the larger circumstances in which the system of science is embedded that further ascent becomes impossible. (This will be sadly the case, e.g., if we succeed in destroying our socio-economic systems through ecological disruption, nuclear warfare, medical pandemic or any other of the many destabilisations of the planetary regulatory system made possible by our technologies.)

Finally, successful refinement will also tend to attract resources to it, for both cognitive and socio-economic reasons, thereby tending to distract from focusing on ascent. Tragically for science, the result is then entrenched cognitive commitment and dogmatism. Then, when recalcitrant experimental results or alternative theoretical ideas emerge, our dogmatic scientists (or philosophers) are unable to integrate them in any meaningful way. The group may disintegrate, ideas be suppressed, or recalcitrant results get explained away by *ad hoc* means. Often, science as a whole must quite literally await the death of the group in order for the institution to be freed of its rigidity, its lack of adaptability.

Thus the inter-relationships between cognitive regulatory refinement and regulatory ascent are complex. No formal logic-based relationship between them will capture the dynamics of superfoliation. Rather the whole process must be theorised within a dynamic regulatory systems context, which provides the right framework for understanding the role and dynamics of methodology within rational science.

Institutional Structure and Dynamics

Now consider the social processes from which science is constituted. A system of the sciencing kind has to have a complex internal organisation to support its regulatory processes. For science this is given in its institutional structure. The design of science's epistemic institutions is crucial to its cognitive capacities and dynamics. The dynamics of those designs is itself an important part of science's cognitive dynamics. The social structure of science is intimately involved in its cognitive characterisation. This position stands in opposition to the standard view on which cognitive and social aspects stand independently, even opposed as the internal and logical versus the external and social. A unification is made possible, and necessary, by the shift to science as a dynamic regulatory process.

The more problem-solving capacity each individual has, the more immediate is the need for cognitively oriented social coordination, i.e., for an 'external nervous system'. Individuals have decidedly finite capacities, thus what must quickly become the focus of nervous system development are not specific individual strategies shaped to particular circumstances and inflexibly retained but general problem solving capacities. General problem solvers have much greater adaptability, because of their behavioural flexibility (behaviour can be conditionalised over more nested conditions) and their greater collective, distributed processing and problem-solving capacities – but only if they can coordinate their individual learning and their criticisms so as to form a systematic investigation of error and organisation of tested information and practice, spread the collective result across the group, focus on it explicitly so as to adapt it, and transmit it from generation to generation. The social processes by which these four crucial cognitive processes occur are structured through social institutions, sets of coordinated behavioural agreement setting up systematic expectations of self and others (cf. Vickers 1968, 1983). In sum, institutions permit cognitive division of labour compatibly with coherence of cognitive strategy. Indeed institutions regulate every important area of scientific activity.

The development of science is a complex process that issues in increased individual and social complexity. The framework for this is epistemic institutional design. To support this dynamic process scientists are required to carry more complex information, even as they are also forced to specialise more and more narrowly within their major discipline. A contemporary molecular biochemist might know less than 0.1% of chemistry as a whole and be an expert on only a few gene structures, yet be familiar with the relevant parts of the whole hierarchy of chemical theory and be sensitive to techniques and results in a dozen different areas in physics, mathematics, biology, neurophysiology, etc. Conversely, chemistry plays intimate roles in biology and geology, while borrowing from physics, engineering and biology, in turn promoting increasing horizontal refinement or adaptation. So, as well as refinement and ascent of regulatory orders, the development of science also leads to increasing cross-theoretical interaction. The result is an intricate set of interrelationships across science, patterned by the context-dependent requirements of local problems and methods.

Effective support and regulation of this intricate and dynamic process requires the development of an increasingly complex institutionalised social structure for the conduct of science. The finitude, flexibility and fallibility of individuals lead, we have seen, to science as a collective enterprise in which the division of labour extends cognitive capacity. Conversely, the socially mediated correlation of labour extends objectivity through systematic intersubjective comparisons. Epistemic institutions, in effect, form the external nervous sub-system organising science.[14] Out of the activities of this larger nervous system grows a larger objective knowledge. Thus institutions are both our necessity and our strength. The institutional complexity of science is only compounded by the rapid increase in scale of scientific operation this century and the increasingly intimate involvement of science in the advance planning of technological development and environmental transformation.

Science is an intrinsically social, institutionalised activity. This is so for deeply embedded regulatory, cognitive reasons. This much may appear more clearly if science is redescribed, so as to situate it more obviously in its biological systems context, as a specialised kind of biological process. Science is a dynamic process in a complex regulatory system which selectively incorporates information from the environment as regulatory structure, accumulating it systematically so as to produce objective knowledge. This system extends from sub-system components of individual humans (e.g. visual system, perhaps the immune system) to individual scientists, institutionalised groups of scientists of various sizes, the sub-ecology of epistemic institutions and the full societal environment. There are complex interactions within and across levels, producing continuous adjustment to environmental signals. Within this regulatory complex there is no principal sub-system boundary which can be drawn, marking out the scientific parts and based solely on intrinsic sub-system properties. Rather, the scientific processes need to be demarcated by their functional consequences, specifically by their contributions to transcending cognitive limitations.

Scientists form social groups of many complex sorts, and the social dimension to science is essential to its nature. The information contained in the abstract regulatory structures is itself transmitted by social communication and is transformed largely during the course of various social exchanges, for example during discussion and debate. Furthermore, the roles which individual scientists adopt in the pursuit of knowledge are heavily influenced by their conception of their own place in the social institutions of science (e.g. whether experimentalist or theoretician, geneticist or ecologist, etc.). With an eye on the relationship to the embedding biological framework, a causal feedback structure for science is given in Diagram 4.

The actions of scientists take place in the four successively more encompassing environments represented in Diagram 4. In order to understand scientific dynamics, we distinguish at least

- the individual scientist,
- the local experimental research/teaching group (includes theoretical research) which forms the locus of daily activity,
- the larger institution of science, and finally
- the wider natural and social environment.

Because of the intricate structure of scientific interactions these environments are not sharply separated from one another in practice; rather they represent convenient abstractions from the complex welter of interactions that is the societal reality. Though the institutional environment of science is internally very complex it is still useful to distinguish within it the group sub-environment which is the immediate locus of testing activities ('experiments'), with the larger scientific environment from which the scientist draws

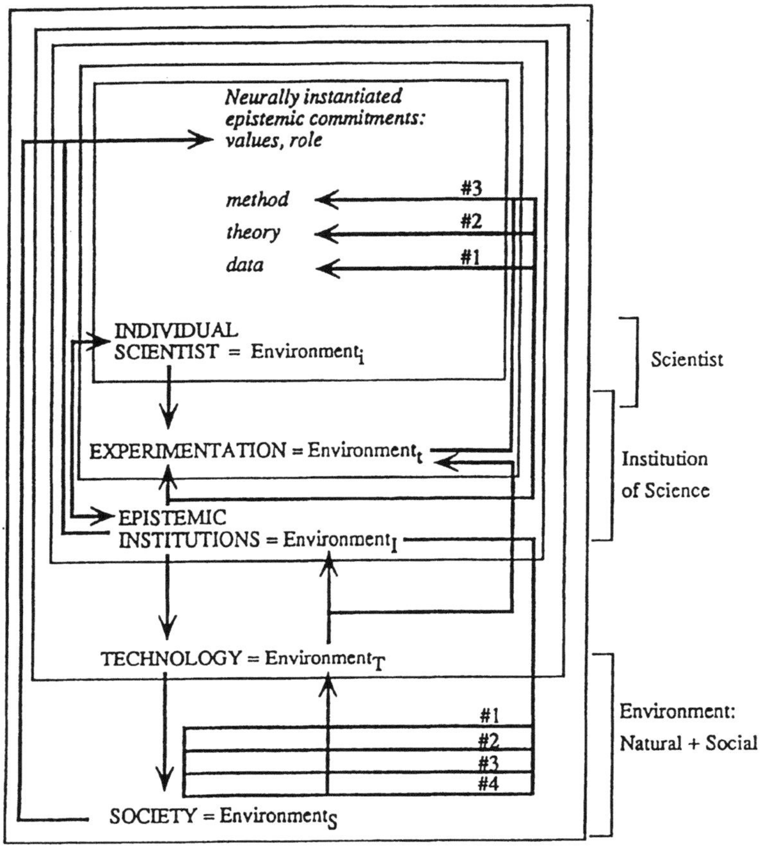

Key:

Science-scientist feedback	*Society-science feedback*
#1 = data shift	#1 = parameter value change
#2 = theory shift	#2 = category or value shift
#3 = method shift, etc.	#3 = institutional design shift
#4 = values, role shift	#4 = science-society relation shift

⟶ : problem-solving/problem-creating process

Diagram 4. Science as a Causal Feedback Structure. Arrows indicate generic kinds of scientist-altering feedback resulting from changes brought about by scientific behavior.

and to which results are submitted. Similarly, though there is tight socio-economic integration in our complex societies, because of the transformative impact of science-based technology on the wider natural and social environment and the feedback to science from technological performance, it is useful to distinguish a specifically technological sub-environment within it. Further discussion of these environments is found in Hooker 1995a. Together they begin the characterisation of science-technology dynamics as a generalised 'genetic algorithm', but one characterised from within the richness of the dynamic systems conception of biological process (rather than a narrow neo-Darwinian selectionism – see *-ed bibliography).

The science-technology system finds its institutional design changed both from within and without. At the same time it acts with self-organising capacity to buffer the epistemic process from distorting change signals and this capacity is an expression of the support for its institutionalised processes which it normally receives from its individual members and its society. Like any other ecology, individuals and the institutions to which they belong undergo a delicate co-evolution. The most convenient picture is probably one in which the science-technology system is thought of as a sub-ecology of a larger ecology of institutions, with individuals participating in several institutions simultaneously (see also Hooker 1987). Understanding this complex and delicate regulatory relationship is the more urgent today as scientific activity becomes more closely entwined with society at large and governments become more active in the transformation of scientific institutions. (It would be good to have to hand, at this point, a theory of the limitations of institutions but we know of no adequate dynamically based theory.)

It follows that both individual and various institutional rationalities should be recognised. But just as various epistemic values may compete, even for an individual scientist, so individual and institutional rationalities may enjoin conflicting actions. But what model of reason to choose for modelling these? The formalist model leads to widespread internal difficulties when pressed and suppresses any individual/institutional distinctions beneath its context-free algorithms (Hooker 1987; 1991; 1995a). The decision-theoretic model of rational action, however, is well able to model the variety of institutional situations involved. Indeed, it provides a natural setting in which to theorise rational institutionalised action because of its specific recognition of context-dependent beliefs and values entering rational action. At all levels of institutionalised decision-making aimed at increasing epistemic utility, methods can be modelled as conjectural and risky resource distributing strategies for theory development and testing.[15]

In sharp contrast to the formalist logic-based conception of rationality, the interleaved patterns of consensus and dissensus which emerge naturally within a decision-theoretical representation of epistemic institutions can be, and typically are, complex. The achievement of collective epistemic gain does require the formation of a consensus, but the critical assessment necessary to objectivity also requires promotion of dissensus. Consensus is of a very specific kind and is compatible with a very large amount of epistemic disagreement. Consider, e.g., an investigation of some theory T. For consensus on what to do, given a background of accepted cognitive commitments C, the participating scientists need to be agreed:

- that it is worthwhile to investigate alternatives to some given theory T,
- that among all of the logically possible alternatives to T a finite set of investigate paths are those which are to be followed first,

- that certain general methodologies (or certain general constraints on specific methodologies) are to apply to formal and experimental exploration and the mutual communication of results, and
- that certain general procedures are to constrain the ways in which the members of the group will each change their own research strategies in the light of the results communicated to them.

Notice, however, that the scientists need not agree on any particular alternative to T as the one most worthy of investigation. Indeed, they need not agree on any ranking of the alternatives in terms of either likelihood of truth or preferability of pursuit. Even more fundamentally, they need not agree on what precisely is at fault with T or even that it definitely has some major flaw. Finally, they need not agree exactly on particular experimental methodologies, on methods of statistical inference, or on the heuristics of theoretical research programs. Indeed, they need not even agree on what the relevant experimental data are at any given point. Despite all of these possible sources of disagreement, it is still reasonable that the scientists conclude, both individually and collectively, that there is a net episemic gain, even a large net epistemic gain, to be had from a cooperative investigation of alternatives to T. It is precisely the purpose of an intelligent epistemic institutional design that it encourages cooperation to proceed, even while also encouraging appropriate critical dissent.

At the individual level the scientists demonstrate their rationality by arranging things so that individuals each follow their own commitments, within institutional constraints, while collectively they exhaust the plausible research strategies. Collectively, the remarkable interdependence of individual epistemic preferences that such arrangements demand requires an institutional explanation focusing on the way in which scientists search for, and adapt to, the institutional roles they perceive themselves to play and the way in which these roles are designed in relation to the overall epistemic enterprise. To the extent that the designs of those roles systematically support the epistemic enterprise we may speak of epistemic institutional rationality.[16]

Epistemic institutional rationality has to do with the generation and reinforcement of patterns of consensus and dissensus of the sort discussed above. Too much emphasis on consensus formation leads to overhasty and uncritical, therefore dogmatic, acceptance; too much emphasis on dissensus leads to paralysis through idiosyncratic speculation and methodological fragmentation unconstrained by critical assessment and synthesis of others' work. In the former case the epistemic pressures emanating from the three enveloping environments in which scientists work (see above) are felt too strongly in one respect (the pressure to agree) and not sufficiently strongly in another (recognition of genuine difference needing investigation and explanation). In the latter case the reverse obtains (or individuals remain indifferent to one another, i.e. feel no pressures). Felt pressure emanating from the social environment to agree with some ideological line is only the worst case of a continuum of epistemic institutional irrationalities here. The institutional framework itself must both give scientists cognitively as well as socially appropriate roles to play and reinforce the values driving those roles through suitable rewards and punishments.[17]

Their support of either consensus or dissensus formation hardly makes epistemic institutions unique among human institutions. Tribal and political rivalries, e.g., generate and reinforce dissensus; group allegiance and coercion generate consensus. Often the two processes will occur together in the same cultural group; but whether separately or together, they are often not systematically related to epistemic development. (Consider, e.g., policy generation processes in democratic political parties.) On occasion they can be more closely related with learning, e.g. when a tribal council seeks consensus on seasonal migration by pooling accumulated experiences and current observations, but not essentially so, and in many institutions, especially where power or profit dominate, truth may be deemed irrelevant or deliberately disavowed.

What is relatively unique about science is the kinds of consensus and dissensus formation it supports; patterns which are systematically relevant to epistemic development. The institutional procedures for submitting a new proposal to public critical assessment through journal publication, conference presentations and the like, e.g., are designed to maximally widen possible dissensus formation and so regulate that formation (the loci of dissensus) and the resultant interactions among dissenters as to focus selectively on the epistemically relevant features. By comparison with most social dissensus processes in our culture, e.g., more and relevant (informed, skilled) actors are drawn into the dispute, yet their interactions are far more complexly and relevantly structured. The feedback structure of this institutional design underpins the construction of the epistemic invariance across methods, scientists and physical contexts which grounds the achievement of objectivity, the product of institutional ration-ality (Hooker 1995a). This is possible, however, only because rational epistemic institutions also so arrange their epistemic activities, especially experimentation and its extension in technological development, that the entities in some domain must, if knowable, play an appropriate causal role in the generation of our representations of them (e.g. Brown 1987, 1988; Campbell 1986).

Rational scientific acceptance requires at least the following:

- The process of acceptance is one for which the relevant features of the world played a relevant causal role in actively bringing about the local consensus on acceptance which is the institutional basis for its wider scientific use as acceptable. This is a necessary condition for being genuinely open to learning the truth (in the surrogate form of our best conjectures thereon). The reason why we object to the intrusion of politics into science is because that intrusion cuts scientists off from causally relevant sensitivity to reality.
- The process of acceptance is one where public, shared epistemic values emerge as a clear counterweight to both sectional and wider social interests and in which what is accepted is critically but fairly assessed. This applies to both local acceptance and wider institutional use; it ensures, e.g., that the evidential bases for local acceptances are always open to retesting.

- The relevant community which is to choose is clearly specified and its role in the choice process is clearly specified.
- The acceptance process itself is continually (in practice: regularly) open to correction. This process should be capable of fundamental reconceptualisation as well as fine tuning. So much is merely prudent in view of the subtleties already raised. But each review has a cost, so that there needs to be an institutional design also for the process of initiating reviews, and this too should satisfy the foregoing clauses.
- If acceptance is to be a rational process, then a minimum requirement is that the institutional design solutions intended to satisfy the first four of these clauses should themselves be argued critically against competing alternatives. That is, these clauses should be meta-consistently applied to their own process of institutional design which they define.

In this context note that the decision-theoretical model of rational agents provides the opportunity to unify the treatment of science methodology and public policies, including science policy, as various cases of institutionalised strategic action. This is not only cognitively valuable but of practical importance in contemporary circumstances. For the practical consequences of science have led to a growing debate about its social control which has struggled with the incoherence generated by the dichotomy between theoretical and practical reason emerging from the formalist conception of reason (Hooker 1987, section 7.9, 1991). The five conditions for methodology sketched above, e.g., have counterparts in a theory of objective public policy contents (see Hooker 1989).

Of course there is operative in all the foregoing design characterisations a large *ceteris paribus* clause to cover the imperfections of human individuals and institutions, the vicissitudes of history, etc. But these are complications; the distinctive design principle is that characteristic of a social learning system, i.e. of a rational epistemic institution. However, the complications run deeper. In science we exclude mental incompetents and the relevantly immoral from the epistemic community, and this already requires careful theoretical design reflection; beyond that the case becomes even more complex. The community for science is increasingly 'soft': very few people can really dispute nuclear physics reasonably, e.g., so science is not in this sense a public enterprise, and this despite the impact of nuclear science on the wider community. Moreover, the competent scientific community is shrinking steadily as scientific complexity increases (cf. the contemporary chemist above who can have read less than 1% of even the important papers in the discipline). Only epistemic institutional design holds the scientific process together, otherwise it would fragment into a thousand specialties, each too small to sustain an objective process of scientific acceptance. One has the sense of some difficult epistemic regulatory problems emerging here for the human species. But at least they again have a principled formulation in this regulatory systems context. Here these design issues are simply noted in order to return to the larger issue of characterising the distinctive epistemic institutional design and dynamics of science.

Progress

The development of the science-technology system over the past three centuries has spurred superfoliation; it has produced greatly increased regulatory refinement but, more importantly, it has also stimulated very significant regulatory ascent. It is the increasing scale, complexity, empirical competence and adaptability of the science-technology system that we associate with the progress of objective knowledge. As the history of major scientific changes or revolutions sharply reminds us, specific cognitive adaptation is not permanent or cumulative. The new principles of Newtonian mechanics contradicted Aristotelian principles at many orders. Yet Aristotelian principles had served humans successfully for two millennia and, as Feyerabend likes to emphasise, precisely because they captured the patterns of the commonsense environment of daily action.[18] The splendid two hundred year history of (largely) successes for Netwonian mechanics, producing a wide array of wonderful refinements, did not prevent the demise of the theory in the changed evidential and theoretical environment that began to support relativity theory instead. Short term explanatory/predictive success is evidently no measure of long term explanatory/predictive success. (Nor, so long as some scientists survive who have access to a theory, is short term failure any guide to long term failure, a favourable environment for renewed application of the ideas may again reappear; see e.g. the fluctuating fortunes of the wave theory of light.)

This evaluation is certainly compelling. But it doesn't seem quite right. To see the primary aim of science being that of empirical adequacy to the current environment, as important as that is, is to nearly completely miss the functional significance of modern scientific knowledge, namely the transition from actuality to possibility. The age-old method of learning has been to observe passively, generalise from what was observed and test by prediction. It is essentially the only survival strategy available to early human societies which had access to relatively little energy and information, and we still find such activities today in areas where we have little information, e.g. in some biological classification, and/or little energy, e.g. astronomy. Its only goal could be refinement, specifically generality and precision, whenever the environmental pattern was stable and simple enough.

But alongside passive observation there has been the equally ancient method of probing to disturb nature and learning from her reaction. Generalised, it is the experimental method. But to properly apply this method really requires information (to know where to direct disturbance and what kind of probe to choose) and energy (to provide the disturbance and to hold other conditions constant); it was only given prominence and systematic application with the Galilean revolution in science at the Renaissance. The secret of its increased epistemological power over passive observation is the ability to create not only regulated conditions, but conditions that would not normally appear (like near-frictionless motion). In this way one passes from knowledge of what is merely actual, is going to happen given present conditions, to knowledge of what is *possible*, even should it never happen in normal conditions. It is this transition which made possible the overthrow of Aristotelian commonsense. Newton's laws of motion tell us not only why as a matter of fact some particular projectile fell where it did, but they also tell us all the possible trajectories which projectiles can have, including ones that didn't exist at Newton's time such as terrestrial satellites.

And the further development of the science-technology system has produced a third, still more powerful, method: that of designed possibility. While experimental method in its classic phase revealed what was naturally possible (instead of just actual), the essence of designed possibility is in turn to obtain possibilities which are not natural but which become possibilities within a deliberately designed system. Study natural chemicals how you will, only under designed conditions can one make possible special-purpose plastics with the tensile strength of steel; study purification and moulding how you will, only in near-zero gravity conditions can you find the possibility of very high purity and sphericity. The scientific relevance of this method is made dramatically obvious by human space travel: there were no facts of humans in space to observe objectively in advance, there were no tentative probings of ways of travelling to the moon to learn from disturbing them; rather we needed to understand in advance all that was possible, though not naturally possible, for us so that we could do it right the first time. The technological relevance is made equally dramatically obvious by the current efforts being put anticipatively into developing fifth generation computers, new designer materials, etc. Any country not able to plan anticipatively on a 10–30 year future time horizon can no longer compete internationally in new technologies but is restricted to producing local variations on existing products (i.e. to horizontal technological progress).

The shift to increasingly wide ranges of possibilities corresponds to conditionalising responses in ever deeper nested layers, i.e. to regulatory ascent or increasing adaptability. An individual or species with sufficient adaptability will not be eliminated by an environmental shift but will be able to adapt to it. Thus it represents increased autonomy. While successful adaptation is fragile, local and non-cumulative, adaptability is, by its functional definition, less local and therefore less fragile. (Note that this is a matter of degree, there is neither guarantee or expectation of either universality or complete resilience.) Moreover, under appropriate conditions (e.g. reinforcement from a heterogeneous environment) and within certain limits, it may be accumulate and hence define a less local direction of progress. Let us call this vertical progress.

The shift to an increasing range of possibilities corresponds quite literally to ascending cognitive regulatory orders. An individual scientist or group of scientists with sufficient cognitive adaptability will not have their cognitive commitments eliminated by introduction of new elements into their scientific environment but will be able to adapt. The degree of adaptability is measured by the number of environments which can be adapted to, i.e. by what possibilities can be encompassed. These can be increased by regulatory ascent, i.e. by vertical progress, which at the same time often also makes possible increasing refinement, i.e. horizontal progress. (The cognitive regulatory framework cuts down possibilities even as it opens them up.) In physics, e.g., very general representation theorems have been proved this century which effectively show that, given only very general assumptions about time and possible physical states, Newton's and Schrodinger's equations are the only possible dynamics for their kinematical structures and that certain symmetries must hold. And Whewellian vertical conciliational ascent has uncovered a number of fundamental constants with tight mathematical inter-relations and consequences for dynamical possibility.[19] Successful adaptability has led us to successful adaptation (not always in the short run) and of course it also requires sufficient adaptive refinement to promote basic functioning. Cognitive adaptability is, by its functional definition, less local and therefore less fragile than local empirical adequacy. Each successively higher regulatory order yields a correspondingly wider notion of (relative) non-local vertical progress. Moreover, under appropriate conditions and within certain limits, it may accumulate and hence define a direction of vertical scientific progress.

Horizontal progress specifies no general or global direction to change because the kinds of changes which count as horizontal progress are specific to each environment. Vertical progress specifies a relatively more global direction to change because it covers a range of environments rather than just one. But the character of the changes constituting vertical progress will also change as one passes outside the adaptability set of environments. And should environmental shift bring this about adaptability will fail, just as horizontal progress fails when its particular environment changes. And yet there is something more to vertical progress than just a less local form of horizontal progress. To see this consider again that each successively higher regulatory order yields a correspondingly wider notion of (relative) vertical progress. Now extrapolate this to the possibility of increasing adaptability until the range of environments encompassed includes all of the possible environments in this cosmos. A creature with this capacity would be the ultimate universal creature, able to maintain its life processes in any possible circumstance. It would be completely autonomous. For cognisers this capacity would involve having a representation of all possibilities. Indeed, just this latter condition suffices to satisfy the goal of science, it is not necessary to actually be able to survive in every possible environment. This notion is a complex one and when thought through in detail may prove difficult to even formulate coherently for various fundamental physics. Nonetheless, it does serve to point out a universal goal for adaptability. It is evidently the idea Piaget had when he claimed that developing a mature cognitive structure was to develop a universal set of operations, ones that could be guaranteed to apply to all possible situations, and hence to truly grasp necessity (see Hooker 1995a, Chapter 5).

The ascent from actuality to wider and wider sets of possibilities connects vertical progress to the achievement of objectivity. The form objective knowledge takes in physics is the representation of each 'viewpoint' as one projection among many in a higher order geometry (Hooker 1992). It is this geometry that specifies the possibilities, including both the dynamical possibilities (manifold of possible solution trajectories) and the various descriptions of these across possible dynamical observers. What is real is what is invariant across these, as determined by available methods, and objective knowledge is roughly a description satisfying these invariances. In a heterogeneous world the search for invariance is driven to a higher order than in a homogeneous world. Sufficiently general patterns must be found that remain invariant across heterogeneous situations (environments); these will only be found at a higher regulatory order than the data. Achieving objectivity will require cognitive adaptability. As heterogeneity increases so will the requirement for adaptability to deliver objective knowledge and (so) successful adaptation. The recent understanding of chaos makes a striking illustration of this (see *-ed bibliography).

Finally, the exfoliations of refinement and ascent require, and lead socially to, the development of increasingly complex and extensive institutions to support the constitutive scientific processes of the kind indicated.

Thus there are complex relationships between the achievement of objectivity, progress and autonomy, and the development of institutional structure, relationships determined, not by logic, but through the cognitive dynamics.

What is, in the last analysis, progressive over the longer term is the superfoliation of the regulatory structure itself. In the prosecution of science we are engaged in a magnificent and potentially unlimited superfoliatory process which takes us 'beyond ourselves' as we increasingly reflect the cosmos in the regulatory order of planetary epistemic organisation. Magnificent as this sciencing process seems to us, at least from our feeble parochial standpoint, as much as we seem here to have caught on to a cosmic process that genuinely transcends us, we close this exposition with a cautionary historical note. Each individual human, and the species as a whole, is in general ignorant (in advance) of the consequences of its own disturbances. The risk carried by inducing change is that it may rebound and eliminate the initiators. Application to the current human situation is all too obvious, with the massive threats to our planetary environment induced by the advent of nuclear warfare, industrial environmental pollutants and industrial resource consumption, and to ourselves with the massive increases in complexity, pace, individual isolation and institutional incoherence induced by our growing Change Machine. Everywhere across the globe, ecologies and personal human relations are being simplified and destabilised, their resiliency decreased, while our economic and scientific institutions complexity. The real issue for us is whether our scientific understanding, and our larger institutional control of our own processes, can develop at a pace fast enough to control our disturbing impact. Recall the Conant/Ashby theorem that a controller must have access to at least as many states as does the system to be controlled (Conant/Ashby 1970; cf. Ashby 1970), and the non-linear increase with complexity of information required for system coherence, and you have the race against ignorance we are in. The secret of sustainable advantage through adaptability is to be able to relate the changes induced to the adaptabilities possessed. It is precisely this relationship which is in doubt for us humans and which at prior times, thanks to our ignorance, we have not even been in a position to sustain. The third and fourth order institutional feedback control loops required here are an urgent developmental priority.

CONCLUSION

Given the historically dynamic nature of even our cognitive processes, are there appropriate naturalist reconstructions for the traditional central notions of scientific method, such as explanation and objectivity? Is there not, after all, considerable danger that the distinctively cognitive dimension of the science-technology process will be dissipated in its wider societal role, so that no distinctive claim on truth can be defended, political abuse being indistinguishable from Platonic ideal? These are indeed present and serious dangers, both practical and theoretical. The practical danger derives from the shaping of research by military and commercial imperatives. The theoretical danger is more subtle but no less devastating. It lies in the possibility that the conception of science-technology as an institutionalised societal sub-system will subvert any attempt to theorise its distinctive cognitive features. Once we have abandoned timeless logic and abstract method for history and institutional process we shall have lost our capacity to theorise rational method, objective knowledge, explanatory power, and so on.

We believe that only a theory which can understand cognition as embedded in a full societal dynamics could produce an articulation of reason in science adequate to its historical complexity. And such an account will be required in order to facilitate intelligent decision making with respect to our still more complex societies in which are embedded increasing interactions among science, technology, commerce, politics and our natural environment.

University of Newcastle, Callaghan, Australia

NOTES

[1] Representatives of these positions are, in order: Bloor 1960; Pickering 1984 (sociological reductionism), Feyerabend 1978a, b (epistemological anarchism), Campbell 1974, 1986; Hahlweg/Hooker 1989 (evolutionary epistemology), Giere 1988; Thagard 1988 (computationalism), Brown 1979, 1988; Churchland/Hooker 1985; Churchland 1989; Hooker 1987, 1995a; Newton-Smith 1981; Shapere 1984 (naturalism/realism), van Fraassen 1980 (neo-empiricism).

[2] We have pursued these theses some distance in Herfel forthcoming a; b and Hooker 1989a, 1991, 1994a, 1995a and further studies are in preparation. Material in section V below is abstracted from Hooker 1995a, which also contains a substantial select bibliography on the relevant dynamical models, now bursting upon science, and from which a brief selection, *-ed, is included in the references at the close of this paper.

[3] For these standard positions in philosophy of science see, e.g., respectively Achinstein/Barker 1969; Neurath *et al.* 1970; Popper 1972, 1979, 1980; Lakatos 1970; Kuhn 1962, Musgrave 1993; cf. Hooker 1987, 1991, 1995a.

[4] See Feyerabend 1978a, b. On Kantian intelligibility frameworks see Friedman 1986; Hooker 1994b.

[5] This inherited knowledge includes various inborn needs and drives (e.g. Popper 1979, pp. 23–4); 'theory-laden perception' (e.g. pp. 35–7); Kantian-like categorical regularities (*a priori* for the individual, as inherited species information, but not *a priori* valid, Popper 1979, pp. 24, 72; Popper 1972, p. 47). The inherited basis not only makes cognitive evolution possible, it provides the initial 'selection' of what aspects of the environment are 'problems' (cf. Popper 1979, pp. 258–9) and also the beginning 'theories' and expectations against which 'problems' appear as mismatches between the expectation and practice.

[6] There is a shift in the locus of selection across levels from the exterior of the organism to its interior. The simplest organisms reactively produce reflex behaviours which result in their being killed or not. All the action, so to speak, is external to them. But adaptable organisms, at least behaviourally adaptable ones, can increasingly run trials within themselves and then anticipatively select the one judged best for the circumstances. Note the similarity of Popper's account here to Piaget's basic principle of an increasing shift from exogenous reaction to endogenous construction (Piaget 1976; cf. Hooker 1995a, Chapter 5) and with Campbell's account of the 'nested hierarchy of inductive achievements', which act to narrow the possible 'search space' for the emission of random trials, cf. Campbell 1974, pp. 421ff, or Campbell 1977, p. 502: 'presumptive procedures' which 'reduce the waste' involved in random trials.

[7] Compare the role of abstractions here to that of reflective abstraction in Piaget, but contrast the attention Piaget give to construction with its neglect in Popper (it is reduced to randomness, or 'psychology'). In Popper this is essentially because it cannot be reduced to formal (logical) operations. Both Popper and Piaget agree about the confinement of reason to formal systems, hence Piaget's attempt to characterise stages formally. But despite the internal tensions it caused, Piaget took the underlying naturalist regulatory model seriously, more seriously than does Popper. See Hooker 1995a, Chapter 5 for further discussion of these themes.

[8] The only exception is the use of the concept of probability, e.g. in corroboration for which Popper provides a measure in probabilistic terms, see Poper 1980, appendix ix. This is still a formal concept and, since classical probabilities are maps from classical deductive logic into the unit real interval (a conception Popper helped pioneer), one not far removed from a purely logical one.

[9] This is argued in detail in Hodges and Hooker 1995.

[10] The idea of the theory-ladenness of observation, and hence of its non-neutrality in relation to theory, is a slippery one, with many meanings. There are many kinds of descriptions of the same situation, differing in their uses of different kinds of theoretical terms (those from instrumental, mathematical and experimental as well as theoretical practices) and in the kinds of epistemic claims made (from vague to precise and experiential to public). See Hooker 1973a, 1974 for one realist view.

[11] In 1974 one of us (CAH) wrote:

> Traditional philosophy of science regards science as an abstract structure . . . entirely separated from the activities of the scientist *qua* human being . . . The consequence has been rational reconstructions of the history of science which ignore large segments of it (discovery is irrational) and in which scientists habitually contradict the alleged canons of *Science* (e.g. by retaining a refuted theory). . . . this leads to the exclusion of all that is importantly human from the account, since most human activities cannot be captured in the combination of basic descriptive statement plus logical machinery. . . .
>
> As against this abstractive approach . . . [Naturalist Realism] proposes placing (or re-placing) the scientist at the centre of the socio-intellectual-ethical complex known as science. . . . then once a rational account of the lives of scientists has been offered we can turn to the

> task of asking in what measure an abstract notion of *Science* can be rescued from the collective participation of *scientists*. The detailed, systematic development of this approach lies in the future; what Feyerbend and Kuhn have done has been to offer us glimpses of the great power of this reorientation where, e.g., what was irrational *vis-a-vis Science* (even though it affected its course profoundly) can now be seen to be rational *vis-a-vis* the *scientist* . . . (Hooker 1974, pp. 411–2)

The period 1974–84 was spent developing a Naturalist Realism which would underpin this account, that of 1985–90 developing an evolutionary dynamical analogy for a population of scientists to provide support for it and 1990+ has been spent developing the insights offered by the dynamics of non-linear complex adaptive systems as the proper framework for a dynamics of science. It is here that WEH, through his work on non-linear dynamical models (Herfel 1990), has made a particular contribution by emphasising the importance of developing a dynamical critique of traditional philosophy of science in these terms, and the fruitfulness of non-linear dynamical concepts, such as constraint, for retheorising science.

[12] The reader is reminded that this is metaphorical functional talk; abstract contents do not literally interact. Rather, the causal description, where interactions actually occur, is concerned with the neural and behavioural causal consequences of the neural states which instantiate the corresponding cognitive commitments. It is a convenient shorthand to speak in this manner, and does no harm where the functional outcomes of the real interactions is characterisable purely logically; but as we have seen, this simplification is only available in suitably confined contexts.

[13] The explosion of new, alternative mathematical forms beginning with non-Euclidean geometries a century ago, and similarly for logic since 1900, bears witness that in these areas also we learn through experience. Cf. Brown 1988; Holdsworth/Hooker 1983; Hooker 1975/79, 1979, 1987. On change in scientific method see e.g. Blake *et al.* 1960; Oldroyd 1989.

Feyerabend is right, no specific transition in the history of science will be quite like any other, any more than any specific change in the evolutionary history of genetic regulatory systems is quite like any other. But he is also wrong: this fact does not prevent our understanding them through theorising them – to the contrary (cf. Hooker 1991). What is truly revolutionary about science is surely not the sheer magnitude of its capacity for collecting information, nor even its capacity for generating valuable theories, important though these are, but its enshrinement of a process which has steadily pushed us to explore ever higher orders of regulatory structure.

[14] See Hooker 1987, sections 7.8, cf. sections 8.8.7, 8.3.9. The notion of an external nervous system is most obvious in the case of the computer, which at present is like a simple of ganglion cluster in a distributed nervous system. Once we have self-programming, self-re-designing computers they will start to look more like us, local quasi-autonomous intelligent information processing centres.

[15] Or some suitable complication thereof. The centrality of risk forces choice among different rationality principles, e.g. between maximum expected utility and minimising maximum expected loss, with the option of various additional constraints, e.g. eliminating possible losses above some ceiling. See e.g. the discussions generally in Brams 1976; Elster 1983; Hooker *et al.* 1977 and McClennen 1990 and specifically Hooker 1994a at note 8 and text. An issue which arises is the nature of the utilities to be attributed to scientists qua scientists; are they purely epistemic, irretrievably mixed epistemic and pragmatic, or purely pragmatic? (Here epistemic utilities are those directly related to the pursuit of truth and pragmatic utilities are those related to other goals, e.g. prestige, salary, institutional power, etc.) While Levi 1967 adopts the first position, Nicholas 1984 the third and Hooker the second (Hooker 1987, Chapter 5), we pursue the issue no further here except to comment that, insofar as pragmatic utilities play an essential role, the explanation of the epistemic character of science must focus on the capacity of its institutional design to so entrain individual behaviour as to collectively pursue truth. So we return to institutional design.

[16] See e.g. Munevar 1989; Hooker 1987, Chapters 7, 8. The discussion of the preceding para-

graph was drawn from Bjerring/Hooker 1979; now see also e.g. Ackermann 1986. On the mutual shaping of individual and institutional roles of relevance here see e.g. Vickers 1968, 1983. Ultimately, we can bring the sociological, political and economic literature on institutional design and functioning to bear on the design of epistemic institutions and also the operations research and cognate literature, especially the viable and inquiring systems literature, see respectively Espejo and Harndon 1989 and Churchman 1972.

17 For example, rejection for fraud, promotion or prestige for successful new theories *and* for successful criticism of extant theories, data or methods. Studies by Galison 1987; Hull 1988; Latour/Woolgar 1979; Latour 1987 and others are beginning to elaborate the extremely complex dynamics of these institutional processes; though we are not yet in a position to subject them to cognitive theorising. Often these authors take themselves to be rejecting scientific reason but, like Feyerabend before them, it is really only the old formalist notion of method as simple logical rules which is rejected; the rest of their studies serve as grist for the mill of the present regulatory systems conception/decision theoretic account. On Feyerabend specifically see Hooker 1991, on the recovery of the cognitive in sociology see Hooker 1995a.

18 See e.g. Feyerabend 1978a. For recent work on Newton, especially the continuing value of his methodological heritage, see e.g. Harper 1989, 1993; Hooker 1994c; Stein 1990 and references.

19 In the first case we refer respectively to the second Stone representation theorem and to the CTP symmetry theorem, see e.g. Bub 1974 or Hooker 1973b and Streater and Wightman 1964. With respect to the physical constants one is referring to the discovery of e, h and c, respectively the electron charge, the quantum constant and the velocity of light; their values are critical to the structure and stability of the universe as we know it. Not only do we not have a plethora of mathematical metaphysics that satisfy all these restraints, despite frequent naive assumptions to the contrary, it is still not obvious that we have even one – see e.g. Hooker 1992.

REFERENCES

* Abraham, R. and Shaw, C. (1982), *Dynamics – The Geometry of Behavior*. 4 vols. Vismath: The Visual Mathematics Library (Ariel Press, Santa Cruz).

Achinstein, P. and Barker, S. (eds.) (1969), *The Legacy of Logical Positivism* (John Hopkins Press, Baltimore).

Ackermann, R. J. (1986), 'Consensus and Dissensus in Science', in *PSA 1986*, A. Fine and P. Machamer (eds.) (Philosophy of Science Association, East Lansing, MI).

* Adams, R. N. (1988), *The Eighth Day: Social Evolution as the Self-Organization of Energy* (University of Texas Press, Austin).

Anderson, P. W. (1988), 'Spin Glass Hamiltonians: A Bridge between Biology, Statistical Mechanics and Computer Science', in Pines, D. (1988).

* Anderson, P. W., Arrow, K. J. and Pines, D. (eds.) (1988), *The Economy as an Evolving Complex System* (Addison Wesley, Redwood City, Cal.).

Ashby, W. R. (1970), *Design for a Brain* (Chapman and Hall Science Paperbacks, London).

Bjerring, A. K. and Hooker, C. A. (1979), 'Process and Progress: The Nature of Systematic Inquiry', in *Perspectives in Metascience*, J. Barmark (ed.) (Berlings, Lund).

Blake, R. M., Ducasse, C. J. and Madden, E. H. (1960), *Theories of Scientific Method* (University of Washington Press, Seattle).

Bloor, D. (1960), *Knowledge and Social Imagery* (Routledge and Kegan Paul, London).

Brams, S. J. (1976), *Paradoxes in Politics* (Free Press, New York).

Brown, H. I. (1979), *Perception, Theory and Commitment: The New Philosophy of Science* (University of Chicago Press, Chicago).

Brown, H. I. (1987), *Observation and Objectivity* (Oxford University Press, Oxford).

Brown, H. I. (1988), *Rationality* (Routledge, London).

Bub, J. (1974), *The Interpretation of Quantum Mechanics*. (Reidel, Dordrecht).

Campbell, D. T. (1974), 'Evolutionary Epistemology', in Schilpp 1974.
Campbell, D. T. (1977), 'Comment on "The Natural Selection Model of Conceptual Evolution" ', *Philosophy of Science* **44**, pp. 502–7.
Campbell, D. T. (1986), 'Science's Social System of Validity-Enhancing Collective Belief Change and the Problems of the Social Sciences', in *Pluralisms and Subjectivities*, D. W. Fiske and R. A. Schwader (eds.) (University of Chicago Press, Chicago).
* Caplan, S. R. and Essig, A. (1983), *Bioenergetics and Linear Nonequilibrium Thermodynamics* (Harvard University Press, Cambridge, Mass.).
Cartwright, N., Shomar, T. and Suarez, M. (1995), 'The Tool Box of Science: Tools for the Building of Models with a Superconductivity Example', in Herfel et al.
Chandrasekhar, S. (1961), *Hydrodynamic and Hydromagnetic Stability* (Clarendon, Oxford).
Churchland, P. M. (1989), *A Neurocomputational Perspective* (Bradford/MIT Press, Cambridge, Mass.).
Churchland, P. M. and Hooker, C. A. (1985), *Images of Science: Essays on Realism and Empiricism* (University of Chicago Press, Chicago).
Churchman, C. W. (1972), *The Design of Enquiring Systems* (Basic Books, New York).
Conant, R. G. and Ashby, W. R. (1970), 'Every Good Regulator of a System must be a Model of that System', *International Journal of Systems Science* **1**, pp. 89–97.
* Diner, S. Fargue, D., and Lochak, G. (1986), *Dynamical Systems: A Renewal of Mechanism* (World Scientific, Singapore).
* Dyke, C. (1988), *The Evolutionary of Dynamics of Complex Systems* (University Press, Oxford).
Elster, J. (1983), *Sour Grapes: Studies in the Subversion of Rationality* (Cambridge University Press, Cambridge).
Elster, J. (ed.) (1986), *Rational Choice* (Blackwell, Oxford).
Espejo, R. and Harnden, R. (eds.) (1989), *The Viable System Model: Interpretations and Applications of Stafford Beer's VSM* (Wiley, Chicester).
Feyerabend, P. K. (1962), 'Explanation, Reduction and Empiricism', in *Minnesota Studies in the Philosophy of Science*, Vol. I, H. Feigl and G. Maxwell (eds.) (University of Minnesota Press, Minneapolis).
Feyerabend, P. K. (1978a), *Against Method* (Verso, London).
Feyerabend, P. K. (1978b), *Science in a Free Society* (New Left Books, London).
* Firrao, S. (1983), *The Theory of Self-organizing Systems in Physics, Biology and Psychology* (CENS, Milano).
* Foster, J. (1987), *Evolutionary Macroeconomics* (Allen and Unwin, London).
Franklin, A. (1986), *The Neglect of Experiment* (Cambridge University Press, Cambridge).
* Frehland, E. (ed.) (1984), *Synergetics – from Microscopic to Macroscopic Order* (Springer-Verlag, New York).
Friedman, M. (1986), 'The Metaphysical Foundations of Newtonian Science', in *Kant's Philosophy of Physical Science*, R. E. Butts (ed.) (Reidel, Dordrecht).
Galison, P. (1987), *How Experiments End* (University of Chicago Press, Chicago).
Galison, P. (1988), 'Philosophy in the Laboratory', *The Journal of Philosophy* **LXXXV**(10), pp. 525–7.
Giere, R. (1988), *Explaining Science: A Cognitive Approach* (University of Chicago Press, Chicago).
* Glass, L. and Mackey, M. (1988), *From Clocks to Chaos* (University Press, Princeton).
Gould, S. J. and Eldredge, N. (1993), 'Punctuated Equilibrium Comes of Age'. *Nature* **366**, pp. 223–7.
Hacking, I. (1983), *Representing and Intervening* (Cambridge University Press, Cambridge).
Hahlweg, K. and Hooker, C. A. (1989), 'Evolutionary Epistemology and Philosophy of Science', in *Issues in Evolutionary Epistemology*, Hahlweg/Hooker (eds.) (State University of New York Press, New York).
* Haken, H. (1977a), *Synergetics: An Introduction* (Springer-Verlag, Berlin).
* Haken, H. (1977b), *Synergetics: A Workshop* (Springer-Verlag, Berlin).

* Hannan, M. T. and Freeman, J. (1989), *Organisational Ecology* (Harvard University Press, Cambridge, Mass.).

* Hao B-L. (1984), *Chaos* (World Scientific, Singapore).

Harper, W. (1989), 'Consilience and Natural Kinds Reasoning in Newton's Argument for Universal Gravitation', in *An Intimate Relation*, J. R. Brown and O. Mittelstrass (eds.) (Kluwer, Boston).

Harper, W. (1993), 'Reasoning from Phenomena: Newton's Argument for Universal Gravitation and the Practice of Science', in *Action and Reaction*, P. Theerman and A. F. Seef (eds.) (University of Delaware Press, Newark, Del.).

Herfel (1990), "Coming Attractions: Chaos and Complexity in Scientific Models', PhD Dissertation, Temple University, Philadelphia, USA.

Herfel (forthcoming a), 'Constructivism and Nonlinear Dynamics: Towards an Integration of the Cognitive and Social.'

Herfel (forthcoming b), 'How Social Constraints Enable Scientific Practice.'

Herfel, W. E. and Hooker, C. A. (1995), 'From Formal Machine to Social Colony: Toward a Complex Dynamical Philosophy of Science', *Preceedings, Xth Congress of Logic, Methodology and Philosophy of Science*, M. L. dalla Chiara and R. Giuntini (eds.) (Kluwer, Boston).

Herfel, W., Niiniluoto, I., Krajewski, W. and Wójcicki, R. (1995), *Theories and Models in Scientific Processes: Poznan Studies in the Philosophy of the Science and the Humanities* **44** (Editions Rodopi, Amsterdam).

Hodges, B. E. and Hooker, C. A. (1995), 'Reason and the Regulation of Decisions: Popper's Evolutionary Epistemology', in Hooker, C. A. 1995a.

Holdsworth, D. G. and Hooker, C. A. (1983), 'A Critical Survey of Quantum Logic', *Logic in the 20th Century, Scientia*, 1–130.

* Holland, J. H. (1992), *Adaptation in Natural and Artificial Systems* (MIT/Bradford, Cambridge, Mass.).

Hooker, C. A. (1972), 'Critical Notice: *Against Method*: P. K. Feyerabend', *Canadian Journal of Philosophy* **II**, pp. 489–509.

Hooker, C. A. (1973a), 'Empiricism, Perception and Conceptual Change', *Canadian Journal of Philosophy* **III** (1973), pp. 59–75.

Hooker, C. A. (1973b), 'Metaphysics and Modern Physics', in *Contemporary Research in the Foundations and Philosophy of Quantum Theory*, Hooker, C. A. (ed.) (Reidel, Dordrecht).

Hooker, C. A. (1974), 'Systematic Realism', *Synthese* **26**, pp. 409–497. Reprinted in Hooker 1987.

Hooker, C. A. (1975), 'Global Theories', *Philosophy of Science* **42**, pp. 152–179. Reprinted in Hooker 1987.

Hooker, C. A. (1976), 'Methodology and Systematic Philosophy', in *Basic Problems in Methodology and Linguistics*, R. E. Butts and J. Hintikka (eds.) (Reidel, Dordrecht).

Hooker, C. A. (1981a), 'Formalist Rationality: The Limitations of Popper's Theory of Reason', *Metaphilosophy* **12**, pp. 247–266.

Hooker, C. A. (1981b), 'Towards a General Theory of Reduction', *Dialogue* XX, Part I, Historical framework, 38–59, Part II, Identity and reduction, 201–36, Part III, Cross-categorial reduction, 496–529.

Hooker, C. A. (1982), 'Understanding and control', *Man-environment Systems* **12**, pp. 121–60. Reprinted in Hooker 1987.

Hooker, C. A. (1987), *A Realistic Theory of Science* (State University of New York Press, Albany N.Y.).

Hooker, C. A. (1988), 'Critical Notice: Churchland, P.S., *Neurophilosophy*', *Australasian Journal of Philosophy* **66**, pp. 240–8.

Hooker, C. A. (1989), 'From Logical Formalism to Control System', in *PSA 1988*, A. Fine and M. Forbes (eds.) (Philosophy of Science Association, East Lansing).

Hooker, C. A. (1991), 'Between Formalism and Anarchism: A Reasonable Middle Way', in

Beyond Reason: Essays on the Philosophy of Paul Feyerabend, G. Munevar (ed.) (Kluwer, Boston), pp. 41–107.

Hooker, C. A. (1992), 'Physical Intelligibility, Projection, Objectivity and Completeness: The Divergent Ideals of Bohr and Einstein', *British Journal for the Philosophy of Science* **42**, pp. 491–511.

Hooker, C. A. (1994a), 'Idealisation, Naturalism, and Rationality: Some Lessons from Minimal Rationality', *Synthese* **99**, pp. 181–231.

Hooker, C. A. (1994b), 'Regulatory Constructivism: On the Relation bewteen Evolutionary Epistemology and Piaget's Genetic Epistemology', *Biology and Philosophy* **9**, pp. 197–244.

Hooker, C. A. (1994c), 'Bohr and the Crisis of Empirical Intelligibility: An Essay on the Depth of Bohr's Thought and our Philosophical Ignorance', in *Niels Bohr and Contemporary Philosophy*, J. Faye and H. J. Folse (eds.) (Kluwer, Dordrecht).

Hooker, C. A. (1994d), 'From Phenomena to Metaphysics', in *Logic and Philosophy of Science in Uppsala*, D. Prawitz and D. Westerstahl (eds.) (Kluwer, Dordrecht).

Hooker, C. A. (1995a), *Reason, Regulation and Realism* (State University of New York Press, Albany).

Hooker, C. A. (1995b), 'Toward a Naturalised Cognitive Science', in *Psychology and Philosophy*, R. Kitchener and W. O'Donohue (eds.) (Sage, London, 1995).

Hooker, C. A. (ed.) (1975/79), *The Logico-Algebraic Approach to Quantum Mechanics*, 2 vols. (Reidel, Dordrehct).

Hooker, C. A. (ed.) (1979), *Physical Theory as Logico-operational Structure* (Reidel, Dordrecht).

Hooker, C. A., Leach, J. and McClennen, E. (eds.) (1977), *Foundations and Applications of Decision Theory*, 2 vols. (Reidel, Dordrecht).

Hoyningen-Huene, P. (1993), *Reconstructing Scientific Revolutions: Thomas S. Kuhn's Philosophy of Science* (University Press, Chicago).

Hull, D. L. (1988), *Science as a Process* (University of Chicago Press, Chicago).

* Jantsch, E. (ed.) (1981), *The Evolutionary Vision: Toward a Unifying Paradigm of Physical, Biological and Sociocultural Evolution* (Westview, Boulder, Colo.)

* Jantsch, E. and Waddington, C. H. (1976), *Evolution and Consciousness: Human Systems in Transition* (Addison-Wesley, London).

* Kauffman, S. A. (1992), *Origins of Order: Self-organisation and Selection in Evolution* (Oxford University Pres, New York).

Kuhn, T. S. (1962), *The Structure of Scientific Revolutions* (University of Chicago Press, Chicago). (Also in Neurath et al. [1970].)

Lakatos, I. (1970), 'Falsification and the Methodology of Scientific Research Programmes', in *Criticism and the Growth of Knowledge*, I. Lakatos and A. Musgrave (eds.) (Cambridge University Press, Cambridge).

Latour, B. (1987), *Science in Action* (Harvard University Press, Cambridge, Mass.).

Latour, B. and Woolgar, S. (1979), *Laboratory Life: The Social Construction of Scientific Facts* (Sage, Beverley Hills).

Leplin, J. (1984), *Scientific Realism* (University of California Press, Berkeley).

Levi, I. (1967), *Gambling with Truth* (A. A. Knopf, New York).

Masterman, M. (1970), 'The Nature of a Paradigm', in *Criticism and the Growth of Knowledge*, I. Lakatos and A. Musgrave (eds.) (University Press, Cambridge).

McClennen, E. (1990), *Rationality and Dynamic Choice: Foundational Explorations* (Cambridge University Press, New York).

Munevar, G. (1989), 'Science as Part of Nature', in *Issues in Evolutionary Epistemology*, Hahlweg/Hooker (eds.) (State University of New York Press, New York).

Musgrave, A. (1993), *Commonsense, Science and Scepticism* (Cambridge University Press, Cambridge).

Neurath, O., Carnap, R. and Morriss, C. (eds.) (1970), *Foundations of the Unity of Science* (University of Chicago Press, Chicago).

Newton-Smith, W. H. (1981), *The Rationality of Science* (Routledge and Kegal Paul, London).

Nicholas, J. M. (1984), 'Scientific and other Interests', in *Scientific Rationality: The Sociological Turn*, J. R. Brown (ed.) (Reidel, Dordrecht).

* Nicolis, J. S. (1986), *Dynamics of Hierarchical Systems: An Evolutionary Approach* (Springer-Verlag, Berlin).

Oldroyd, D. (1989), *The Arch of Knowledge* (University of NSW Press, Kensington).

Piaget, J. (1976), 'The Possible, the Impossible and the Necessary' (trans. G. Voyat). *The Genetic Epistemologist* **6**, pp. 1–12.

Pickering, A. (1984), *Constructing Quarks* (University of Chicago Press, Chicago).

* Pines, D. (1988), *Emerging Syntheses in Science* (Addison West Publishing Company, Redwood City, Calif.).

Polikarov, A. (1973), *Science and Philosophy* (Bulgarian Academy of Sciences, Sophia).

Polikarov, A. (1983), *Methodological Problems of Science* (Bulgarian Academy of Sciences, Sophia).

Popper, K. R. (1972), *Conjectures and Refutations* (Routledge and Kegan Paul, London).

Popper, K. R. (1979), *Objective Knowledge: An Evolutionary Approach* (rev. ed.) (Oxford University Press, Oxford).

Popper, K. R. (1980), *The Logic of Scientific Discovery* (Hutchinson, London). (First published as *Logik der Forschung*, Wien, 1934.)

Popper, K. R. (1987), 'Natural Selection and the Emergence of Mind', in *Evolutionary Epistemology, Theory of Rationality and Sociology of Knowledge*, G. Radnitzky and W. W. Bartley, III (eds.) (Open Court, La Salle).

* Prigogine, I. and Stengers, I. (1984), *Order Out of Chaos* (Shambhala, Boulder, Colorado).

Reichenbach, H. (1961), *Experience and Prediction* (University of Chicago Press, Chicago).

* Saviotti, P. P. and Metcalfe, S. (eds.) (1991), *Evolutionary Theories of Economic and Technological Change* (Hartwood Press, London).

Schilpp, P. A. (ed.) (1974), *The Philosophy of Karl Popper* (Open Court, La Salle).

Sellars, W. (1965), 'Scientific Realism or Irenic Instrumentalism', in *Boston Studies in the Philosophy of Science*, Vol. II. R. S. Cohen and M. W. Wartofsky (eds.) (Humanities Press, New York).

Shapere, D. (1984), *Reason and the Search for Knowledge* (Reidel, Dordrecht).

Solomon, M. (1992), 'Scientific Rationality and Human Reasoning', *Philosophy of Science* **59**, 439–55.

Stein, D. L. (1989), 'Spin Glasses', *Scientific American*, July 1989, 52–9.

Stein, H. (1990), 'From the Phenomena of Motion, to the Forces of Nature: Hypothesis or Deduction?', in *PSA 1990*, vol. 2, A. Fine, M. Forbes and I. Wessels (eds.) (Philosophy of Science Association, East Lansing, Mich.).

Streater, R. F. and Wightman, A. S. (1964), *PCT, Spin, Statistics and All That* (Benjamin, New York).

Suppe, F. (ed.) (1974), *The Structure of Scientific Theories* (University of Illinois Press, Urbana, Ill.).

Thagard, P. (1988), *Computational Philosophy of Science* (MIT Press, Cambridge, Mass.).

* Trappl, R. (ed.) (1986), *Power, Autonomy, Utopia: New Approaches toward Complex Systems* (Plenum, New York).

van Fraassen, B. C. (1980), *The Scientific Image* (Clarendon Press, Oxford).

Vickers, G. (1968), *Value Systems and Social Process* (Penguin, London).

Vickers, Sir G. (1983), *Human Systems are Different* (Harper and Row, London).

* Weber, B., Depew, D. and Smith, J. (eds.) (1988), *Entropy, Information, and Evolution* (MIT, Cambridge, Mass.).

* Yates, F. E. (ed.) (1987), *Self-organising Systems: The Emergence of Order* (Plenum, New York).

* Zurek, W. H. (ed.) (1990), *Complexity, Entropy and the Physics of Information* (Addison-Wesley, Redwood City, Cal.).

PETER JANICH

METHODICAL CONSTRUCTIVISM

PRELIMINARY TERMINOLOGICAL REMARK

In recent years, an inflation of "constructivisms" has been restricting the information value of the label "constructivist" – the more so since (the younger) "Radical Constructivism" developed by biologists or from biological approaches to the role of the brain is a naturalistic position. The claims advanced as well as the means used by this type of constructivism are totally different from the methodical approach.

The object of the following text is "Methodical Constructivism" as it grew from the early works of Hugo Dingler (1881–1954) and was developed mainly by the "Erlangen Schule" founded by Wilhelm Kamlah (1905–76) and Paul Lorenzen (1915–94). The first and the second generation, of this school of German philosophers included (in alphabetical order) Peter Janich, Kuno Lorenz, Jürgen Mittelstrass, Oswald Schwemmer and Christian Thiel, and was continued by the Konstanz group (Janich, Mittelstrass and Friedrich Kambartel). Affiliated were Carl Friedrich Gethmann, Franz Koppe, Mathias Gatzemeier, and others.

Still historically speaking, the Erlangen period lasted from the early sixties to the late seventies, the Konstanz period through the seventies. Beginning with the eighties, Methodical Constructivism was worked on almost exclusively in Marburg and Göttingen (Lorenzen). A comparison with the different branches of Radical Constructivism a few years ago[1] showed that, in terms of professorships, published books and articles, Methodical Constructivism is not only much older and, with respect to the philosophical fields treated, much broader than Radical Constructivism, but also much more elaborate, for instance with regard to philosophy of language, logic, philosophy of science and ethics.[2] It was, however, less recognized in the English-speaking countries which – as prominent examples of present-day Anglo-American philosophy illustrate – may be explained by the abstinence of Anglo-American philosophers from reading non-English texts. Lacking Anglo-American recognition of Methodical Constructivism also has deep roots in the fact that it was in opposition to the mainstream philosophy of the Vienna Circle, Critical Rationalism of the Popperian type, to analytical philosophy of language, both in the form of ordinary-language philosophy and the philosophy of scientific language in the Carnap-Hempel tradition, and other forms of Anglo-American mainstream philosophies.[3]

Whereas, in Germany, the opposition between the two philosophical orientations, often exemplified by the Erlangen and the "Stegmüller-Schule", was known to everybody interested in philosophy, even a series of English

D. Ginev and R. S. Cohen (eds.), Issues and Images in the Philosophy of Science, 173–190.

publications by constructivists did not bring this opposition to the attention of the philosophical public in English-speaking countries.

Therefore, the main traits of Methodical Constructivism shall be presented here as opposed to competing approaches at a given time – approaches, to be sure, which are not only those of academic philosophers but also develop in the form of scientists' philosophies – in the beginning mainly mathematicians' and physicists' philosophies.

GEOMETRY: CONSTRUCTIVISM AS OPPOSED TO FORMALISM, AXIOMATICISM AND EMPIRICISM

The history of geometry from the early days of Eudoxos and Euclid (synthetic geometry) via Descartes (analytic geometry) all the way up to David Hilbert (formal-axiomatic geometry) leaves open some foundational questions: not to mention the well-known problematic character of both the definitions (mainly of basic concepts such as "point", (straight) "line" and (plane) "surface") and the postulates and axioms (mainly the parallel-postulate) in Euclid's *Elements*. It is questionable in the ancient form of geometry from which Euclid starts out with planimetry, the geometry of plane figures, to come to stereometry by rotation of plane figures only at a later stage (beginning with from book 11). The definitions of, say, the circle and the sphere clearly show that the background of geometrical concept formation lies in a practice of drawing and sculpturing.[4] However the constitutive poietic practice which would have to explain the technical availability of the drawing plane, a ruler and a pair of compasses was not reflected by ancient geometers. The philosophers of Greek antiquity did not see the epistemological bearing of bringing about the objects of geometrical reflection by non-linguistic or non-mental activities (even though the specification of types of human actions by Aristotle had included "poietic" actions as opposed to practical ones).

The "Euclidean gap" of geometry was a burden for its further history, the failure to clarify the relation between "ideal" mathematical geometry on the one hand and the geometrical properties of real bodies on the other (Dingler later would call this the *Anwendungsproblem* – the problem of application). The logical deficiencies of Euclidean geometric language can be traced back to this gap as well as to the unclarified epistemological status of the axioms, above all, of the axioms of orthogonals and of parallels and the particular "Euclidean" character of this geometry.

Descartes protracted the defect by importing it into his ingenious idea of translating into calculations problems of the construction of figures by means of rulers and compasses. He did not ask what it means that the plane of coordinates is "plane" and that the axes are "straight". He simply presupposed that this is known or unproblematic. And he added to the problems by omitting any definition or explanation as to how points and numbers have to be correlated on the axes. He did not see that only the investment of an implicit act of measurement attributes numbers to steps of the same geometrical length

– but how can the congruence of equal length at different places on the axes be defined, and how can it be controlled in practice (both in the sense of the concept of congruence and of real measurement in any case of application of analytic geometry to physical descriptions)? This may be called the "Descartes Gap", and it continues to characterize all analytical geometry up to its modern forms.

The third foundational gap of geometry may be termed the "Hilbert Gap". David Hilbert (in his famous *Grundlagen der Geometrie*, originally published in 1899) gave up any efforts to define basic geometrical terms. Instead, he regarded his axioms as mere forms of propositions. They remained open for so-called "interpretations" – mainly by empirical physics: If one replaces the Hilbertian sorts of variables by traditional words such as "point", "straight line" and "plane", and then interprets, say, "point" by "tiny hole in a screen" and "straight line" by "light ray", then the formal axioms provide a model the validity of which is considered an empirical question.

As far as the empiristic understanding of (applied) geometry is concerned, the famous empiricist Hermann v. Helmholtz contributed an additional gap to the tradition: he held that measurement of length (as well as of other physical parameters) should build up operationally in a way parallel to the building of the body of numbers by mathematicians, beginning with simple counting in units by integers. Consequently, "to measure a magnitude" was consequently understood as looking for a (rational) number to express the relation of a given magnitude to a measuring unit. The gap lies in the fact that this proposal cannot serve as an operational definition of magnitudes since the reproducibility of measuring units itself is accessible only on the basis of empirical knowledge based on successful measurements.

THE CONSTRUCTIVE ANSWER: METHODICAL OPERATIVISM

This sketch of the historical background of current mainstream philosophy which is shared by mathematicians, physicists and many philosophers, characterizes the formalism, axiomaticism and empiricism which challenged constructivists beginning with Dingler. He saw that the problem of the fitting of geometry to the physical objects was not clarified, that the validity of axioms was open, that the art of measurement actually practised in science and technology was burdened by circular or missing definitions, and that questions, such as the one concerning the character of geometry, remained unanswered. The dogma of Logical Empiricism had no choice but to regard geometry as merely logically consistent or empirically valid; and both understandings revealed more mistakes and open questions than it provided answers. In this situation, Dingler was deeply impressed by the fact that common scientific practice deals with instruments in which geometrical forms are technically realized. In Aschaffenburg, where he lived at that time, he came into contact with the proper industry for the production of planes (*Richtplatten*) which, for instance in optical industry or measuring-tool production, were used as standards of planarity. He developed his idea of an operational foundation of geometry (1913) which centers around the idea that geometric forms can be produced (and reproduced) methodically before any measurement of length is technically available.[5]

This work of Dingler's – more than other important insights, e.g. in causality (as defined by a pragmatic theory of experiment), or in the programmatic character of theories (50 years before very similar ideas were developed by I. Lakatos) – was continued by Lorenzen's proposal of "protophysics" (the term however was already used by F. R. Lipsius in 1937, to describe Dingler's approach). In 1961, Lorenzen reinterpreted Dingler's approach to geometry by Leibnitz's "principium identitatis indiscernibilium" as a principle of homogeneity: the craftsman producing a geometrical plane tries to make it homogeneous i. e. to make its points indistinguishable. Later (1984), he gave up this idea in favor of a "principle of form" according to which all figures are equal which follow the same steps of construction so that geometrical similarity (and, consequently, an axiom of parallels) is established by a norm.

METHODICAL PHILOSOPHY OF LANGUAGE

The publication of the book, *Logische Propädeutik* (1967), by Wilhelm Kamlah and Paul Lorenzen, may be regarded as the historical starting point of the "Erlangen School". It not only revealed some influence by the phenomenological background of Kamlah but, above all, introduced an alternative position in the philosophy of language as compared to the mainstream philosophy in English-speaking countries. Whereas analytical philosophy of language was done in a merely descriptive attitude with regard to both ordinary language and language of science, the *Logische Propädeutik* recommended a far-reaching critical attitude towards any form of historically grown language. "Methodical reconstruction" was considered the remedy against the burden of a history of thinking in unclear concepts in everyday language, language of science and traditional philosophy. Consequently, the core of the *Logische Propädeutik* was a theory of how to perform such reconstructions, beginning with a kind of ostensive introduction of first words (named "predicators", according to the Carnapian proposal), guided by examples and counterexamples. This proposal included consideration of situations of teaching and learning language, which are always genuine practical situations, and it imported the Dinglerian idea of a "full foundation free of circles and gaps" into the philosophy of language.

This type of philosophy of language, to be sure, was influenced by the linguistic turn of analytical philosophy, but developed proposals for many fields of philosophical discussion which differ from its precursors. For instance, logic was no longer treated in an axiomatic or formalist way but was "founded", by incorporating an intuitionist tradition, as dialogical logic by a system of rules that guide a dialog between two partners, a proponent and a opponent, and test logically composed sentences by resolving, step by step, their logical composition. In Germany, where constructive logic was prominently developed and defended by Lorenzen and by Kuno Lorenz, a disciple of Paul Lorenzen, there was a deeply rooted controversy about the foundation of logic and its role in the philosophy of mathematics. Lorenzen worked out a constructive philosophy of mathematics and of metamathematics to some extent. "Constructive mathematics" was not understood as being based on recursive definitions but rather on a step-wise construction of the objects of mathematics where, instead of adhering to Peano's axioms, the construction of integers followed a (schematic) calculus for the production of counting symbols which was subsequently connected with a logical process of "abstraction" (a type of definition by certain invariances, following a basic idea of Gottlob Frege) and eventually proposed explicit definitions of "set", "function", "differential" and "integral".

The philosophy of language of the Erlangen School shared with analytical and empiristic traditions a skeptical attitude towards metaphysics. (Kamlah became famous for a public proclamation of his personal aversion to his former teacher, Martin Heidegger.)

PROTOPHYSICS

Another crucial issue of the constructivism of the *Logische Propädeutik* was the understanding of geometry, as mentioned before, and of protophysics. Where the tradition of logical empiricism of the Vienna Circle allowed only for either the logical or the empirical truth of scientific propositions, Methodical Constructivism turned its attention to the non-linguistic, technical side of physical research.

It was mainly in the writings of Peter Janich that the idea of a norm of undisturbedness, guiding production and use of scientific instruments, was used to elaborate protophysics. The first partial theory consequently finished was his "Chronometry".[6] Following the insight that so-called "natural laws", said to guide the function of clocks, cannot distinguish between well-functioning and disturbed (and therefore useless) clocks, an operational definition was provided for the uniform motion of a clock-pointer by a principle of homogeneity which guides the construction of clocks as well as the logical properties of a metrical concept, "time". Janich first postulated that such definitions in protophysics have to be unequivocal in the following sense (and that their uniqueness must be explicitly proven): The norm guiding the construction, production and use of measuring instruments must logically imply that different realizations of the same recipe lead to instruments of predictable relative properties. So, for instance, two different clocks (in the same place) which fulfill the principle of homogeneity must, at any time, run at a constant ratio of the velocities of their clockhands.

Corresponding proofs were provided by Janich also for the uniqueness of operational definitions of the geometrical basic forms "plane", "right angle" (Hence, interpreting Euclid's axiom that all right angles are equal) and (in 1992) of the "parallelity" of planes and of straight lines.[7]

The third partial theory of protophysics concerning mass measurement was treated in varying ways by constructivist philosophers. Bruno Thüring[8] used the Newtonian law of gravitation for the definition of mass since, being an astronomer, he was interested in an unequivocal definition of "motion" relative to a cosmological reference system. Lorenzen used the proposal, initiated by Galileo and Hermann Weyl, that the equality of mass should be defined by an (ideal) inelastic impact.[9] Because the Lorenzenian definition could not solve the problem of a suitable reference system for mass definition, and was therefore deficient, Janich proposed to define homogeneous density in an operational way which then allows for an unequivocal concept of equality of mass independent of reference systems – thus also allowing for a definition of inertial reference frame and a proof of the uniqueness of mass definition.[10] (These proofs function as reasons for accepting the transsubjectivity of measurement results, i.e. the independence from the particular measuring tool.)

The discussion between protophysicists and physicists mainly concerned problems raised by special and general relativity theories relative to the corpus of operational definitions given in protophysics. To some extent, this discussion is still open, since an empiristic position cannot evaluate the normative character of constructive philosophy. Empiristic critics of protophysics conveniently disregard the fact that it is mainly metatheory or philosophy of relativistic physics which is criticized by constructivist philosophers.

FROM PHILOSOPHY OF SCIENCE TO ETHICS

Constructivism exerted an important influence on German philosophy in other areas as well. Jürgen Mittelstrass, in his Habilitationsschrift *Neuzeit und Aufklärung*, entered into a discourse with history of philosophy and, in particular, with epistemology. Also Friedrich Kambartel, not a constructivist teacher's disciple, joined the group of constructivist philosophers and dealt with epistemology, logic, philosophy of mathematics and fundamental problems of economy. Oswald Schwemmer, a disciple of Paul Lorenzen's, developed (in collaboration with Lorenzen) a constructivist ethics and published it in a book which, to some extent, replaced the *Logische Propädeutik: Konstruktive Logik, Ethik und Wissenschaftstheorie* (second edition, 1975). Although both authors later turned away from their approach, this book is important inasmuch as it shows that, after the linguistic turn, a pragmatic turn has taken place to some extent, even though not in full consequence, in constructivist philosophy. First parts of a constructive theory of action are found in that book. It overcame remnants of a heritage from logical atomism which still persisted in *Logische Propädeutik*. On the other hand, the pragmatic turn – already Dingler was a pragmatist to some extent – was carried out somewhat half-heartedly: It mainly dealt with, on the one hand, the poietic actions that provide the technical basis of empirical physics and, on the other hand, the practical actions with regard to ethics. But it did not entail philosophy of language in the sense that speaking is consequently regarded as acting in terms of means-and-ends rationality.

After the Lorenzen's death (1994), after Schwemmer had turned his back on constructivism and after some changes in the philosophies of Kambartel, Lorenz, and Mittelstrass, the present situation of constructivism may be characterized as follows:

Christian Thiel, who succeeded Paul Lorenzen on his philosophical chair at Erlangen, published constructive writings on logic and, only recently, on the philosophy of mathematics (1995).[11] In Konstanz, the place where Mittelstrass teaches philosophy, there has remained the editorial staff of the four-volume *Enzyklopädie Philosophie und Wissenschaftstheorie* as virtually the only left-over of the constructivist tradition at the University of Konstanz. This well-reputed encyclopedia is very liberal with respect to its own philosophical position, and it involves many authors in addition to constructivists.

A separate development is pushed forward by Carl Friedrich Gethmann whose background is, in part, phenomenology. He could demonstrate the many influences that phenomenological types of philosophy of science had on constructivist authors, and he contributed to the discussion of logic, philosophy of language, and ethics. In a sense, he continues with a slightly modified program of constructivist ethics as it was advanced by Lorenzen and Schwemmer in the book mentioned previously.

FROM CONSTRUCTIVISM TO METHODICAL CULTURALISM

A comparatively large group of philosophers, however, tries to continue constructive ideas at the University of Marburg. Holm Tetens, a disciple of Janich's, submitted a Habilitationsschrift on the problem of experiment in physics in connection with questions of concept formation.[12] This book, as did Dingler's book, *Das Experiment, sein Wesen und seine Geschichte* (München 1928) and Klaus Holzkamp's *Wissenschaft als Handlung* (Berlin 1964), remained completely unconsidered in the entire recent discussion on experiment in the English speaking world that was initiated by I. Hacking. In the meantime, the theory of experiment is carried on by a group of Marburg philosophers (D. Hartmann, P. Janich, R. Lange).

As major recent developments of Methodical Constructivism we can regard:

- The consequent pragmatic turn, which now reascribes the methodical importance that seemed to belong to philosophy of language to the theory of action: Since speaking is acting and, for instance, also is subject to some constraints of the rationality of means and ends, philosophy of language has to be rewritten in terms of the theory of action.
- Constructivism, opposed as it still is to some mainstream philosophies

focuses on various forms of naturalism prevalent in philosophical developments in the English-speaking world (as well as the strong naturalism of most sciéntists), and it stresses the cultural character of science and philosophy.

On the other hand, it abstains from different forms of relativism, from the Kuhnian and Feyerabendian type to R. Rorty's hinging on factual agreements of groups of persons for acknowledging scientific results. So the history of constructivism up till now might be summarized by three turns: the linguistic turn in the *Logische Propädeutik*, the pragmatic turn beginning in the seventies, and the culturalist turn during the last past few years. Consequently, the present form of Methodical Constructivism – to avoid its being mixed up, as usual, with Radical Constructivism – nowadays prefers to be referred to as "Methodical Culturalism". It can be characterized by the following traits:

The central idea of constructive philosophy was to take seriously the fact that any science and any discipline in the humanities is brought about by people. It is human acting which constitutes the objects of any scientific effort. Although initially, the concept of "Constructivism" was created with regard to the objects of mathematics and logic, it was still possible to speak of the construction of scientific objects.

As mentioned in the historical review, the means used by that classical form of Methodical Constructivism, were mainly philosophy of language and theory of action. Means-and-ends rationality, the foundation of any assertoric statement in philosophy and science, and the justification of any prescriptive sentence were held to be indispensable. These main traits of constructive philosophy are maintained in the recent development toward methodical culturalism. However, the critical reconstruction of linguistic and pragmatic means concerns achievements brought about during a long cultural history which submits both real acting and philosophical reconstruction to some constraints. Therefore, according to the conviction of their authors, constructive philosophy has to be expanded in the following respects.

A THEORY OF EVERY-DAY LIFE

All approaches, from the early one of Hugo Dingler through the writings of Paul Lorenzen up to the most recent books and articles by members of the methodical constructive school, distinguish between pre-scientific and scientific knowledge. Science, of course, grows out of everyday life in a twofold sense: There were times when modern sciences (and humanities) did not exist; in some way, they must have developed from every-day practices by a process of *Hochstilisierung*, i.e. by a qualitative change in methods and claims. Therefore any methodical reconstruction has to consider this passage from every-day practice to scientific theory. On the other hand, every form of science or humanities is still embedded in a historical and social practice. Not only because all members of scientific communities have to be competent participants in every-day-life communities in order to be able to become scientists by education in the first place; science and humanities are embedded in social practice also inasmuch as they support or make more effective the modern complex way of living – from support of technologies by natural sciences through any application of mathematics e.g. in economy to the support of social and legal organization of states received from social sciences and the humanities.

However none of the previous approaches mentioning the life-world (*Lebenswelt*, a term adopted from phenomenologists) or the world of every-day life as basic or even fundamental for the sciences and humanities, developed a theory about it. Since it can be seen that scientific goals and means grow out from every-day practice and every-day life, where skills and human abilities are needed to bring about science, there cannot be a successful philosophy of science without an epistemology of knowledge and truth in every-day life. Constructive philosophy up to now has considered only the methodical steps necessary to support universal claims of validity for scientific results. Consequently, it has discussed methods as schemata of human actions which underly certain norms of rationality. The entire remainder of human investment into science and humanities, in the sense of pragmatic and linguistic human abilities, were only mentioned but never analyzed or founded. All constructive, normative foundations served, as it were, as conditioned norms: Provided that a human being decided to share the process of scientific research (and consequently to share its aims), then the constructive philosophy of science had to offer some normative rules as to how to bring about the objects of logic, mathematics, protophysics, and almost any other field of human culture. But neither the constraints for reasonable aims nor the constraints of accessible means imposed by the respective cultural situation were seriously taken into account.

THE CULTURAL BACKGROUND

Also, Methodical Culturalism – like constructivism – is consistently concerned only with "universal" problems and solutions, i.e. fields where the foundation of knowledge and the justification of norms are at stake. The term "culturalism" should not lend support to the interpretation that the nonuniversal, the singular – e.g. the particular creative action of an artist or a scientist – is the object of a methodical culturalistic philosophy. It only concerns what can be well-founded and justified, and therefore forms a counterposition to culturalistic relativism. On the other hand, it is of course also a counterposition to any naturalistic philosophy which, originating from realist or evolutionist views of natural science, regards any achievements of human culture only as a particular form of natural history. Methodical culturalism maintains, in other words, the constructivist claim to found propositions upon complete, non-circular chains of arguments and to justify any prescription, be it a methodological rule, a moral maxim or an ethical norm. It seriously attempts, on the other hand, to embed this business of foundation and justification in communities which share forms of a prediscursive consensus by historical processes of (cultural) selection.

Particular examples may serve to demonstrate advances of culturalistic over constructive approaches: For instance, the foundation of geometry as offered by Lorenzen (and by his disciple, Rüdiger Inhetveen)[13] abandoned the earlier approach of defining basic geometric concepts by means of homogeneity principles. And they immediately sought the reconstruction of historically given systems of axioms, for instance of the geometry of David Hilbert. In this approach, however, it was disregarded that the object of reconstruction has itself to be selected by means of a justified choice. Lorenzen and Inhetveen do not provide such justification. They are content with the historical fact that mathematicians like these systems of axioms. On the other hand, the "form-principle" advanced by Lorenzen and Inhetveen has no basis in the practice of either mathematicians or technicians or physicists.

It can be seen from Euclid (as mentioned before) that his geometry is based on practices of drawing and sculpturing. And still today, geometry deals with the technical shaping of bodies for the production of instruments for drawing, measuring, experimenting, and so on. Hence, the methodical reconstruction in the sense of Lorenzen seems to be a "top-down reconstruction" which takes the most advanced modern theories as the top from where an analysis by reconstruction leads down to certain *ad hoc* principles.

By contrast, the culturalistic alternative takes into account the bringing about of basic spatial forms by procedures which were already guiding for the philosophy of Hugo Dingler, for instance, the grinding of surfaces by a triple of stones until they fit in pairs. Geometric theory is understood as the field in which (sometimes even counterfactual) arguments about the technical goals of the production of forms of real bodies (machines, instruments, sheets of glass for windows and mirrors, and so on) are developed and discussed. In other words, the purpose of geometric theory is taken as a complex aim which, on the one hand, has to be justified within a certain historical practice and which, on the other hand, has to be achieved by technical means which entail operational definitions and theoretical elements for the geometric theory.

THE PRINCIPLE OF METHODICAL ORDER – REVISITED

The older constructivist and the more recent culturalist views both hinge on one principle of rationality in theoretical philosophy which distinguishes them from the empiristic mainstream philosophy of science: the principle of methodical order. Since objects of our technological civilization, no matter whether they belong to every-day life or to scientific research, are brought about by complex chains of human actions, it must be taken into account that a change in the order of sub-actions will cause failure. Everyone knows from every-day life that certain chains of actions are defined by their purposes: if water is to be poured from a bottle into a glass, it is well-known that the bottom must be opened before it is tilted. If someone wants to enter a closed door he knows well that he first has to insert the key into the lock, then to turn the key, then to push the handle and, finally, to move the door. In scientific theories, however, for syntactic or didactic purposes, the order of the descriptions of these steps is frequently changed. Thus, theories that begin with the geometry of Euclid and with the classical Newtonian mechanics offer an order of definitions, or of axioms, that does not follow the pragmatic order in which the actions to which they refer are performed. (Of course, such theories simply neglect that they are referring to particular actions at all.) Consequently, the scientific world is full of theories which represent the practice and the results of science in an order which hides rather than reveals the practice (and its success) on which it is based. Therefore a "principle of methodical order" is formulated that forbids the use of any description or prescription of chains of actions that differs from the order which leads to success in obtaining the purposes of the entire enterprise. Thus far the recent culturalistic philosophy follows the same paths as did the previous constructivist one. However, the ends of scientific achievements are not seen in the recent textbooks. Hence, methodical reconstruction follows the maxim, "What steps are needed to found theory t? (Lorenzen-type reconstruction) instead of, "What means are needed to reach the purposes actually achieved by science".

EPISTEMOLOGY – CULTURALIZED

Recent developments in analytic-empiristic or descriptive philosophy, mainly in the fields of theories of truth and of cognitive sciences, reveal a tendency to "naturalize" any process of gaining knowledge. If they are delegated from a philosophical epistemology to empirical cognitive sciences (psychology, neuroscience) – as is the case in Quine's program of "epistemology naturalized" – or if they are described by means of systems theory (as in the biologistic approaches of H. Maturana's "Radical Constructivism") the anti-realistic and anti-naturalistic criticism of constructive philosophy has to be extended by the distinction between the "observer" and the "participant". The introduction of the concept of observer, historically beginning with relativistic physics and ending in systems theoretic approaches of Radical Constructivism (or in Niklas Luhmann's sociology), cannot cope with the problem of claiming truth. Systems, be they natural or artificial ones, just work, and, according to these approaches, the philosophical or scientific descriptor observes them. Irrespective of whether the observer is held to see the world from the outside or from an inside position, the observer cannot regain the perspective of the participant who acts in the pursuance of ends, and who claims to succeed in reaching his aims, goals or ends.

The culturalistic view which postulates the non-exclusion of the perspective of the committed participant overcomes the restriction according to which norms can only be dealt with in descriptive terms, and it achieves the possibility of actually sharing norms. The act of acknowledgment of normative orientations is fundamental both to theoretical and practical (in the sense of moral or ethical) attitudes. Any scientist not only has to describe but to follow methodical norms. Acting persons cannot restrict themselves to considering or even judging normative orientations from the distance stance of an uninvolved observer. Rather, they must decide whether to accept or to refuse a norm, to realize it or to refrain from its realization.

The direct consequence of these extensions of constructive into culturalistic philosophy, on the one hand, lies in regained interest in traditional branches of philosophy such as philosophy of nature. Where the empiristic and formalistic tradition of analytical philosophy and even methodical constructivism shared the view that philosophy of nature is replaced by philosophy of the natural sciences, culturalistic philosophy takes into account that sciences do not generate an explicit concept of nature, even though, for instance, physicists, chemists or biologists often speak about nature and do so in ways specific for their disciplines. Those implicit concepts of nature have far-reaching influence on the development of sciences. Therefore, they are in need of philosophical clarification. In addition, the relations people have towards nature are not restricted to the management of nature by natural sciences and technology. Aesthetic categories as well as ecological or moral needs have to be considered even though the preservation of nature for mankind, of course has to be based on sound knowledge provided by natural science. The attitude towards nature cannot dispense with normative orientations for handling and preserving nature in terms of their technological availability.

Epistemology forms another field of philosophy which both the (earlier) empiristic tradition and methodical constructivism counted among metaphysical errors of traditional philosophy. Apparently, however, it is not only the field of founding and justifying sciences and humanities where we have and need universal every-day knowledge. In every-day life – independent of its foundational role towards science as was discussed above – we also depend on the possibility of gaining and communicating universal knowledge. And, in every-day life, we practically distinguish many forms of knowledge and truth through different ways of judging linguistic representations and validity of knowledge. Whether sensual perception, the result of a simple calculation, some kind of statistical judgement about, say, the weather of the last summer, or any other example is considered, we can see that, already in every-day life, almost all different types of knowledge are present which later turn up in the sciences and humanities.

Another reason for caring about epistemology, again, lies in the fact that the empirical cognitive sciences (which were meant to replace epistemology) need to know how to distinguish knowledge from error. A scientific explanation of how knowledge comes about and how the relations in the outer world, be they natural or cultural ones, are represented in our consciousness (or brain), must distinguish, in the first place, between real knowledge as opposed to deception, error or even mere personal opinion. Cognitive sciences would not have an object at all if they could not apply a normative distinction in the traditional epistemological sense, that is, if they were unable to tell true from false.

PROTOSCIENCES

In the field of philosophy of science, the older program of Methodical Constructivism that was to develop a prototheory of physics is extended to other sciences and, as mentioned above, is interpreted as the task to methodically reconstruct the bases of empirical sciences from their historical practices. Since the art of measurement of length, time and weight (and even of inert mass, force, and other mechanical magnitudes) was established long before classical mechanics took it to be a central method in experiments, these arts were invented by craftsmen and technicians. Also, experimenting is no more than a scientifically stylized way of constructing machines and apparatuses by craftsmen, engineers and technicians. Protophysics, as a normative theory, only explicates the norms that guide the practice of making measuring tools in order to arrive at, for instance, operative definitions of the logical and mathematical structure of scientific concepts. Theories of physics are instrumental for technical applications, and they possess an inherent way of confirmation: technical success. Instrumentalism in methodical culturalistic philosophy, therefore, is not restricted to linguistic activities (such as successful predictions of natural events). It rather aims at the successful construction of machines. Only where this technological success is reached in laboratory research, are models available for the description and explanation of natural events.

In a way similar to that of protophysics, protochemistry is built up by culturalistic philosophers (Peter Janich, Nikos Psarros, Gerd Hanekamp).[14] Whereas all current textbooks of chemistry nowadays present chemical knowledge in a top-down version – the periodic table of elements and the most recent results of chemical research form the top from which all chemical methods are considered, e.g., in terms of the atom/molecule-schema, – Methodical Culturalism tries to find the way that leads from reproducible measures to change substantial properties to the distinction of chemical compounds and, eventually elements. Recognized "laws" or "principles" are not considered empirical results but rather definitions and technical principles for gaining basic chemical concepts. The law of the constant ratio of reaction partners, for instance, is not based on empirical observation but rather stipulates possible definitions and distinctions of substances. It is not really open for empirical falsification, but fundamental for concept formation. Also, in protochemistry, processes of making substances homogeneous are involved so that our knowledge about "natural" properties of substances or processes (like metabolism in organisms) can only be researched in and known by way of making artificial properties reproducible in laboratories. Therefore, the historical development of technical goals for mastering substantial properties has to be taken into account.

The results of protochemistry, as elaborated up to now, provide a survey of basic concepts and fundamental methods, and they methodically reconstruct chemistry up to state-of-the-art theories. In a sense this issue formulates a counterposition to reductionist attempts to reduce chemistry to quantum physics. According to the principle of methodical order, however, chemistry precedes physics. Chemistry is not based on microphysics. Rather microphysics methodically depends on the availability of identified substances – identified by chemical procedures, to be sure.

Another field of culturalistic philosophy of science is protobiology.[15] The Darwinian idea of explaining natural history and evolution by mutation and selection in terms of the model of breeding animals and plants has to be connected with the theory of the construction of organisms. The field in which a human breeder (and, by analogy, nature) can support or prevent the development of organisms and of their particular properties is restricted by the constraint that the organism has to be able to live at all. It has to avail itself of a certain construction which is successful with respect not only to metabolism but even, e.g., to its own mechanical constructive coherence and stability.

Protobiology, therefore, tries to make clear what an organism is in terms of models that help to describe and explain its functions. From this concept of organism, the shortcomings of the concept of species as developed by the traditional "synthetic" theory of evolution can be overcome. Again, in protobiology the program starts from every-day knowledge and every-day practice in coping with animals and plants. There are the practices of the butcher (as the prescientific precursor of the anatomist) and of the farmer who breeds animals, or of the gardener who breeds new forms of roses and who develops rational means for his breeding purposes. And, yet, in prescientific practice, the behavior of animals, in some respect even the behavior of animals towards men, is important for breeding success. Hence, there is some sort of pre-Darwinian selection or an internal selection of properties of organisms before they are submitted to the struggle for life in the Darwinian sense. As protophysics and protochemistry, protobiology also methodically reconstructs the basic concepts and fundamental methods of a natural science by taking into account the fact that it is embedded into every-day life both in its historical and present form. Also, biology needs (for instance for modern branches of microbiology and genetics) criteria for the validity of its results, which – according to the culturalistic interpretation – depend on the epistemic goals of this science. And, of course, the means again are to be judged relative to those goals.

A fourth field of the prototheoretical approach is protopsychology.[16] As the writings of Dirk Hartmann[17] show, two forms or parts of psychology have to be distinguished which may be referred to as the natural-scientific and the cultural-scientific parts of psychology. The first one pursues the task of bringing about technologies for, say, the support of successful education, training of pilots, advertising goods, preventing traffic accidents, making more effective, administration and so on. The entire research for those techniques concerns means towards certain ends which themselves are not questioned anymore. The "natural-scientific" part of psychology, therefore, is instrumental in supporting practices, the goals or purposes of which are presupposed as being justified.

On the other hand, the traditional practice of psychologists, e.g. in psychoanalysis, in theories of education, marital counseling, and so on, is not restricted to techniques for overcoming certain problems relative to maintained ends, but have to enter into transformations of needs and ends themselves. There has to be a kind of counseling in (often: morally) questioning and justifying ends in order to overcome, to solve or at least to reduce conflicts. This part of psychology needs different methods and a different protopsychology.

Also, the theories of protopsychology are elaborated to some extent, and they follow the culturalistic program, in order to grasp science as a support for everyday-life that is based on scientific knowledge.

CONCLUSION

Other prototheories are still programmatic. They demand, as once theories did which have matured, at least fundamental competence in the particular science as well as in philosophy. Therefore, they depend on "manpower", on persons willing to turn to philosophy after having finished an academic education in a scientific discipline. But philosophy also has to be advanced in its own right, e.g., in the fields of the theory of action, of epistemology, of philosophy of language, of nature, of art, of history etc., and in ethics. Some initial steps have already been taken and published so that the intentions of traditional Methodical Constructivism, again, seem to be fertile for a philosophical enterprise of our time. It might serve to advance that enterprise not least, because it plays an alternative, challenging role vis-a-vis the mainstream philosophies of analytical persuasion as it is predominantly pursued in the English-speaking countries. Not to take refuge in various forms of relativisms, and not to claim too much in the form of absolute, last or eternal foundations (or to impose the responsibility for human actions and their results on "nature") this always deserves a new, methodical effort.

University of Marburg

NOTES

[1] Cf. 'Die methodische Ordnung von Konstruktionen. Der Radikale aus der Sicht des Erlanger Konstruktivismus', in S. J. Schmidt (ed.), *Kognition und Gesellschaft*. Der Diskurs des Radikalen Konstruktivismus II, Frankfurt 1992, S. 24–41.

[2] Cf. the attached list of a selection of constructivist writings.

[3] This may be stated as many books and articles by authors of the Erlangen School have been translated into English and appeared in reputable places.

[4] The Euclidian definition of the sphere runs over a rotation of a semi-circle which is the method to control the shape of a stone sphere by sculptors, and is not analogous to the definition of the circle (in book 1) which is taken from the drawing by means of a pair of compasses.

[5] The operativism already of Dingler, the more so of Janich, is therefore basically different from that of P. W. Bridgman who also allows "paper- and pencil-operations". The core of constructive operativism lies in the non-linguistic, poietic actions as a foundational source for linguistic entities like definitions. Cf. P. Janich, H. Tetens, 'Protophysik. Eine Einführung', in P. Janich (ed.), *Protophysik heute*; Sonderheft *Philosophia naturalis* 1985/1, pp. 3–21.

[6] Cf. P. Janich, Protophysics of Time, in *Boston Studies in the Philosophy of Science*, Volume 30 (Dordrecht, Boston, Lancaster, 1985).

[7] Cf. P. Janich, 'Die technische Erzwingbarkeit der Euklidizität', in P. Janich (ed.), *Entwicklungen der Methodischen Philosophie* (Frankfurt a.M., 1992), pp. 68–84.

[8] Cf. B. Thüring, *Die Gravitation und die philosophischen Grundlagen der Physik* (Berlin, 1967).

[9] Cf. P. Lorenzen, *Lehrbuch der Konstruktiven Wissenschaftstheorie* (Mannheim/Wien/Zürich, 1987), pp. 206–9.

[10] Cf. P. Janich, 'The Concept of Mass', in R. E. Butts and J. R. Brown (eds.), *Constructivism and Science. Essays in Recent German Philosophy* (Dordrecht, 1989), pp. 145–62.

[11] Ch. Thiel, *Philosophie und Mathematik* (Darmstadt, 1995).

[12] H. Tetens, *Experimentelle Erfahrung* (Hamburg, 1987).

[13] R. Inhetveen, *Konstruktive Geometrie. Eine formentheoretische Begründung der Euklidischen Geometrie* (Mannheim/Wien/Zürich 1983); P. Lorenzen, *Elementargeometrie. Das Fundament der analytischen Geometrie* (Mannheim/Wien/Zürich, 1984).

[14] Cf. P. Janich, 'Protochomie. Programm einer konstruktiven Chemiebegründung', in *J. for General Philosophy of Science* **25**, 1994, pp. 71–87; P. Janich, *Philosophische Perspektiven der Chemie*, 1. Erlenmeyer-Kolloquium zur Philosophie der Chemie, with contributions by M. Eisvogel, G. Hanekamp, P. Janich, C. M. Liegener, N. Psarros, K. Ruthenberg, J. Schummer (Mannheim/Leipzig/Wien/Zürich, 1994).

[15] P. Janich, 'Naturgeschichte und Naturgesetz', in O. Schwemmer (ed.), *Über Natur. Philosophische Beiträge zum Naturverständnis*, Frankfurt 1987, pp. 105–22; M. Weingarten, *Organismuslehre und Evolutionstheorie* (Hamburg, 1992); M. Weingarten, *Organismen – Objekte oder Subjekte der Evolution*? (Darmstadt, 1993); M. Gutmann and M. Weingarten, *Veränderungen in der evolutionstheoretischen Diskussion: Die Aufhebung des Atomismus in der Genetik*, in *Natur und Museum*, Berichte der Senckenbergischen naturforschenden Gesellschaft, Vol. 124, 6 (1994), pp. 189–95.

[16] Cf. D. Hartmann, *Naturwissenschaftliche Theorien. Wissenschaftstheoretische Grundlagen am Beispiel der Psychologie* (Mannheim, Leipzig, Wien, Zürich, 1993); D. Hartmann, Psychologie: Natur- oder Kulturwissenschaft? in E. Jelden (ed.), *Prototheorien – Praxis und Erkenntnis*? (Leipzig, 1995), pp. 177–90; P. Janich, Naturwissenschaft kulturalistisch verstehen: ein Angebot an die Psychologie? in G. Jüttemann (ed.), *Regelgeleitetes Handeln. Zur Wiederbegründung einer geisteswissenschaftlichen Psychologie* (Heidelberg, 1991), pp. 1–9; P. Janich, Der erkenntnistheoretische Status von Prototheorien, in E. Jelden (ed.), *Prototheorien – Praxis und Erkenntnis?*, (*Leipziger Schriften zur Philosophie* 1) (Leipzig, 1995), pp. 31–40; P. Janich, Das Experiment in der Psychologie, in H. P. Lengfeldt and R. Lutz (eds.), *Sein, Sollen und Handeln.* Beiträge zur Pädagogischen Psychologie und ihren Grundlagen (Göttingen/Bern/Toronto/Seattle, 1995), pp. 41–51.

[17] Cf. note 16. D. Hartmann also developed a constructive logic of questions, an approach which is important also because of its criticism of Lorenzen's "Dialogical Logic".

CONSTRUCTIVIST PUBLICATIONS

Gethmann, Carl Friedrich (1979), *Protologik* (Frankfurt a.M.).

———. (1980), *Theorie des wissenschaftlichen Argumentierens* (Frankfurt a.M.).

Hartmann, Dirk (1990), *Konstruktive Fragelogik. Vom Elementarsatz zur Logik von Frage und Antwort* (Mannheim/Wien/Zürich).

———. (1993), *Naturwissenschaftliche Theorien. Wissenschaftstheoretische Grundlagen am Beispiel der Psychologie* (Mannheim/Leipzig/Wien/Zürich).

Inhetveen, Rüdiger (1983), *Konstruktive Geometrie. Eine formentheoretische Begründung der Euklidischen Geometrie* (Mannheim/Wien/Zürich).

Janich, Peter (1969), *Die Protophysik der Zeit* (Mannheim).

———. (1973), 'Zweck und Methode der Physik aus philosophische Sicht', G. Hess (ed.), Konstanzer Universitätsreden Nr. 65 (Konstanz).

———. (ed.) (1981), *Wissenschaftstheorie und Wissenschaftsforschung*. Contributions by C. F. Gethmann, P. Janich, Th. Luckmann, J. Mittelstraß und O. Schwemmer (München).

———. (ed.) (1984), *Methodische Philosophie*. Beiträge zum Begründungsproblem der exakten Wissenschaften in Auseinandersetzung mit Hugo Dingler, Contributions by J. Mittelstraß, F. Kambartel, J. Willer, W. Krampf, G. Wolters, R. Inhetveen, H. Tetens, P. Lorenzen, P. Janich (Mannheim, Wien, Zürich).

———. (1980), *Die Protophysik der Zeit. Konstruktive Begründung und Geschichte der Zeitmessung* (Frankfurt).

———. (1985), 'Protophysics of Time', in *Boston Studies in the Philosophy of Science*, Volume 30, R. S. Cohen and M. W. Wartofsky (eds.) (Dordrecht, Boston, Lancaster).

———. (ed.) (1985), *Protophysik heute*. Sonderheft von *Philosophia Naturalis* 1985/1, Contributions by P. Hinst, R. Inhetveen, P. Lorenzen, B. Thüring, H. Tetens und P. Janich.

———. (1989), *Euklids Erbe. Ist der Raum dreidimensional?* (München).

———. (ed.) (1992), *Entwicklungen der Methodischen Philosophie*. Contributions by: C. F. Gethmann, R. Inhetveen, P. Janich, F. Kambartel, R. Kötter, K. Lorenz, P. Lorenzen, J. Mittelstraß, H. J. Schneider, O. Schwemmer, H. Tetens, Ch. Thiel, U. Weiß, J. Willer and G. Wolters (Frankfurt a.M.).

———. (1992), *Grenzen der Naturwissenschaft. Erkennen als Handeln* (München).

———. (1992), 'Euclid's Heritage: Is Space Three-Dimensional?', in *The University of Westers Ontario Series in Philosophy of Science*, Vol. 52, R. E. Butts (ed.) (Dordrecht, Boston, London).

———. (1993), *Erkennen als Handeln. Von der konstruktiven Wissenschaftstheorie zur Erkenntnistheorie*. Jenaer Philosophische Vorträge und Studien, Wolfram Hogrebe (ed.) (Erlangen und Jena).

———. (ed.) (1994), *Philosophische Perspektiven der Chemie*. 1. Erlenmeyer-Kolloquium zur Philosophie der Chemie, Contributions by: M. Eisvogel, G. Hanekamp, P. Janich, C. M. Liegener, N. Psarros, K. Ruthenberg and J. Schummer (Mannheim, Leipzig, Wien, Zürich).

———. (ed.) (1994), *Die Konstruktive Wissenschaftstheorie. Einführung in die allgemeine und spezielle Konstruktive Wissenschaftstheorie* (Hagen).

Janich, Peter, Friedrich Kambartel und Jürgen Mittelstraß (1974), *Wissenschaftstheorie als Wissenschaftskritik* (Frankfurt).

Kambartel, Friedrich (1908), *Erfahrung und Struktur. Bausteine zu einer Kritik des Empirismus und Formalismus* (Frankfurt a.M.).

———. (1989), *Philosophie der humanen Welt* (Frankfurt a.M.).

Kamlah, Wilhelm (1972), *Philosophische Anthropologie. Sprachkritische Grundlegung und Ethik* (Mannheim).

Kamlah, Wilhem und Paul Lorenzen (1990), *Logische Propädeutik. Vorschule des vernünftigen Redens*, 2nd ed. (Mannheim, Leipzig, Wien, Zürich).

Lorenzen, Paul (1970), *Formale Logik*, 4th ed. (Berlin).

———. (1965), *Differential und Integral. Eine konstruktive Einführung in die klassische Analysis* (Frankfurt).

———. (1974), *Methodisches Denken* [1st ed. 1968] (Frankfurt).

———. (1969), *Normative Logic and Ethics* (Mannheim/Zürich).

———. (1974), *Konstruktive Wissenschaftstheorie* (Frankfurt).

———. (1978), *Theorie der technischen und politischen Vernunft* (Stuttgart).
———. (1980), *Metamathematik*, 2nd ed. (Mannheim/Wien/Zürich).
———. (1984), *Elementargeometrie. Das Fundament der analytischen Geometrie* (Mannheim/ Wien/Zürich).
———. (1987), *Constructive Philosophy* (Amherst).
———. (1987), *Lehrbuch der konstruktiven Wissenschaftstheorie* (Mannheim/Wien/Zürich).
Lorenzen, Paul und Kuno Lorenz (1978), *Dialogische Logik* (Darmstadt).
Lorenzen, Paul und Oswald Schwemmer (1975), *Konstruktive Logik, Ethik und Wissenschaftstheorie*, 2nd ed. (Mannheim/Wien/Zürich).
Mittelstraß, Jürgen (1970), *Neuzeit und Aufklärung. Studien zur Entstehung der neuzeitlichen Wissenschaft und Philosophie* (Berlin/New York).
———. (1974), *Die Möglichkeit von Wissenschaft* (Frankfurt a.M.).
Schwemmer, Oswald (1980), *Philosophie der Praxis. Versuch zur Grundlegung einer Lehre vom moralischen Argumentieren* (Frankfurt a.M.).
Thiel, Christian (1995), *Philosophie und Mathematik. Eine Einführung in ihre Wechselwirkungen und in die Philosophie der Mathematik* (Darmstadt).
Weingarten, Michael (1992), *Organismuslehre und Evolutionstheorie* (Hamburg).
———. (1993), *Organismen – Objekte oder Subjekte der Evolution*? *Philosophische Studien zum Paradigmawechsel in der Evolutionsbiologie* (Darmstadt).

JOSEPH J. KOCKELMANS

HERMENEUTIC VS. EMPIRICIST PHILOSOPHY OF SCIENCE

When I started my studies in the philosophy of science in the early fifties, it was quite clear to me that logical positivism was the leading school in the philosophy of the natural sciences. This conception was strongly promoted by the Vienna Circle between 1923 and 1938, as well as later when the events leading to World War II saw the Circle members leaving Vienna and establishing themselves all over the world, particularly in the United States. In the United States, Carnap would soon become a leading figure, whereas in England, Popper eventually would become the main proponent.

From the very beginning I did not feel at home in this way of thinking in view of the fact that it seemed to me that such a view is inherently empiricistic and thus skeptic. Yet from the start I have immersed myself in the philosophy of science promulgated by logical positivism. In the Netherlands, I encountered this kind of philosophy of science in many publications, but most importantly in the works of Evert Willem Beth and later in the United States in the works of Carnap. Due to my association with the University of Pittsburgh I received first hand knowledge of logical positivism through my contacts with Grünbaum and Hempel and later in the form of logical empiricism as defended by Wilfrid Sellars. There I also first encountered Feyerabend who for some time lectured at the University of Pittsburgh during those years. Later I also became fully familiar with the philosophy of logical empiricism in and through the works of Stegmüller, Popper, and Lakatos. My conviction that this philosophy is inherently skeptical has not changed over the years. Yet I have gradually realized the positive merits of this way of thinking. Thus I do accept that in the sciences there is the dimension that logical positivism and logical empiricism make the central point of their attention. The sciences are preoccupied with methodical, orderly, and "logical" procedures and scientists are convinced that experience has an essential part to play in the development of any science.

As I have mentioned already, my first contact with logical positivism was in the Netherlands and later in the United States, mainly in Pittsburgh. Yet my most important immediate contact with logical positivism and logical empiricism came about through my contacts with Stegmüller, with whom I stayed in close contact until the day of his untimely death. I have not only discussed some of his ideas in articles and review articles, but I have also discussed a number of issues pertinent to philosophy of science with him personally as well as in our correspondence. I am particularly impressed by his work on scientific explanation and foundation and his work on the dynamics of scientific theories.

On a still more personal level I became involved in a discussion about

D. Ginev and R. S. Cohen (eds.), Issues and Images in the Philosophy of Science, 191–215.

the "foundations" of logical empiricism and related forms of empiricism, such as scientific realism, in my discussions with van Fraassen who at one time took some of my courses. Obviously I do not take any credit for the ideas which van Fraassen would later develop and promulgate, because they were undoubtedly his own and partly inspired by Sellars, Reichenbach, Grünbaum and others, but I merely suggest that my contact with him has helped me considerably to determine for myself why any empiricist conception of the sciences would be unacceptable to me.

With respect to van Fraassen's book on constructive empiricism I have tried to formulate my opinion about this important book, particularly in my Presidential Address to the members of the Eastern Division of the American Philosophical Association. In that lecture I have tried to explain that any type of empiricist philosophy of science is incapable of effectively dealing with the problem of truth in the sciences, partly because such a view is inherently skeptical and, thus, prohibits any access to the truth, but also because of the fact that it works with several basic unclarified notions, experience and truth being the most important of them.[1]

Over the last 40 years I therefore tried to find another approach to the philosophy of the natural sciences. At first I called this kind of thinking phenomenology or existential phenomenology; later I have learned to call it hermeneutic phenomenology of the natural sciences. I have explained my position first, very inadequately indeed, in my book *Phenomenology and Physical Science*, but later more effectively in a number of publications from the eighties in articles as well as in book form. I find my *Hermeneutic Phenomenology of the Natural Sciences*[2] the most successful of these efforts, not only because in that book I was able to lay out my basic position in outline, but also because I was able to relate these ideas to the views of Carnap, Stegmüller, Lakatos, Kuhn, and Popper.

During the past two years, van Fraassen has undertaken the task of trying to convince me and many others of the view that a constructivist conception of empiricism is still by far superior to any other conception, including my hermeneutic position, and that there are even a number of reasons why an empiricist philosopher is unable to accept a transcendental attitude in philosophical reflections on the sciences. In this essay I shall take the opportunity to explain once more why I hold this position to be unacceptable because it remains skeptical and dogmatic.[3]

In what follows I shall first briefly explain van Fraassen's latest position and then present a number of "arguments," or better reflections, which are meant to substantiate what I call the hermeneutic phenomenological conception of the natural sciences. Finally, I shall indicate why van Fraassen's position is still unacceptable to me.

I. VAN FRAASSEN'S EFFORTS TO DEFEND EMPIRICISM

Van Fraassen was introduced to the philosophy of science from the perspective of logical empiricism. Within the domain of logical empiricism there has been a long and ardent debate about the problem of truth in the natural sciences. This debate is led mainly by realists and antirealists. For a realist a scientific theory can be said to be true; an antirealist denies this. Some antirealists are conventionalists who claim that any set of phenomena ideally can be explained equally well by more than one scientific theory. Thus when a community of scientists prefers one theory over another, then this is not because of the belief that the former theory is true and the latter false, but rather because of an agreement among them, by convention, to describe the relevant phenomena in terms of the one theory rather than the other. And this choice is always made on pragmatic grounds. Instrumentalism, on the other hand, stresses the notion that a scientific theory is no more than some "instrument" which enables us to make correct predictions about the course of nature. Theories should not be evaluated in terms of being true or false, but rather in terms of being effective or ineffective with respect to further research. Radical instrumentalists simply deny that scientific theories are true; on the other hand, moderate instrumentalists hold that scientific theories perhaps may be true; yet in view of the fact that we will never know whether they are actually true, one must limit the debate about the value of theories to questions about their usefulness or effectiveness.

Van Fraassen, too, is opposed to scientific realism; yet he does not think that the antirealist views just mentioned can be uncritically accepted.[4] This is the reason why he has set out to develop a constructive alternative to scientific realism, which he calls constructive empiricism. In his view it is difficult to explain in a few statements what is meant by scientific realism, because here, too, several positions must be distinguished. After a brief discussion of some descriptions of scientific realism presented by some of the leading scientific realists, van Fraassen suggests that we define scientific realism with the following two theses:

1. in its theories science aims to give us a literally true story of what the natural world is like; and
2. acceptance of a scientific theory involves the belief that the theory is true.

Constructive empiricism states, in opposition to these two theses, that science aims to give us theories which are empirically adequate, and that acceptance of a scientific theory involves a belief only that it is *empirically adequate*. The basic difference between scientific realism and constructive empiricism is thus the difference between *literal truth* and *empirical adequacy*, which in these two positions appear in the definition of the goal of science, and that of the criterion for theory acceptance. For constructive empiricism, a theory is empirically adequate if, in what it says about the observable things and events in the world, it "saves the phenomena." Formulated more carefully, a theory is empirically adequate, if and only if such a theory has at least one model into which all the actual (and not just the actually observed) phenomena fit. For constructive empiricism the assertion of empirical adequacy is a great deal weaker than the assertion of truth by scientific realism.

In addition to the epistemological dimension just touched on, constructive empiricism also mentions the pragmatic dimension of science, which involves explanation, the commitment to a certain research program, etc. In the view of constructive empiricism, however, the pragmatic aspects of theory acceptance cannot contribute to our epistemic judgments about the truth of theories. One theory may be explanatorily more powerful than its rival; yet this fact does not warrant our inferring that the theory is true, or even that it is more likely to be true than another theory.

It is clear that constructive empiricism proposes a *new* picture of scientific theories, one which is notably different from the one proposed by logical positivism in which scientific theories were represented one-sidedly from a predominantly linguistic point of view. For logical positivism to present a theory was to specify an exact language and a set of axioms, to which a partial dictionary was to be added in order to relate the theoretical terms to the observed phenomena. Constructive empiricism, on the other hand, states that to present a theory is to specify a family of structures, called models, and to specify certain parts of these models, namely, the empirical substructures, as candidates for direct representation of observable phenomena. The structures which can be described in reports of experiments and measurements are called appearances. The theory is said to be empirically adequate if it has a model such that all appearances are isomorphic to the empirical structures of that model. Obviously much more should be said about this new and intriguing proposal; yet what has been said may suffice for the discussion to follow.[5]

If we compare these theories and carefully examine their presuppositions from the ontological perspective from which they have been formulated, we shall find, I think, that the position of scientific realism is too strong because in some of its formulations it would seem to leave no room for the historical dimension of the sciences, whereas in others, the account of the historical dimension encounters insurmountable difficulties, as both Laudan and van Fraassen have shown independently in a convincing manner.[6] Furthermore, realism is committed to a physicalist ontology the precise nature of which has never been spelled out in detail. Most importantly, the scientific realist accepts as true the statement that if one has good reasons for holding a theory, one also has good reasons for holding that the entities implied by the theory really exist. But this statement is true only under the following conditions:

1. that "holding a theory" means "taking the theory to be true"; and
2. that the expression "true" is taken in the sense of the classical correspondence theory of truth.

But these two conditions can be questioned.[7]

Finally, the uncritical but common conception of the correspondence theory of truth seems to force the scientific realist into the position that things in the real world actually are exactly as the scientists claim them to be *and that they cannot be in any other way except in the one brought to light by the sciences.* I would like to call this last claim the positivist thesis. In the final analysis, I thus cannot accept scientific realism because of the fact that in most of its forms it remains ontologically undetermined and, furthermore, because in the end it appears to be a form of positivism.

On the other hand, the positions of conventionalism and instrumentalism, taken in their original forms, are too weak. Conventionalism and instrumentalism seem to lead to a radical form of skepticism; the logical consequence of such a form of skepticism would be that the engagement in science becomes an irrational endeavor. Conventionalism and instrumentalism are unacceptable also insofar as they are unable to account for the fact that scientific theories can be falsified. They are thus unacceptable because scientific theories are not just human fabrications but are indeed *theories of what is*. Finally, it may be the case that there are some theories which seem to rest on one or more "mere" conventions; yet it is difficult to see how the atomic theory, for instance, could be called "just a convention or an instrument."[8]

I find myself in agreement with constructive empiricism when it states that in the past one has formulated many problems of philosophy of science in a form that was too one-sidedly linguistic. This is true particularly for the position of logical positivism, which, in addition, rests on a theory of meaning and language which is now quite generally rejected by most philosophers of science.

I also agree with constructive empiricism where it states that it is very difficult to subscribe to the view that scientific theories are true without further qualification.[9] If one considers contemporary theories in physics, such as quantum mechanics, then it is clear that it cannot be true that there is always a one-to-one correspondence between the elements of the model and the elements of real things in nature. For there are numerous elements in the model to which nothing corresponds in the real world; furthermore, there may also be elements in the real world about which the theory does not (yet) purport to make any claims. The scientific model provides us with an extremely vast and complex relational framework of a mathematical nature, in which certain structures and elements of the real subatomic world can be embedded. Thus we cannot meaningfully claim that the theory is true without further qualification.[10]

And so, as long as we remain within the context of the basic philosophical assumptions which are shared by both scientific realism and constructive empiricism, we can say, I feel, that science aims at theories that are empirically adequate, provided one understands here by theory not just a set of propositions, but rather a scientific model. It seems clear that, of scientific models, one may legitimately demand that they be adequate to explain the relevant phenomena that are known by observation and experiment, or that perhaps will show up in further research. It seems rational to conclude from this that accepting such a model implies the belief only that the model is empirically adequate. For our scientific models are always theoretical, usually mathematical, models which contain nothing but well-defined entities which have only those characteristics which the theorist has exactly specified. The question of how such theoretical models relate to real physical systems cannot be answered simply. One cannot possibly defend the view that these models are faithful replicas, in all details, of what there is in our physical world. The maximum claim one could perhaps make is that the designated "real" system is similar to the proposed model in specified respects and to specified degrees.[11]

In my discussion of scientific realism, conventionalism, and constructive empiricism I have thus far limited myself to reflections in which one can engage without yet questioning the assumptions made by realists, instrumentalists, and constructive empiricists, alike. I wish to turn now to one of these assumptions, namely the one that constitutes the very subject matter of this paper.

As I see it, many philosophers who today concern themselves with philosophy of science accept the basic ontological theses of empiricism. This basic empiricist perspective is usually not explicitly examined critically in the publications of these philosophers. Usually one assumes this perspective to be the correct one. Yet some of the basic ontological assumptions underlying all forms of empiricism are questionable, and this is the case also for some of the assumptions which empiricism shares with classical rationalism.[12]

Van Fraassen at first sight is an exception to this "rule," for he states explicitly what he understands by being an empiricist. For him to be an empiricist implies the acceptance of at least three basic commitments:

1. to withhold belief in anything that goes beyond the actual, observable phenomena and to recognize no objective modality;
2. to develop an account of science which depicts science as involving a search for truth only about the empirical world, about what is actual and observable; and
3. to reject resolutely the demand for an explanation of the regularities in the observable course of nature by means of truths concerning a reality beyond what is actual and observable, even though a full account of science must be accompanied by auxiliary theories about explanation, conceptual commitment, modal language, and perhaps other issues.[13]

Yet it seems to me that this description of what it means to be an empiricist is really a description of what *from an empiricist point of view it means to be an empirical scientist.* The precise nature of the underlying empiricist perspective remains here, too, undetermined. The basic empiricist ontology is here not made explicit but is merely posited dogmatically. Yet, the history of philosophy since Locke and Hume has shown convincingly that empiricism rests on unacceptable ontological assumptions. Thus it seems to be mandatory that one follow the critical discussion of empiricism systematically from beginning to end, if one is to understand contemporary interpretations of the empiricist perspective completely.[14] It will then be clear that a passing reference to Ockham, Mach, and Poincaré will just not do here, and particularly that Kant and German Idealism on the one hand, and modern philosophies such as the philosophy of the later Wittgenstein, Husserl's phenomenology, and the hermeneutic tradition must also make a substantial contribution to the debate about empiricism. I feel that both Hooker and Giere, too, have correctly criticized van Fraassen's uncritical adherence to "classical" empiricism.[15]

In a recent essay, entitled "Against Transcendental Empiricism"[16] van Fraassen has responded to my criticism of his position and rejected my suggestion that no form of empiricism is acceptable, as long as it does not allow for some transcendental dimension. In making this claim I had made it clear that the term "transcendental" in this case need not be taken in the sense of Kant.

In part I of his essay van Fraassen critically discusses a number of empiricist positions which he, too, rejects and about which we can be in agreement. He begins by stating that the fact that he wants to be an empiricist is intimately connected with his preference for certain philosophical moves or themes developed in the past as against some others. When he reports characteristics of empiricism he includes only those he endorses and neglects what he regards as common errors that past empiricists have fallen into. For a critique of historical empiricism by a contemporary empiricist, he refers to an essay by Hans Reichenbach, "Rationalism and Empiricism: An Inquiry into the Roots of Philosophical Error," which appeared in *Modern Philosophy of Science* in 1959.[17] In so doing his goal is to develop a philosophical position, which has two basic characteristics: a) it contains what is admirable in past versions of empiricism, and b) it must still be feasible today.

The first question to be raised therefore is the following: what is a philosophical position? To have a philosophical position is to believe something. We can therefore say that for each philosophical position X there exists a statement X+ such that to have, or take position X, is to believe, or decide to believe, that X+ (*Principle Zero*). Applied to empiricism this would lead to the following principle: For empiricism there must be some statement E+ such that to be an empiricist is to believe that E+. This is the basic position of a naive empiricist. In the remainder of part I, van Fraassen now shows that there cannot be any statement that could possibly play the role of E+ in naive empiricism. He concludes from this that there is only one way out of the dilemma with which naive empiricism is confronted, namely that any (E+) would have two contradictory functions: a) to constitute a respectable belief for the empiricist's ideal rational believer, and b) a basis for the empiricist's critiques of metaphysics. Thus we must deny Principle Zero as a characterization of philosophical positions *überhaupt*. This means that philosophical positions cannot be captured in dogmas that ask for beliefs. Van Fraassen now presents the following radical proposal: a philosophical position can exist in something other than a belief in what the world is like.

Taking the empiricists' *attitude* toward science, rather than his or her beliefs about it, as the more crucial characteristic, we are led to the suggestion that an alternative to Principle Zero is that a philosophical position consists in a stance (attitude, commitment, approach). Such a stance can be expressed, and may involve or presuppose, some beliefs as well, but cannot be simply equated with having beliefs.[18]

In part II of his essay van Fraassen critically discusses Husserl's distinction between two basically different attitudes: the natural and the philosophical. He makes it quite clear that for him such a distinction does not hold. All philosophical questions are scientific questions or they are irrelevant questions. All legitimate questions can be treated by empirical and logico-mathematical research. Having rejected Husserl's distinction van Fraassen remains within logical empiricism, and instead of Husserl's distinction he proposes a distinction between "having the belief that X" and "adopting a certain attitude." Van Fraassen can now reject the view that committing oneself to a method would imply the belief that this method indeed is adequate. Already in his essay "Empiricism in the Philosophy of Science," which appeared in *Images of Science*, van Fraassen had shown that empiricism will remain inadequate and unacceptable as long as one insists that some belief is involved that would not be either a logical tautology or an empirically significant statement.[19]

Finally in Part III van Fraassen tries to identify the empiricist stance and to isolate those attitudes that characterize empiricist philosophical positions. For van Fraassen a certain attitude toward empirical science is central to empiricism. To spell out this central attitude toward empirical science one must make use of concepts and also articulate beliefs. In spelling out the central attitude toward science, there must be left some leeway, because there are different varieties of empiricism, predicated on variations in this conception of science. Furthermore, one can adopt different attitudes toward science as so conceived. The crucial question now is one of how much of science, and which aspects of science must be selected in setting an ideal for practical and theoretical reason. Then how much of actual science bears out this ideal? Where is this ideal to be left behind as no longer offering adequate or relevant guidance? In trying to answer these three basic questions one must keep in mind that empiricism does not defend scientism; rather it entails a criticism of scientism in the name of science.

Van Fraassen first points out that in his view attitudes need not be justified by beliefs. It is usually held that reason demands that one can justify one's attitudes. Thus it is often said that attitudes are to be justified by beliefs. In van Fraassen's view, anyone who holds this view immediately falls into the fact/value problem. The only way to avoid this is to deny the initial bias of reason: rationality does not require that our attitudes be justified by beliefs.

Yet it must be added immediately that attitudes do involve beliefs. We must expect, therefore, that if we isolate those attitudes that characterize empiricism, we will find them inextricably involved with certain beliefs. Since van Fraassen takes it that the attitude toward empirical science is central to empiricism, it is to be expected that the centrally involved belief must be a view of what science is.

Van Fraassen has made an effort to answer the essential questions to be asked here in a book-length study, *The Scientific Image*, and does not here make an effort to condense the content of that book into a few statements. Instead he addresses two subsidiary questions. The first question is one of how we should approach the subject of attitudes involving beliefs; the second is one of what exactly is believed, in holding a certain view as to what science is.

One cannot answer the questions of how beliefs are involved in attitudes and of what the content is of the involved beliefs, except by appealing to a very rich psychological idiom. This most certainly cannot be done by limiting oneself to the reductionist belief-desire psychology of Hume and other 18th century empiricists. The psychological idiom required contrasts sharply with the idea that an empiricist position might be identifiable with a belief to the effect that there is a certain purely semantic relation between scientific theories and facts, or between scientific models and the empirical world.[20]

As van Fraassen sees it, for empiricism science is the paradigm of human rationality. Empiricism advocates the scientific method as the guide to life. Science is seen as the ideal. Van Fraassen is aware of the fact that if the attitude of empiricism is defined in such bold terms, it seems to imply two terrible corollaries:

1. Empiricism appears to remain in the natural attitude and from the perspective of almost all philosophers that would be the death of philosophy.
2. The thesis advocated here seems to imply intellectual submission to actual science which always was, is, and will be imperfect.

But for van Fraassen this is not actually the case. For being truly scientific necessarily involves a skeptical attitude toward extant science, rejection of scientism, and refusal to allow any intellectual scheme to keep us "pinned and wriggling on the wall." Science does not advance through submission to its previous generations. The scientific method is the skeptical method, with one difference: it allows that it is also rationally permitted at any point to lay to rest doubts one cannot disprove, and to commit oneself to a theory that goes infinitely far beyond the evidence we have for it, or even ever could have for it. But in so doing rational inquiry is not replaced by blind dogma, provided the step taken is acknowledged as such.[21]

What is proposed here requires the possibility of a certain detachment, of simultaneously having convictions, while being capable of standing back and assessing them critically. Van Fraassen thinks that this may suffice for the second corollary, but what about the first? Husserl claimed that in the natural attitude questions about the very possibility of cognition are absent. To make the discussion of such questions possible Husserl distinguished between the natural and the philosophical attitude. For van Fraassen, what Husserl described here are really two extreme poles, both impossible: the consciousness so immersed in things that it cannot be troubled by preconditions of cognition, and the opposite consciousness so philosophically free that its inquiry is presuppositionless. In actuality we are never at either extreme. Furthermore it is not the intention of empiricism to replace our everyday belief-structure by a parallel scientific belief-structure. What we learn in our everyday life is real and, indeed, is not among the deliverances of scientifically disciplined experience. The right advocacy of the scientific orientation is to advocate it only against its direct rivals: approaches that suggest to acquire the same information about the world by consulting superstition, spirits of the dead, metaphysical arguments, innate ideas, and the comforts of explanation and other intellectual indulgences, where these go beyond the test of experience.[22]

Van Fraassen finally turns to a last observation which he regards as a serious research problem. In the preceding he stated that what empiricism admires in science is the ideal of an epistemic attitude, which is indeed belief, but belief that is incompatible with dogmatism. The scientist's acceptance of theories and hypotheses involves belief, but an undogmatic belief. Scientists hold these beliefs but always with a certain detachment. The question is: is this really possible?

As stated earlier, van Fraassen holds that acceptance of a scientific theory involves not the belief that this theory is true, but only that it is empirically adequate; that it saves the phenomena. But in each case our opinions go very far beyond the evidence we have. These opinions involve many beliefs about observable facts elsewhere and at other times. Such beliefs will be equally hostage to the fortunes of future experience. The empiricist attitude which van Fraassen admires is the following: with respect to all these beliefs, I must display a certain *amor fati*. If they become very unlikely in light of new experience, I shall not hold on to them at all costs, but revise my opinion.[23]

This resolve to allow one's beliefs to be hostage to the fortunes of future experience, empiricism holds to be crucial and central to scientific method. Moreover, empiricism holds this up as an ideal for us with respect to all our beliefs and opinions and even indirectly to our attitudes. But how can this sort of resolve possibly be anything more than empty posturing?[24]

Van Fraassen does not discuss this last issue in detail. Yet it is clear that in his view such a resolve is indeed a constitutive component of the empiricist attitude in regard to science.[25]

Before turning to a brief description of my own position in regard to the basic issues raised here, let me state first that I am practically in complete agreement with van Fraassen on what he has stated about the empirical sciences. I, too, have a great admiration for the sciences. I have at several times stated that if the term "rationalism" has any meaning at all, it should be used to characterize the work of the scientists. I also subscribe to the notion that a good scientist must give up his theories and hypotheses, when they become unlikely in light of new experience. I could mention several other points in which I would agree with van Fraassen, even though in some instances I might have used other expressions to state my opinions or beliefs. Yet I should immediately add to this that I share this view only if the sciences in question are natural sciences. In the case of the sciences that concern themselves with human beings and their individual, social, and political behavior, I take it that this position would only be true in part. As I see it, in these sciences descriptive and interpretive methods will have to be added to the empirical method. Finally, for the historical, linguistic, and juridical disciplines I take the model of the natural sciences to be irrelevant in most cases. Since I have stated my opinion about these sciences elsewhere in detail, I shall no longer dwell on them here.[26]

II. REASONS FOR WHY A TRANSCENDENTAL DIMENSION IN THE PHILOSOPHY OF THE NATURAL SCIENCES IS NECESSARY

Before going into details let me first state briefly the most important reasons for my reluctance to go along with the position outlined by van Fraassen. The prime reason for my disagreeing with him is to be found in the fact that he is unwilling to admit the possibility of Kant's transcendental logic, or Husserl's transcendental phenomenology, or Heidegger's fundamental ontology, or some other transcendental dimension. I see this as the main reason why his position remains dogmatic, because unexamined.

It seems to me that anyone who denies every transcendental dimension is doomed to skepticism and unable to speak of truth.

Because every form of empiricism rejects a transcendental dimension, it is unable to define an *a priori* that can function as the condition of the possibility of what empiricism calls experience. As a matter of fact, the notion of "experience" remains unexamined in empiricism; yet some conception of experience is held and taken for granted; experience is taken as a given that is unproblematic, something active that can be structured by certain processes of association. Experience is certainly not taken as something that is in need of critical analysis and explanation. It may be true that contemporary empiricist theories reject Hume's conception of experience as totally inadequate; yet one has not been able to substitute a defendable conception for it, either. I do not know of any empiricist study that explicitly asks what the necessary conditions are that must be fulfilled if human beings are to have experiences. It is not easy to determine precisely what is meant by experience, what it is in our cognitive life that we call an experience, and what having an experience really means. Empiricists never explain the large contexts of meaning within which people have experiences; I would not know how to determine what is meant by an experience if first the world has not been revealed in which alone such experiences can take place. It is clear to me that what most empiricists and scientists call experience is really a derivative notion; the original conception of experience is the one Hegel tried to elaborate in the *Phenomenology of Spirit*,[27] or perhaps the hermeneutic conception of experience critically discussed by Gadamer in *Truth and Method*.[28]

Also, van Fraassen often speaks of "saving the phenomena". Yet he never explains what is to be understood by "phenomenon", nor does he ever explain how phenomena are to be related to appearances, and most importantly how phenomena are constituted.

Another basic point of disagreement consists in the fact that in his reflections van Fraassen begins with theories and hypotheses, but never turns to the complex activities of leading scientists since the 16th century, in which these theories and hypotheses were developed and tested. Thus van Fraassen never discusses the process of thematization in and through which scientists make natural things become objects of research. No reference is made to the fact that scientific thematization always involves objectification, formalization, functionalization, and mathematization. If one were to have paid attention to the activities in and through which science becomes constituted it would have been possible to show why the empiricist conception of induction is unacceptable, also.

Van Fraassen states that pragmatic reflections cannot have any direct bearing on the epistemic validity of a theory. This may be so from the perspective of his analytic conception of language. Yet it certainly is the case that the theoretical reflections and the scientific research activities made possible by them have a direct bearing on the meaning and validity of the statements in which the result of these reflections and research activities has been articulated. As a matter of fact, I would like to state that no one could possibly have an idea of the meaning and validity of scientific theories and statements derived from them, except by projecting them upon the research activities in and through which they were generated as well as the entire broad background of meaning or world that is necessarily presupposed in this scientific work.

Finally, by stating that a philosophical position consists in a stance or commitment which cannot be equated by beliefs, van Fraassen reduces each philosophy really to its method. He says that each philosopher should begin by saying that he has no preconceived notions about what the world is like, but that he certainly knows how to go about it in an effort to find out. Thus each philosophy must begin with a reduction of all doxic beliefs about what is, as well as with the demand always to apply no method other than the empirical method. But this last requirement leads to complete skepticism and nihilism. It seems to me that Descartes through his metaphysical reflections and Kant through his first *Critique* have tried to avoid this disastrous consequence. Now I do not suggest that we follow either Descartes or Kant; yet I do suggest that it is all-important to heed their warning. For an empiricist there is no method other than the empirical one because by *fiat* he decrees that the empirical method is the only one that needs no justification and does not presuppose anything beyond itself. On the other hand, all non-empiricist and non-nihilist philosophers are convinced that the empirical method is in need of justification and that the meaningful application of this method presupposes precisely all the elements needed to avoid skepticism and nihilism.

It seems to me that all other points of disagreement in some way or other follow from the basic points of disagreement just mentioned.

Let me now turn to a more detailed discussion of some of the issues raised. But before doing so I would like to make some preliminary observations. First of all, in what follows I shall restrict my reflections and claims to the philosophy of the *natural* sciences for reasons mentioned above. I would like to note secondly that even though van Fraassen explicitly rejects a transcendental component in his empiricist philosophy, in the essays discussed above, he nonetheless *in some sense* is engaged in transcendental reflections. Let me explain.

In an essay, entitled "On the Meaning of the Transcendental Dimension in Philosophy", I have explained that most 19th and 20th century philosophers following Kant admit a transcendental dimension, even though they do not all conceive of this dimension in the same manner.[29] Yet in most cases these conceptions continue to refer back to Kant's conception of the transcendental. As we all know, in the *Critique of Pure Reason*, Kant defined transcendental knowledge as follows: "I entitle transcendental all knowledge which is occupied not so much with objects as with the mode of our knowledge of objects in so far as this mode of knowledge is to be possible *a priori*" (CPR, B, 26). Now insofar as the statements made by Van Fraassen did not directly state anything about things, events, or "objects," but rather stated directly something about our scientific knowledge of such objects, one could say that his reflections were indeed transcendental.

Yet van Fraassen will obviously immediately respond by saying that he rejects a transcendental dimension on the ground that he rejects any *a priori*, i.e., any knowledge one might appeal to that does not have its origin in experience. But if we again carefully look at the statements which van Fraassen actually made about the sciences, then they, too, do not have their origin in the kind of experience that the scientists rely on, and empiricist philosophy is willing to accept. Now it is obvious at once that the "*a priori*" of van Fraassen is completely different from the *a priori* defined by Kant by means of categories and principles; yet it is true, nonetheless, that van Fraassen, too, appeals "unconsciously" to some *a priori*, and this *a priori* remains underdetermined in his reflections, if not undetermined altogether.

I have already observed that in contemporary philosophy there is no common agreement on the use of the expression "transcendental." Yet in all cases the term is used to refer to the conditions of the possibility of all finite knowledge, to which human beings can have access. Kant defined the transcendental in terms of the *a priori* forms of man's sensibility, the *a priori* concepts of the understanding, and certain *a priori* principles of all finite understanding. Kant in addition claimed that forms, categories, and principles are given *a priori*, i.e. independent of any form of experience. Husserl, on the other hand, defined the transcendental in terms of an absolute, transcendental, constituting subjectivity. Finally, in the case of Heidegger, the transcendental is defined in terms of *Dasein's* standing-out and transcending in the direction of the world, and in and through the world, to "being" as the totality of all possible meaning. In other words, Heidegger defines the transcendental dimension of philosophy in reference to the intimate relation between *Dasein*, taken as transcendence, and being, taken in the form in which it manifests itself today as the concrete world in which we live, i.e. the actual world in which scientific work is being done. In Heidegger's view a systematic investigation of the kind presented in *Being and Time* is needed to bring this transcendental dimension to light in systematic articulation. Only on the basis of such a "fundamental ontology" can a philosophy of science be developed systematically. It is also only on the basis of a fundamental ontology that a meaningful conception of finite and historical truth can be developed.[30] In view of the fact that I have presented these ideas systematically elsewhere in great detail, I shall not repeat them here.[31]

Once empiricism rejects the idea of developing some form of "fundamental ontology" it deprives itself of the possibility of developing a philosophically justified theory of experience. The same can be said for a similar conception of (finite and historical) truth. In avoiding the term "truth" altogether it seems as if an empiricist identifies the notion of truth with eternal and perfect correspondence between what is stated and what is the case. It is obvious that such a conception of truth, which could apply probably only to an absolute, eternal, and divine being, is not applicable to any human form of behavior, not even to our scientific knowledge, of which van Fraassen admits that for humans it is the ideal of knowledge and its most perfected form of rationality.[32] I find it very strange that humans would have to admit that even for the result of their most rational behavior the term "truth" would not be applicable.

What makes things worse is the fact that empiricist philosophers of science usually begin with theories and hypotheses and never critically analyze the activities which finally led to these theories and hypotheses. Let us assume for a moment that we have already succeeded in developing some fundamental ontology of the kind Heidegger has given us in *Being and Time*, so that from this very broad perspective we can now turn to the question of the meaning of science. It seems to me that anyone who wishes to focus on the problem of the meaning of science must begin by conceiving of the sciences in a truly "active" sense and by starting from the entire scientific practice, i.e., from what scientists actually do when they engage in scientific research.[33] To prevent misunderstanding and to explain the point which is of prime importance here let us recall for a moment how a novice is introduced to one of the natural sciences.

When students receive their introduction to a science they first learn how one makes natural things objects of investigation and how one articulates them and delineates the domain of research. This is done by introducing the student to the manner in which things are thematized by the discipline on the basis of an ontological projection which is typical for each science. In this projection a certain domain of natural things is staked out, the approach to the domain is given its particular methodical direction, the structure of the conceptual and discursive explanation receives its orientation, and a specific "language" is constituted. The purpose of the thematization is to free the relevant natural things or a particular group of them in such a manner that they can become the objects of a purely theoretical discovery and, therefore, can be examined objectively.

To clarify what is meant here by thematization, it is perhaps helpful to refer to the manner in which Galileo and Newton laid the foundations of modern mechanics. They selected the domain of inanimate natural things as the subject matter of their investigations and determined that these would be viewed only from the perspective of carefully defined characteristics: mass, force, space, and time determinations. Furthermore, these things were to be examined on the basis of certain principles or axioms, namely, Newton's laws of motion. It is on the basis of these definitions and axioms that actual research in the domain of classical mechanics was performed.

Be this as it may, the entire scientific enterprise is to be taken into consideration in the discussion of ontological problems, i.e., problems of meaning and truth, which the sciences pose to us. One cannot just limit oneself to problems of experience and observation on the one hand, or problems of a linguistic nature on the other. Now the entire scientific enterprise taken as a whole is geared toward getting a better understanding of a certain domain of nature. We must thus focus first on all the activities that are being performed daily in large communities of scholars, all of which ultimately lead to the models and the linguistic structures mentioned. It may be true that some or even most linguistic structures as such do not seem to have a direct relationship to a real state of affairs; yet the entire active involvement of the community of scholars was oriented precisely toward the possibility of making claims about natural phenomena. The theories with the help of which they approach these phenomena cannot be evaluated on their truth value as long as they are not seen as an essential part of this ongoing process. And the same is true for scientific laws and propositions.

In other words, it seems to me that in a discussion of problems of meaning and truth we must not only move away from the domain of narrow linguistic analysis of logical empiricism, but we must even move beyond the domain of scientific models, because even those are still "linguistic" structures.[34] We must move beyond semantics, mathematics, logic, and epistemology, to an ontological domain where issues concerning meaning and truth can be discussed. It should be stressed here that this ontological realm is no longer a linguistic realm even though it is and remains true that there is meaning and truth only *within* the domain of language. Yet to say that meaning is found only within the domain of language is not tantamount to reducing meaning to something that is essentially linguistic only. After all, meaning is the meaning of things and events, and all meaning is meaning for human beings; otherwise it would be the meaning of nothing and for nobody and thus no meaning at all.

One will observe that anyone who moves into the domain of meaning and truth along these lines will inevitably be confronted with the problems on which Kant focused in his first *Critique* and which he failed to solve there in a satisfactory manner. I admit that we are indeed led into the problematic of Kant's critical philosophy, and I also grant that Kant's effort to solve the basic problem to some degree has failed. Yet the fact that Kant did not fully succeed does not entail that we are forced to return to an empiricist position, classical or otherwise.[35] For as I see it, Kant may have failed; yet his effort was not a complete failure. There is something extremely important to be found in the positive dimension of Kant's work. I agree with Kant that claims about natural entities and events can be said to be true only in regard to the phenomena. Yet to say this is not tantamount to accepting Kant's philosophy as a whole without reservation.

Thus, we must examine Kant's conception of phenomenon more carefully. A phenomenon is, as the word says, that which shows itself; it is the thing insofar as it manifests itself. But what shows itself, manifests itself always in a context of meaning. Confronted with this fact it makes no sense to think that one is facing here an unavoidable dilemma: either to demand access to absolute truth or to accept radical skepticism. For humans there is nothing wrong in accepting that to know always means to understand within a context of meaning that is already given in advance. This acceptance brings indeed an interpretive dimension into the picture. But what is called for here is not skepticism but understanding of, and respect for, the limits of human understanding.

When we make a claim we must compare our claim with the thing about which the claim is made to the degree that the thing in question has manifested itself to us already, independent of that claim. The fact that all phenomena manifest themselves as part of a world and, thus, that all human claims are context-bound makes our claims indeed non-absolute claims, i.e., they are not the kinds of "claims" an absolute Knower could make Who is not bound by contexts. Yet this fact itself does not make our claims arbitrary, nor does it make it impossible for us to hold certain claims to be true. It merely means that we never can make claims without having to examine critically the contexts from which our claims were made.[36]

It is indeed the case that in all contexts of meaning in which things manifest themselves to us, there are elements that are not necessary. This is so for claims we make from a pre-scientific point of view, where the context of meaning usually is our Western world today. It is true also for all philosophical claims. Even in the most sophisticated, critical reflection it is impossible for us to go beyond at least assuming some context of meaning. Such contexts always contain prejudgments which need further analysis and critical examination from ever larger perspectives of meaning. Yet it can always be shown that such prejudgments, even though they are not true in an absolute sense, nonetheless are often not arbitrary insofar as they rest on very rational and reasonable grounds. After these preliminary reflections let us now turn directly to the problem of truth.

In discussions about the problem of truth in the sciences, virtually all authors define truth as correspondence; thus they adhere to some interpretation of the correspondence theory of truth. But this does not mean that there is no room for the coherence and the pragmatic theories of truth, also. For these latter conceptions, too, have their place in the discussion; yet they are not so much concerned with the nature of truth as with a criterion that must be fulfilled if in certain contexts one is to be able to speak about truth in a meaningful manner.[37]

I share the view of those who define truth as correspondence. Yet I feel that most authors who are involved in the debate about the problem of truth in the sciences more or less dogmatically select a certain *interpretation* of the classical correspondence theory and do not explicitly reflect on the conditions under which this theory can be meaningfully applied. Others, such as Tarski, Field, Putnam, Boyd, and Ellis do ask questions about the nature of truth as correspondence and about the conditions under which this notion is to be applied, but limit themselves to semantic or linguistic reflections. Since they do not raise the issue on a strictly ontological level, they end up, I feel, with a theory which is both epistemically and ontologically underdetermined, or they find themselves forced to accept the view that what is true *is* no more than a limit of reasonable belief. Let us therefore reflect for a moment on the nature of truth.[38]

In the tradition truth was usually defined as the adequation of intellect and thing. It was assumed in this conception that the proper place of truth is the judgment and that the essence of truth lies in the correspondence between judgment and object. Although this conception has been defended by most philosophers between Plato and Aristotle on the one hand and Husserl and Wittgenstein on the other, and although this conception, as we have just said, is undoubtedly correct, it is still a conception that because of its formal emptiness is affected with serious difficulties. Furthermore, the theory is still unfounded.

One of the first issues to be raised here is what one is to understand by agreement. For in view of the fact that judgment and thing or object are in no way similar to one another, one must ask how and in what sense judgment and thing can indeed correspond to each other. The judgment must state the thing "as it is," as Aristotle says;[39] yet it can never become identical with the thing. Furthermore, how is the thing in question to be understood? It is obvious, as we have said already in our reflections on what is to be understood by phenomenon, that if one interprets the expression "the thing" to mean "the thing as it is in itself independent of our understanding of it," we shall again encounter the problem with which Kant struggled in his first *Critique*. It is clear also that the agreement we are looking for cannot be one between a real psychical process and a physical thing, nor one between a psychical process and an ideal object or idea which somehow relates act and thing, standing between them, as it were, nor finally one between a representation and a "real" thing. The agreement is rather one between the content of a claim about a thing and the thing insofar as this can be discovered by us independently of our claim, by means of some process of confirmation which is intersubjectively acceptable. Thus to say that a statement is true really means that this statement reveals the thing as it manifests itself to be in some process of confirmation that logically antecedes the claim. Thus how the thing is, we know independently of the statement by the manner in which the thing reveals itself in other activities of man, for instance, in perception, or in observation and experiment.

It is understood here that in order for the thing to manifest itself to man it has to manifest itself time and again in a certain context of meaning. These contexts of meaning belong ultimately to the totality of meaning of which we can now conceive. Now the limits which we have to impose on our judgments and on the possible confirming activities, as well as the limits of the contexts of meaning within which things manifest themselves, make it impossible for us to claim that in our judgments we state how things are "in themselves," comprehensively, exhaustively, definitively, and absolutely. We can claim only that our judgments state how things are as seen from some limited context of meaning or, in the final analysis, from the perspective of the whole of meaning of which we can conceive. Thus, every form of revealment implies for us also some form of concealment. By revealing the one thing, I conceal the others; by focusing on one aspect, I omit others; by considering the matter from one perspective, I forego other similar perspectives. But if every human effort to reveal things just as they are implies various forms of concealment, then for us truth is in principle always connected with untruth and meaning with lack of meaning. And yet we can still maintain that our claims are true to the degree that they reveal things just as they manifest themselves to be.

Those who find this conception of truth strange and unacceptable should realize that the problem of truth would not have to be asked if human beings could make truth claims totally independently of contexts of meaning which are limited and historical. Most philosophers since Parmenides and Heraclitus who have thought about the problem of truth have tried to get rid of the real problem by an appeal to an absolute Being or to an ideal order; those who refused to do so have often fallen into an unacceptable form of skepticism and intellectual anarchy. However, the fact that all human truth claims are claims within limited contexts of meaning does not yet by itself doom us to skepticism and anarchy; it merely implies that for human beings all truths are and will be finite and limited truths which in principle will always be accompanied by some lack of truth and which thus always remain open to revision.[40]

Let us now see what this interpretation of the correspondence theory of truth means for a systematic discussion of the problem of truth in the sciences. It is clear that even the sciences are no exception to the general condition outlined here. Scientific claims are also context-bound, and constructive empiricism is correct, I think, in stating that these contexts, which include our scientific models, contain elements that are not necessary. Furthermore, one cannot deny either that our models are formed by means of processes of abstraction and idealization, that at a certain moment mathematical relationships are substituted for empirical generalizations, and that principles are formulated which may very well, by themselves, not yet refer to any concrete set of phenomena. Thus one cannot deny that there always remains a gap between the model and the empirically observed phenomena, whose observation itself again is theory-laden. Yet all of this does not change the fact that all suppositions made in scientific theories can be shown to be rational and reasonable assumptions, not mere arbitrary ones.[41] Making such assumptions makes it possible for us to engage in research projects in which processes of verification play an essential role. It is through these processes that we try to close the gap between model and observed phenomena, even though we will never succeed completely.[42]

Yet all of this does not entail that there will be no longer any reason why one cannot call these statements true. By calling a statement true we merely mean to state that everyone at any time under similar assumptions and circumstances can approach the given phenomena from the same context of meaning and will find the same insights as we have come to, even though at a later time one may discover a model that is even better or more adequate empirically.

Thus far I have tried to explain why and in what sense I take well-established scientific claims to be true. One may be inclined to observe that all of this is rather trivial because no one really denies this thesis, except perhaps the most extreme skeptic. It is indeed true that my position is basically in harmony with that of many other philosophers of science: scientific statements for which we have adequate empirical evidence can legitimately be called true. Yet if this claim is taken in the context which was outlined briefly in the preceding pages then it no longer means that making this claim implies some "one-to-one correspondence" between what the proposition states and the thing taken independently of all claims we can make about it, the thing-in-itself. The claim that scientific statements are true, taken in the interpretation that I have given of it, does not imply that the thing is, and is nothing but what my scientific statements posit it to be. My interpretation of the truth of scientific claims does not imply either that things genuinely are what science claims them to be and that our prescientific conceptions are false or less likely to be true. For me, defending the truth of scientific statements is tantamount to maintaining that things indeed manifest themselves independently of my actual claims as having the properties, characteristics, or relationships which my scientific claims assert them to have. Thus my interpretation of the truth of scientific statements is not endangered by the realization that all observations and experiments are theory-laden, because I have already granted that all human understanding is context-bound, and I have shown, I believe, why the fact that all knowledge and understanding is finite does not make it impossible to speak about truth.

Yet the difference between the view presented here and that held by realists and constructive empiricists will become even clearer if we now turn once more to the truth of scientific theories. I have already indicated that I find the position of scientific realism too strong and that of classical conventionalism too weak. I have also indicated in what sense I here favor the position of constructive empiricism. I shall now try to explain to what degree I would also like to go beyond this position.

But before doing so a few introductory remarks must be made. It seems to me that it would be difficult to defend the view that all the theoretical assumptions made in scientific theories should be taken to be true. Many theoretical assumptions have the character of an "invitation" to do, see, or approach the relevant phenomena in this way rather than in another way. It seems to me, difficult also, to defend the view that all elements of a theory must correspond with certain elements of the real phenomena. For there are always some assumptions about elements that were meant to be claims not about what is but rather about how one should go about dealing with the phenomena. Thus I take it to be difficult to disagree with van Fraassen that we "do not need to postulate that there are elements of reality corresponding to all elements of the [theoretical] model." In other words, there may be elements in the model to which nothing corresponds in reality. In my view this remark is of great importance in view of the fact that in the theory of relativity and certainly in quantum mechanics many such assumptions must be made.[43]

To recognize that scientific theories contain elements that are not essential and can be explained only historically, and that each scientific theory in the final analysis is only a possible interpretation of the meaning of natural phenomena, is not to dispute the fact that scientific theories obviously are and remain theories of what is real and can legitimately be said to be true to the degree that they do indeed make an essential contribution to the revealment of a certain aspect of natural things. In other words, the moment the truth question is raised on an ontological level, and the moment truth is defined so that the primary *locus* of truth is no longer the isolated statement but rather the entire set of intentional relations that constitute our scientific investigation, of which the formulation of scientific models is only an essential part, we realize that this set of intentional relations taken as a whole does indeed reveal the natural phenomena in the manner in which they manifest themselves to be. At that moment one realizes that scientific theories can be said to be true, not because they present us with a precise replica of the relevant natural phenomena, but because they make it possible for us to give a rational account of the natural entities and their states, relationships, and interactions and thus to reveal them in a manner which appears to be in harmony with the manner in which they manifest themselves to be independently of that theory in observation and experiment. Obviously, it is assumed here that the theory in question is a theory that is accepted by the entire scientific community and adequately verified by observation and experiment. I also grant that the borderline between theoretically postulated entities and entities to which something corresponds in the real world of nature is difficult to define and cannot be drawn once and for all. Finally, in my view there is also truth in the conviction of Churchland and others that it is not correct to take descriptive excellence at the observational level to be the only genuine measure of a theory's truth. Empirical adequacy or observational excellence is only one criterion among others of great importance, which together constitute the overall excellence of a theory. I also agree with Glymour that the explanatory power of a theory remains a very important criterion of the truth of scientific theories.[44]

It seems to me that the moment we rethink our conception of truth as correspondence totally independently of reflections on an *intellectus archetypus* and also independently of any conception of an absolute Knower, then one can say legitimately that scientific claims made on the basis of universally accepted scientific theories and adequately supported by the relevant scientific empirical evidence are indeed true. This statement means to express that anyone who is ever willing to conceive of the respective phenomena (things, states, events, occurrences, relationships, structures) from the viewpoint of the scientific conceptual framework that we for good reasons now employ, will find the state of affairs to be the way we claim it to be, regardless of how deeply theory-laden the empirical evidence involved may be. This statement obviously does not imply that our conception as now formulated will be eternally true; nor does it imply that our claims are true independent of the conceptual framework from which the claims are made. In the statement it is implied that human beings as finite knowers can never make claims that are totally independent of the conceptual framework which we must employ because of our intrinsic limitations. Should such limitations not exist we would be able to have absolute knowledge and thus would no longer need any science.

The moment one has made room for a more relaxed, strictly human interpretation of the correspondence theory of truth, leaving room for an established correspondence which nonetheless is not an absolute, definitive, comprehensive, and creative correspondence, there no longer seems to be a strong objection to calling scientific theories true to the degree that they form a constitutive element of our complete involvement with the relevant natural phenomena, an involvement which was oriented toward an effort to reveal these phenomena as they are from the perspective of our basic scientific thematization. When we actively engage in our scientific practice and when we thus no longer limit ourselves to *speaking or writing* about natural phenomena (which obviously is an essential part of "doing science"), when we do scientific work and engage in an intentional inter-action with the phenomena we have selected for investigation, then our entire engagement is true to the degree that it reveals the relevant phenomena in the way they manifest themselves to be independent of the particular claim we now make about them (even though not independent of all human involvement). If on this basis we now call adequately verified scientific statements true, I see no urgent reason why we cannot also call true the scientific theories which are a necessary condition for the establishment of these true statements.

III. CONCLUSION: WHY IS EMPIRICISM UNACCEPTABLE TO ME AS A PHILOSOPHICAL POSITION

In the preceding I have mentioned a number of reasons for my position that every form of empiricism is philosophically inadequate. The most important of these reasons is the denial of a transcendental dimension in philosophy. For me, as for most other philosophers, there is an essential difference between philosophy and the sciences. This difference precisely consists in the fact that philosophy, contrary to the sciences, contains a reflective and critical part that uses a kind of knowledge, transcendental knowledge, to examine the meaning and limits of all human knowledge, the empirically scientific knowledge included. Connected with this rejection of a transcendental dimension is the radical rejection of every form of metaphysics, even though in all transcendental kinds of philosophy the rejection of certain forms of metaphysics does not exclude other forms of metaphysics. Connected with the rejection of any transcendental dimension is also the denial of any *a priori*. I take it to have been the great merit of hermeneutic phenomenology to have eliminated unacceptable forms of metaphysics, unacceptable *a priori's*, and unacceptable assumptions and opinions or beliefs, without having to return to the philosophy of Kant and his conception of the transcendental, the *a priori*, and his idea of a transcendental logic. As I see it, a fundamental ontology is necessary to take the place of Kant's transcendental logic in any effort to determine both the meaning and the limits of all human knowledge.

Secondly, empiricists refuse to take into consideration the development of scientific theories and the application of the insights gained by these theories in the daily research activities when it comes to determining what science really is and what scientific statements precisely mean. For hermeneutic phenomenologists a careful examination of how the objects of a science become constituted in the relevant scientific thematization is precisely the decisive factor in the determination of the meaning and the truth of scientific theories and statements.

A third important reason, connected with the first, is empiricism's inherent positivism, insofar as it adopts the point of view that only those problems and questions are meaningful that can be answered or considered by one of the empirical sciences.

Fourthly, there is the rejection of the notion of truth and the skepticism that is its consequence. For me the notion of truth is essential both for our conception of philosophy and that of the sciences.

Fifthly, there is the notion that only sciences can guide human beings in all theoretical and practical problems.

Finally, because of its naive and dogmatic approach empiricism must leave a number of basic issues unexamined. In the preceding I mentioned in this connection the concept of experience and the concept of phenomenon, as obvious examples.

These basic sources of disagreement between a hermeneutic phenomenologist and an empiricist philosopher, do not at all entail that much of what van Fraassen has to say about the empirical sciences would not remain important for the phenomenologist. The disagreement has to do mainly with the kind of knowledge that van Fraassen uses to make the relevant claims.

Pennsylvania State University

NOTES

[1] Joseph J. Kockelmans, 'Stegmüller on the Relationship Between Theory and Experience', in *Philosophy of Science* **39** (1972), pp. 397–420; 'Reflections on Lakatos' Methodology of Scientific Research Programs', in *The Structure and Development of Science* [*Boston Studies*, vol. 59], Gerard Radnitzky and Gunnar Andersson (eds.) (D. Reidel, Dordrecht, 1979), pp. 187–203; *Heidegger and Science* (Center for Advanced Research in Phenomenology and University Press of America, Washington D.C., 1987), pp. 1–18, 117–89 (*passim*).

[2] *Phenomenology and Physical Science: An Introduction to the Philosophy of Physical Science* (Duquesne University Press, Pittsburgh, 1966); *Ideas for a Hermeneutic Phenomenology of the Natural Sciences* (Kluwer Academic Publishers, Dordrecht, 1993).

[3] Bas van Fraassen, *The Scientific Image* (Clarendon Press, Oxford, 1980); 'Empiricism in the Philosophy of Science', in *Images of Science*, Paul M. Churchland and Clifford Hooker (eds.) (The University of Chicago Press, Chicago, 1985), pp. 268–76; 'Against Transcendental Empiricism', in *The Question of Hermeneutics*, Timothy Stapleton (ed.) (Kluwer Academic Publishers, Dordrecht, 1994), pp. 309–35.

"Against Transcendental Empiricism" was provoked by some critical remarks in my 1986 Presidential Address to the American Philosophical Association, entitled: 'On the Problem of Truth in the Sciences'. This paper was later reprinted in *Ideas for a Hermeneutic Phenomenology of the Natural Sciences*, pp. 127–50. Van Fraassen writes in his response that "Against Transcendental Empiricism" has as companion "Against Naturalized Empiricism," which appeared in *On Quine*, P. Leonardi and M. Santambrogio (eds.) (Cambridge University Press, Cambridge, 1993).

[4] Bas van Fraassen, *The Scientific Image*, pp. 6–7.

[5] Bas van Fraassen, *The Scientific Image*, pp. 1–5, 11–9, 64–9, 201–3, and *passim*.

[6] *Ibid.*, pp. 39–40 and *passim*; Larry Laudan, 'A Confutation of Convergent Realism', in *Philosophy of Science* **48** (1981), pp. 19–49.

[7] Brian Ellis, 'What Science Aims to Do', in *Images of Science*, pp. 52–4; Gary Gutting, 'Scientific Realism versus Constructive Empiricism: A Dialogue', in *Images of Science*, pp. 118–31 (*passim*).

[8] Imre Lakatos, 'Falsification and the Methodology of Scientific Research Programmes', in *Criticism and the Growth of Knowledge*, Imre Lakatos and Alan Musgrave (eds.), pp. 95–138; Mary Hesse, *Revolutions and Reconstructions in the Philosophy of Science* (Indiana University Press, Bloomington, 1980), pp. 135–64; 173–86; 'Laws and Theories', in *Encyclopedia of Philosophy*, Volume IV, P. Edwards (ed.) (Macmillan, New York, 1967), pp. 404–10; Kurt Hübner, *Critique of Scientific Reason*, trans. Paul R. Dixon, Jr. and Hollis M. Dixon (University of Chicago Press, Chicago, 1983), Part 2.

[9] Bas van Fraassen, *The Scientific Image*, pp. 3–4.

To prevent misunderstanding here it is important to make a distinction between different types of scientific theories and to realize that the discussion is not about models or systematic theories, for instance, but merely about so-called causal theories, of which scientific realists hold that they are genuinely true. (Cf. Brian Ellis, 'What Science Aims to Do', pp. 55–6). One should note also that there are theories which seem to bear out the position of constructive empiricism (quantum mechanics, for example), whereas other theories *prima facie* seem to substantiate the position of scientific realism. If one makes a study of chemistry and biology it seems difficult not to become a realist. Giere writes that reading the history of molecular biology "it is difficult to find evidence of an overriding concern with saving the phenomena. The whole profession acted as if it were after the real molecular structure of real molecules." (Ron Giere, 'Constructive Realism', in *Images of Science*, p. 96).

[10] Bas C. van Fraassen, 'Empiricism in the Philosophy of Science', in *Images of Science*, pp. 268–76.

[11] Ron Giere, 'Constructive Realism', in *Images of Science*, pp. 75–82.

[12] *Ibid.*, p. 82.

[13] Bas van Fraassen, *The Scientific Image*, pp. 202–3.

[14] Bas van Fraassen acknowledges that he feels most akin to French and German writers on science in the half century before logical positivism and to the nominalists of the fourteenth century (*Images of Science*, p. 300).

[15] Ron Giere, 'Constructive Realism', p. 82; Clifford Hooker, 'Surface Dazzle, Ghostly Depth', in *Images of Science*, pp. 154ff.

[16] See note 3 above.

[17] 'Against Transcendental Empiricism', pp. 309–10. Reichenbach's book was published in London by Routledge and Kegan Paul.

[18] *Ibid.*, p. 318.

[19] *Ibid.*, p. 325.

[20] *Ibid.*, p. 328.

[21] *Ibid.*, p. 329.

[22] *Ibid.*, p. 330.

[23] *Ibid.*, p. 331.

[24] *Ibid.*, p. 331.

[25] *Ibid.*, p. 323–33.

[26] Joseph J. Kockelmans, 'Phenomenology and Psychology: Theoretical Problems in Phenomenological Psychology', in *Phenomenology and the Social Sciences*, Maurice Natanson (ed.) (Northwestern University Press, Evanston, 1973), pp. 225–80; 'Empirische, phänomenologische und hermeneutische Psychologie. Gedanken zu einer mehrdimensionalen Bestimmung des Problems der Psychologie', in *Versuche über Erfahrung*, A. Métraux (ed.) (Hans Huber, Bern, 1976), pp. 35–49; 'Some Reflections on Empirical Psychology: Toward an Interpretive Psychology', in *Reconsidering Psychology: Perspectives from Continental Philosophy*, James E. Faulconer and Richard N. Williams (eds.) (Duquesne University Press, Pittsburgh, 1990), pp. 75–91; 'Reflection on Social Theory', *Human Studies. A Journal for Philosophy and the Social Sciences* **1** (1978), pp. 1–15; 'Some Reflections on the Meaning and Function of Interpretive Sociology', *Tijdschrift voor Filosofie* **42** (1980), pp. 294–324; 'Hermeneutic Phenomenology and the Science of History', *Phänomenologische Forschungen* **2** (1976), pp. 130–79; 'Toward a Hermeneutic Theory of the History of the Natural Sciences', in *Ideas for a Hermeneutic Phenomenology of the Natural Sciences*, pp. 253–82. Cf. *Heidegger and Science*, pp. 190–248.

[27] G. W. F. Hegel, *Phenomenology of Spirit*, trans. A. V. Miller (Oxford University Press, Oxford, 1993). *Introduction*, pp. 46–57, particularly pp. 55ff.

[28] Hans-Georg Gadamer, *Truth and Method* (The Seabury Press, New York, 1975), pp. 235–341.

[29] 'On the Meaning of the Transcendental Dimension in Philosophy', in *Perspektiven transzendentaler Reflexion*, Gisela Müller and Thomas Seebohm (eds.) (Bouvier, Bonn, 1989), pp. 27–49.

[30] *Ibid.*, pp. 32–43.

[31] *Ideas for a Hermeneutic Phenomenology of the Natural Sciences*, pp. 60–126.

[32] 'Against Transcendental Empiricism', pp. 328–30.

[33] Ron Giere, 'Constructive Realism', p. 95; Richard N. Boyd, 'Lex Orandi est Lex Credendi', in *Images of Science*, p. 32.

[34] Ron Giere, 'Constructive Realism', p. 75.

[35] Imre Lakatos, *op. cit.*, p. 94.

[36] Cf. Martin Heidegger, *Being and Time*, trans. John Macquarrie and Edward Robinson (SCM Press, London, 1962), sect. 44.
[37] Nicholas Rescher, *The Coherence Theory of Truth* (Oxford University Press, Oxford, 1973).
[38] Cf. Brian Ellis, 'What Science Aims to Do', pp. 62–73. For what follows see Martin Heidegger, *On the Essence of Truth*, trans. John Sallis in Martin Heidegger, *Basic Writings*, David Farrell Krell (ed.) (Harper & Row, New York, 1977), pp. 117–41.
[39] Aristotle, *Metaphysics* **IX**, 10, 1051b6–9.
[40] These reflections on truth should obviously be developed in a more refined and systematic fashion and explained in greater detail. It is impossible to do this here because of lack of space. Yet what has been said may suffice at least to give an idea of the direction in which I would like to reflect on the problem of truth in the sciences.
[41] Kurt Hübner, *Critique of Scientific Reason*, pp. 42–7, 82–9, 184–203.
[42] Cf. Joseph J. Kockelmans, *The World in Science and Philosophy* (The Bruce Publishing Company, Milwaukee, 1969), pp. 123–33.
[43] Bas C. van Fraassen, 'Empiricism in the Philosophy of Science', p. 276.
[44] Paul M. Churchland, 'The Ontological Status of Observables', in *Images of Science*, p. 35; Clark Glymour, 'Explanation and Realism', in *Images of Science*, pp. 99–116; Clifford Hooker, 'Surface Dazzle, Ghostly Depth', in *Images of Science*, pp. 166–7.

WŁADYSŁAW KRAJEWSKI

MUST THE EXPLANANS BE TRUE?

As is known, the classical Hempelian scheme of scientific explanation requires the truth of the *explanans*. Some contemporary philosophers of science claim that this requirement should be abandoned. The aim of this paper is to discuss their arguments.

1. THE 'EXPLANATION' BY FALSE THEORIES

Historians of science often say that a theory, which is now held to be false, explained some phenomena. Larry Laudan (1984) adduces many examples of 'successful' theories which we hold to be not true and even not referential. Indeed, we say that Ptolemy's system 'explained' the observed motions of planets (their 'loops'), the caloric theory 'explained' many heat phenomena, the phlogiston theory 'explained', to some extent, combustion and other chemical reactions. These theories also predicted, to the same extent, future events. Otherwise, they could not be admitted or even discussed by the scientific community.

To use Wiliam Seager's expression, we may say that reasonings based on the above-mentioned theories had only "the form of an explanation" (Seager 1987b, p. 311). However, they were not genuine explanations. A false theory gives a spurious or illusionary explanation, an 'explanation' in quotation marks.

Consider the three above-mentioned examples. Only the Copernican heliocentric system gave a genuine explanation of the motion of planets, Lavoisier's oxygen theory – of combustion, molecular kinetic theory – of heat phenomena. In other words, only the scientific revolutions in astronomy, in chemistry and in heat physics, which led to the true (at least, approximately true) theories, made genuine explanations possible.

Hence, we distinguish genuine explanations from illusory 'explanations'. Should we make an analogous distinction with respect to prediction? Rather, no. When Ptolemy predicted motions of planets and his predictions came true, we cannot speak about illusory predictions. His predictions were genuine ones – of course, with a limited accuracy – whereas his 'explanations' were spurious. False theories have no explanatory force but they may have a predictive force. Many philosophers of science speak about so-called 'Kuhn losses' entailed by a scientific revolution. They claim that some, or even all, scientific revolutions lead not only to new successes but also to some explanatory losses.

A typical example is the transition from the Cartesian to the Newtonian theory of planetary motion. Descartes 'explained' many features of the Sun system by means of his vortex theory, not only the revolution of planets around

D. Ginev and R. S. Cohen (eds.), Issues and Images in the Philosophy of Science, 217–222

the sun but also their motion in the same (approximately) plane, in the same direction, the motion of moons in the same direction, etc. Newton explained much less: only the revolutions of planets and moons along the ellipses. Other above mentioned features of the Sun system were not explained by Newton. Only later various cosmogonic hypotheses tried to explain them.

Have we here to do with genuine losses? Certainly, no. Cartesian theory was false, hence its 'explanations' were spurious. The Newtonian theory is (approximately) true. Its explanation is genuine. Hence, the transition from Descartes to Newton entailed no losses of explanation but losses of illusions of explanation.

In general, 'Kuhn losses' are spurious. They are losses of illusions.

2. THE EXPLANATION BY NOT YET TESTED HYPOTHESES

Scientists often say that a hypothesis explains some events although they do not yet know whether it is true. Bas van Fraassen (1980) draws therefrom a conclusion that we don't assume the truth of the *explanans* when we speak about explanation.

Indeed, when we search for an explanation of an anomalous event E, we create a hypothesis H. When the conjunction of H and other premisses (known initial conditions, etc.) entails E, we usually say that H explains E. Does this mean, however, that we don't bother about the truth of H? Of course, no. We always test H and only in the case of positive results of the tests (confirmation of H) do we say that H indeed explains E. In the case of negative results of the tests (falsification of H) we abandon H and speak no more about its explanatory force.

We see that the ascription of explanatory force to H is conditional: when we say that H explains E we mean that H will explain E if H is true. This reservation is not explicitly expressed but it is always implicitly assumed. We may speak about 'tentative explanation' (cf. Seager 1987a, p. 194).

Hence, the Hempelian requirement of the truth of the *explanans* is not violated by the fact that we often speak about the explanatory force of a not yet tested hypothesis. Van Fraassen is wrong. However, his criticism was useful: we now realise that we often speak of the explanatory force conditionally, tentatively.

3. EXPLANATION BY PARTIALLY TRUE THEORIES

We now pass to a more difficult problem. An explanation of a phenomenon may be partial. Sometimes we say that a theory explains some features of an object, another theory – other features. We shall now consider an example adduced by van Fraassen as one more argument aiming to show that truth is not necessary for the explanation.

The corpuscular theory of light, created by Newton, explained reflection well and refraction, but hardly interference and diffraction. The undulatory theory of light, especially in the form created by Young and Fresnel, explained the latter processes well. However, it failed to explain the photo-electric effect. Only the 'dualistic' quantum theory, created by Planck and Einstein, explained all these phenomena, all the known features of light.

Both previous theories were one-sided, therefore each of them explained only some features of light. Were these theories true or false?

We shall not solve this problem if we admit the absolutistic conception of truth according to which a statement is either simply true or simply false. We must admit a more elastic or 'dialectical' conception which takes into account the concept of relative truth, and even in two senses. Firstly, the truth may be approximate, more or less accurate. Secondly, the truth may be partial, may grasp one side, one aspect of the given object. When a theory rightly grasps an existing aspect of the object, it is referential and partially true.

Both corpuscular and undulatory theories of light were relatively true in the sense of partial truth. Hence, they had a partial explanatory force. The dualistic quantum theory of light is wholly true and has a full explanatory force.

In general, a partially true theory has an explanatory force, but a partial »one.

4. EXPLANATION BY APPROXIMATELY TRUE THEORIES

We mentioned approximate truth. Probably, all quantitative scientific theories are only approximate. Hence, we should consider all of them to be false if we confined ourselves to the two-valued classical conception of truth. If we, however, admit the concept of relative truth as approximation we shall consider scientific theories to be relatively true, approximately true.

The explanatory force of approximately true theories is approximative. There are various degree of approximation, various degrees of relative truth. We may speak about the truth-content of an approximately true statement. This concept is close to Popper's concept of verisimilitude. As is well-known, Popper's definition of verisimilitude is not satisfactory. In general, it is not easy to find a good measure of the relative truth or verisimilitude. I tried to sketch a project of such a measure in some cases in my (1977). There are other proposals. Anyhow, we may hope that in the course of the growth of science a new theory has a greater truth-content and is closer to the absolute truth than the previous one.

Like relative truth (verisimilitude), explanation by its means has different degrees. The more true a theory is the better is the explanation by its means (cf. Tuomela 1985). E.g., Copernicus gave a rough approximation of the structure of the Sun system, hence his explanation of planetary motion was rough, too. Kepler gave a better approximation, hence his explanation was better, too. Newton gave a further improvement of the approximation (taking perturbations into account) and explanation of planetary motion. Einstein in his General Relativity gave the best known approximation and the deepest explanation of this motion. We have here a sequence of better and better explanations.

5. EXPLANATION BY IDEALIZATIONAL LAWS

We pass to another controversial problem, the problem of idealization. Fundamental laws of physics usually have an idealization nature. They are exactly fulfilled only in ideal models. We create the models assuming that some factors (lateral factors) do not exist or, in quantitative language, relevant parameters take zero values. Idealizational laws are approximately fulfilled in some real systems, namely, in those in which lateral factors play no great role, i.e. the values of relevant parameters are close to zero. In order to apply these laws to other real systems we must gradually 'factualise' them, obtaining factual laws which have a large scope of application.

Idealizational laws are false in the classical correspondence sense. They are approximately true within a narrow scope. We may, however say that they are true (absolutely true) in a 'model sense' since they are fulfilled in ideal models. Hence, we must introduce, besides classical (absolute and relative) correspondence truth, a concept of model truth, correspondence to an ideal model (cf. Krajewski 1977).

The role of idealization was, for a long time, ignored in the philosophical literature. Nowadays, some philosophers of science notice it. One of them is Nancy Cartwright. She writes (1983) that fundamental laws of physics are false; they are fulfilled only in simplified models. Nevertheless, they have a strong explanatory force. Phenomenological laws are true but their explanatory force is poor (the truth "does not explain much"). Cartwright comes to a conclusion that we should abandon the requirement of the truth of the *explanans*.

I agree that idealizational laws are more fundamental and have stronger explanatory force than factual laws. However, this does not mean that we may resign from the truth of the *explanans*. We should not abandon but amend the Hempelian conception of explanation. In order to explain a complicated phenomenon, we must create its simplified, ideal model, discover its laws (idealizational laws), then gradually take into account lateral factors and obtain factual laws, then apply these laws according to the Hempelian scheme.

We cannot start with the complicated factual laws which are true in the classical sense. Only theories which contain idealizational laws give a deep explanation. They explain factual laws which in turn explain phenomena. Notice that we may formally reduce idealizational laws to their factualizations but, in this case, reduction is not identical with explanation. On the contrary, they have opposite directions (see Krajewski 1984).

Why do idealizational laws give a deep explanation? It seems the reason is as follows: An ideal model grasps an essence of the given phenomenon. Rather *an* essence than *the* essence because there are various layers and sides to the essence of each phenomenon. They may be grasped by different idealizational laws. In the course of the growth of science, we penetrate through successive layers, we "probe deeper and deeper into the structure of our world" (Popper 1973, p. 196). If we admit this form of essentialism, we shall be able to say that model truth is a sophisticated kind of correspondence truth, which deals not directly with phenomena but with their essence. Hence, idealization brings us in fact closer to the deep truth.

CONCLUSIONS

We discussed various arguments against the requirement of the truth of the explanans. We come to the conclusion that this requirement withstands all objections. Nevertheless, the criticism was useful. It has turned our attention to various circumstances which show that the situation is not so simple as Hempel and other proponents of the classical model of explanation supposed.

- We often speak about the 'explanation' of some phenomena by false theories. We must distinguish their spurious 'explanation' from genuine explanation by means of true theories.
- We often say that a not yet tested hypothesis explains a phenomenon. This claim should be understood conditionally: it will explain if it is true. Its explanatory force is tentative.
- Theories used for explanation need not be wholly true. They may be partially true. Their explanatory force is then partial.
- Approximately true theories have an approximate explanatory force. The closer they are to the absolute truth the better they explain.
- Theories which contain idealizational laws give a deep explanation. These laws are false in the simple classical sense but true in a model sense. They grasp an essence of phenomena. However, they need a factualization. Hence, the explanation by idealizational laws is indirect.

The general conclusion is as follows. The *explanans* must be always true but the concept of truth is not as simple as it is usually supposed. The Hempelian conception of explanation should be not abandoned but amended.

University of Warsaw, Poland

REFERENCES

Cartwright, N. (1983), *How the Laws of Physics Lie* (Clarendon Press, Oxford).

van Fraassen, B. (1980), *The Scientific Image* (Clarendon Press, Oxford).

Krajewski, W. (1977), *Correspondence Principle and the Growth of Science, Episteme 4* (D. Reidel, Dordrecht).

———. (1984), 'May We Identify Reduction and Explanation of Theories?', in *Reduction in Science*, W. Balzer *et al.* (eds.) (D. Reidel, Dordrecht).

Laudan, L. (1984), *Science and Values* (University of California Press).
Popper, K. R. (1973), *Objective Knowledge* (Oxford University Press).
Seager, W. E. (1987a), 'Scientific Explanation and the Trial of Galileo', *International Studies in the Philosophy of Science, The Dubrovnik Papers* **1**(2).
———. (1987b), 'Credibility, Confirmation and Explanation', *British J Phil. Sci.* **38**.
Tuomela, R. (1985), 'Truth and the Best Explanation', *Erkenntnis* **22**.

DAVID LAMB

COMMUNICATION WITH EXTRATERRESTRIAL INTELLIGENCE: SETI AND SCIENTIFIC METHODOLOGY

1. THE SEARCH FOR AN EXTRATERRESTRIAL MESSAGE

Whilst beliefs in extraterrestrial intelligence have been expressed for centuries it is only recently that the technological means of conducting a search have been available. The discovery of electromagnetic radiation in the nineteenth century and the invention of the radio telescope in this century appears to have moved SETI from the realm of speculation to scientific enquiry. The early modern pioneers of SETI in the late 1950's were Guiseppi Cocconi, Philip Morrison, Frank Drake and Carl Sagan, who marshalled arguments based on what we know about terrestrial life and its origins, and applied them to astronomical data on conditions elsewhere in the universe. In 1959 SETI achieved scientific respectability when the scientific journal, *Nature*, published a paper by Cocconi, a cosmic ray specialist, and Morrison, a physicist. This paper addressed the question of radio communication from ET sources. The problem of communicating with ET's is that first there has to be an agreement on a medium, which in the 1950's was radio, and then agreement on a frequency. If it is assumed that ETs want to communicate by radio, then the problem of which noise-free frequency out of millions has to be considered. Cocconi and Morrison proposed a frequency which was based on the rate at which hydrogen atoms emit radiation when the spin axis of the electron orbiting the nucleus flips over from being parallel to the nucleus's spin to being opposite to it. They argued that the hydrogen frequency 'has a universal uniqueness, not set by anthropocentric considerations, that fits it as the outstanding choice for potential communicators who have not had the opportunity to agree on a frequency' (Cocconi and Morrison 1959, p. 844). Hydrogen is one of the most abundant elements in the Universe; about 999 out of every 1000 atoms that exist are of this substance. The characteristic frequency for hydrogen is 1,420, 405,752 times per second, which can be expressed as 1,420 megahertz – a band in the noise-free area. Another band in the noise-free area represents the hydroxyl radial (OH) and is known as the 'Waterhole'. The hydroxyl radial emits radiation when it changes its rate of spin. Because hydrogen and hydroxyl combine to make water the frequency corresponding to its radiation emission is known as the 'Waterhole'. The Waterhole has several lines near 1.65 GHz. Because these frequencies have associations with life it is assumed that intelligent ETs would choose to communicate through them.

Throughout the 1960's Cocconi and Morrison, at Cornell University, employed radio telescopes tuned to the hydrogen frequency to search for an

D. Ginev and R. S. Cohen (eds.), Issues and Images in the Philosophy of Science, 223–251.

intelligent signal. A similar search was conducted by Frank Drake, head of Project Ozma, which began in 1960, and used a 26 meter Tatel radio telescope in the National Radio Observatory (NRAO) in West Virginia. Project Ozma searched for radio signals near the hydrogen frequency from two nearby Sun-like stars, Epsilon Eridani (10.2 light years away) and Tau Ceti (11.9 light years away). There were no results after 400 hours listening. Since then over 60 searches, accumulating over 200.000 search hours, in the USA, the former Soviet Union, Australia, Canada, France, Holland, England, Japan and Argentina, have been conducted at various 'magic' frequencies. Of these about 90% have been conducted on the hydrogen line. One negative search during the 1970's of ten stars on the narrow band 21-cm wavelength was reported by Gerrit L. Verschuur (1980) of the NRAO, who suggested that advanced extraterrestrials might have instituted a series of protected bands for radio astronomy such as the 21-cm wavelength. Whilst several searches have been funded by governments others have been financed by private means. Some, the 'dedicated' searchers, have diverted radio telescopes to pure SETI work, whilst other 'parasitic' searches have examined data supplied by conventional astronomic research; some have hooked up to disused early warning systems for detecting Soviet missile attacks, whilst others have relied upon portable receivers which they connect to radio telescopes whenever they can beg a few hours. None have reported contact.

On Columbus Day in 1992 NASA scientists began a search for ET radio signals. The search was originally scheduled to run for ten years with an upper limit budget of $100 million dollars. It was described as part of the USA's commemoration of the discovery of America, looking towards its future as well as its past. The search was to be conducted primarily by means of radio telescopes and involved two complementary teams. The first team, based at the Ames Research Center, were to employ a 'Targeted Search' to examine at high sensitivity about 800 stars like our Sun which are located within 80 light years from Earth. It began with a concentrated search targeted on a star in the Constellation of Opiuchus. Using the largest antennas available, including the 305 metre dish at Arecibo Observatory in Puerto Rico, each star was to be carefully scrutinised on frequencies between 1 GHZ to 3 GHz in the microwave band, which would include the Waterhole. Radio telescopes at various sites throughout the world were to be programmed to follow each designated star for several minutes at each frequency whilst the 'targeted search analyzer' was to search for complex signal patterns. The assumption underpinning the 'Targeted Search' was that intelligent civilizations are so common throughout the Universe that our nearest neighbours may be within 80 light years of the Sun.

The second search was based at the Jet Propulsion Laboratories in Pasadena, California, employing a 34 metre radio telescope at the Goldstone complex of telescopes in the Mojave Desert, California. This was known as the 'All Sky Survey'. It was to be less sensitive than the 'Targeted Search' but the sacrifice of sensitivity allowed greater sky coverage. It assumed a smaller number of communicating intelligences, although Frank Drake saw this as the most likely to succeed. Whilst sacrificing sensitivity, the All Sky Survey was intended to employ a search covering 99% of the heavens ignored by the Targeted Search. It was intended to tune into frequencies that cover almost the entire ground-based (1–10 GHz) microwave window. Unfortunately the project suffered a serious set-back in the 1993 US Budget cuts which killed off the All Sky Survey. However, the Targeted Search managed to escape the cuts as the SETI Institute, which is responsible for it, is independent of the government and has been able to solicit outside funds. Its current name is Project Phoenix.

Two assumptions can be discerned in these radio searches. First, the sending civilization will be more advanced than us; if they were not then they would not have the appropriate technology. And if they have the technology they are very likely to be more advanced as our communicative history is so short. Radio communication on Earth is less than 100 years old and has almost reached technical perfection – a state in which further technical refinements will not improve results. If radio communication can be as good as it can get within such a short time it is likely that other civilizations have either got it perfect or not at all. Second, the sending civilization, being more advanced than ours, would do their best to make it easy for us to listen. This is why they are likely to choose one of the various 'magic' frequencies in the noise free area. This assumption provides a powerful heuristic limitation on the scope of the search.

Current proposals for radio searches are committed to receiving information rather than broadcasting messages. This is bound up with the belief that the sending civilization will be technologically superior. They will discover us, just as Columbus discovered America; the native Americans did not discover Europe. So instead of sending them signals we should await theirs. It is also more economical with resources to receive signals rather than broadcast. But if they are so much more advanced what is their point in communicating? And could they communicate even if they are intelligent? They may be like the ancient Greeks; intelligent but lacking the technological means to communicate over long distances. Extraterrestrials outside of our technology threshold may be intelligent but we can have no way of confirming it if they have no access to a radio transmitter.

2. SETI AND PSEUDO-SCIENCE

From the start SETI scientists have sought to distance themselves from Ufology and other branches of inquiry deemed to be pseudo-scientific. There are clear methodological rules regarding the recognition of ETI: for example, the rule that every object must be assumed to be of natural origin unless it is decidedly proven to be unnatural places the burden of proof on SETI researchers to provide adequate evidence for their theories. This is in sharp contrast to many exponents of 'pseudo-science' where the onus of proof is directed to the skeptic. SETI scientists work within the limits of existing theory and levels of technology and eschew appeals to as yet unheard of theories when offered as solutions to today's insoluble problems. The case for SETI, it must be stressed, rests on its appeal to plausibility which is based on knowledge derived from state of the art technology. Recent confidence in SETI's chance of success is predicated on developments in astronomy, space-exploration, information technology, and the life-sciences which offer a sound empirical foundation for SETI research.

There are, however, several logical pitfalls in the arguments which frequently appear in SETI literature and philosophers of science have drawn attention to them. Ernan McMullan (1971) suggests that two types of probability may be sometimes confused in arguments concerning the likelihood of extraterrestrial life. The first type is *inductive probability* which is based upon a frequency count of outcomes: this might involve a count of heads or tails in a coin tossing experiment or predictions based on a correlation between traffic accidents and teenage motorists. Conclusions drawn from inductive probability need not require either an understanding or an explanation of the processes involved. The second type of probability, *theoretical probability*, does require at least a partial understanding of the causes. Thus if one has a reasonable theoretical explanation of the psychological factors affecting teenage motorists one could attempt a calculation of the percentage of likely accidents without observing numerous instances. Now appeals to inductive probability are of little value to SETI as we only know of one planet where intelligent life has evolved. But appeals to theoretical probability are not without methodological problems: although theoretical progress is under way, we are not yet in a position to advance a definitive theory as to how planets, life and then intelligence, came about. This lack of a firm theoretical base, argues McMullan, is fundamentally damaging to the SETI enterprise and partly accounts for four fallacies which he detects in SETI literature.

The first fallacy, according to McMullan, is the belief that given enough time "the probability of any natural outcome dependent on universal natural processes increases to virtual certainty" (*ibid*: 292). For example, given enough time any environment containing the constituents associated with life will inevitably produce life. The problem with this argument is that, lacking a theory of how life came about, it appeals to nothing more than a random juxtaposition of elements. But this is unlikely, as what we do know about life is that it emerges out of a gradual biochemical process involving, as McMullan says, "a host of unknown interdependent environmental factors" (*ibid*: 292). The problem is not resolved by arbitrarily lengthening the timescale, but rather by knowing which of the essential factors will be present and what forces will be operating upon them.

The second fallacy McMullan identifies is the 'uniformitarian fallacy', according to which it is believed that if life can develop in one place it must develop in another place with similar conditions. This argument also appeals to general scientific beliefs such as the belief that science does not admit uniqueness. Again, McMullan identifies the fallacy with reference to our lack of a theory as to how life came about: for unless we have this theory for one environment we can have no idea how life will emerge elsewhere. The same appeal to the lack of a theoretical account of the emergence of life underpins McMullan's depiction of the third SETI fallacy, the appeal to large numbers. For unless we have a theory of some sort we cannot attach any theoretical probability to any predicted outcome. It might be said that it is logically possible that life can emerge elsewhere, but expressions like 'logically possible' cannot be assigned a numerical rating unless there is a theory to support it.

It might be countered that evolutionary theory provides a theoretical grounding for speculations about extraterrestrial life. But McMullan cites SETI's appeal to evolutionary theory as an example of the fourth fallacy; namely that it is fallacious to use the theory of evolution as a predictive theory. SETI scientists often argue that, once life originates, the operation of natural selection will inevitably lead to more complex life forms and ultimately to consciousness and intelligence. Yet biologists constantly point out that no predictions can be made on the basis of evolutionary theory; intelligence may bestow selective advantage, but its emergence cannot be guaranteed as other features may turn out to be more advantageous.

McMullan's criticism suggests that SETI's methodological status is not as sure-footed as it sometimes sounds. The appeal to an almost infinite length of time and large numbers, without some theoretical direction, is certainly problematic. But these appeals are not in conflict with well-established scientific theories and SETI researchers attempt to work within such a framework, acknowledging that more data and more theorizing will enrich the program.

SETI is still a young science and therefore, it is argued, should be given enough breathing space to grow. But how does it compare with other young sciences? It is approximately the same age as molecular biology and older than chaos theory. It differs fundamentally from these two because its subject-matter might not even exist. On the other hand, just one authentic detection would yield massive rewards and a place in posterity alongside Columbus, Galileo and Darwin. SETI has also focused attention on what we mean by 'intelligence' in a way that may turn out to be more philosophically rewarding than inquiries into artificial intelligence systems. It offers a cosmic standpoint for the understanding of intelligence and forces a critical assessment of hidden assumptions concerning its origin and function.

SETI's richness renders it non-falsifiable. If not on the Moon, then extraterrestrial intelligence is to be found on Mars. If not in the Solar System, then on other stellar systems, somewhere in the Galaxy or other galaxies. This is not to say that SETI scientists tenaciously cling to refuted theories, but simply that even when refuted, hypotheses concerning the possibility of life in certain sites, can be re-applied to other sites. The avoidance of falsification is not a ruse to protect a research programme, but simply reflects the fact that only a small fraction of space has been searched. Although, strictly speaking the hypothesis that there is extraterrestrial life cannot be falsified, SETI scientists base their research on empirical statements and observations which can be and are falsified. Moreover, it is unscientific to conclude that an ETI hypothesis has been tested and failed unless a really extensive search has been carried out. There are also strict rules governing confirmation, which in SETI research is not vacuous as in spiritualism where almost any experience may be deemed confirmatory. SETI research has strict regulations over confirmability, such that no purported signal has yet had an acknowledged confirmation. Repeated failure, however, might not be seen as decisive falsification, but it would certainly contribute to a loss of interest and a reduction in resources.

Is SETI a paradigm in the Kuhnian sense? Apart from the optimism of its exponents it is not committed to any revolutionary epistemological rupture with the past nor does it reflect a crisis in any contemporary branch of science. Yet it does exhibit a set of ground rules and doctrines characteristic of a Kuhnian paradigm. Its scientific credentials are underpinned by three fundamental features of the Universe. First, the Universe is believed to be uniform with the same essential building materials, the same chemical elements, functioning according to universal natural laws. The basic chemicals of life are widely distributed throughout the Universe. This is expressed as the cosmological principle which states that the Universe is the same everywhere, except for local irregularities. This assumption that the Universe is homogenous and isotropic simplifies models used to describe it. Second, the two major revolutions affecting modern science lend support to SETI: the Copernican revolution revealed that we are not privileged in our location and the Darwinian revolution showed that we are not a privileged species. There is no unique spot, or centre, in the Universe; there is nothing unique about our Sun, and no reason to assume that events on our planet could not happen elsewhere. Thus modern cosmology supports SETI's plausibility; the Universe has no centre, no outer limit or surface, no privileged place, but continuous expansion. Third, the Universe is massive: within telescopic range are over a hundred billion galaxies, each containing hundreds of billions of stars.

In addition SETI researchers are committed to two sets of beliefs. The first is that intelligent alien life forms will have similar thought patterns to those displayed by intelligent life on Earth. This does not have to commit SETI scientists to a search for other human beings. Carl Sagan has argued that, given different evolutionary programs and different environments, it would be astonishing if ET life resembled ours. But whilst biological diversity is expected there is a belief among SETI scientists in a similar technology. This is because of the universality of physical laws. A NASA sponsored design study for SETI, Project Cyclops, reported in 1971 that:

> Regardless of the morphology of other intelligent beings, their microscopes, telescopes, communication systems, and power plants must have been at some time in their history, almost indistinguishable in working principles from ours. To be sure there will be differences in the order of invention and application of techniques and machines, but technological systems are

shaped more by the physical laws of optics, thermodynamics, electromagnetics, or atomic reactions on which they are based, than by the nature of beings that design them (Cyclops 1971, p. 4).

With so much in common, it is presumed that there would be few problems in constructing a dialogue based upon technology, science and mathematics.

The second set of beliefs are found in SETI's: commitment to the Drake Equation.

The Drake Equation

The Drake Equation was first presented by Frank Drake at a meeting of scientists gathered to discuss 'Intelligent Extraterrestrial Life' at the Green Bank Observatory under the auspices of the Space Science Board of the US National Academy of Science in November 1961. The equation has consequently become a SETI creed, its logical justification. Reflecting on the equation some thirty years later Drake said that its basic premise 'is that what happens here will happen with a large fraction of the stars as they are created, one after another, in the Milky Way Galaxy and other galaxies' (Drake 1990, p. 151). This is the equation:

$$N = R^* \times f_p \times n_e \times f_l \times f_i \times f_c \times L$$

It is not a definitive equation, nor is it inviolable like $E = MC^2$. Its components can be broken down as follows:

N = the number of technically advanced civilizations in the Galaxy that are currently capable of communicating with other solar systems.
R^* = the number of new stars formed in the Galaxy each year
f_p = the fraction of those stars that have planetary systems
n_e = the average number of planets in each such system that can support life
f_l = the fraction of such planets on which life actually exists
f_i = the fraction of life-sustaining planets on which intelligent life evolves
f_c = the fraction of intelligent life-bearing planets on which beings develop the means and the will to communicate over interstellar distances
L = the average life-time of such a technological civilization.

R The Rate of Star Formation*

There are held to be over 40 billion stars in the Milky Way Galaxy with enough loose gas and dust for millions more. Estimates put the rate of star formation at 10 each year. This, however, is a theory-laden guess in that any attempt to assess the rate of star formation must be derived from a theory as to how stars are formed. The dominant theory is the Big Bang theory, the moment when the Universe began. According to this theory the first stage was when the elements flew off in rotating clouds. Then they cooled and formed galaxies. The early stars were massive combinations of hydrogen and helium, and when they exploded heavy elements were created and flung about the Universe. Eventually other generations of stars condensed from this matter, leaving their rocky substances that formed planets in their orbiting paths. There are countless of billions of stars in over a billion galaxies. A star's colour is a clue to its surface temperature and the scale for classifying them is indicated by the letters O B A F G K M, running from the hot blue stars to the cool red ones. A star's life span is measured in millions of years. Our Sun, for example, is intermediate and is classified as a G type. It has existed for about 4 billion years and supports life on Earth. It is likely to survive for another 8 billion. Similar stars to our Sun can be found in the F to K groups. Whether they can support solar systems similar to our own is a matter for speculation. Application of the Goldilocks criterion eliminates the remaining groups as either too hot or too cold. It is estimated that about 25% of the total number of stars are found in the intermediate F-K groups. From this it is concluded that a quarter of all the stars in the heavens are capable of supporting planetary systems with life.

f_p *The Fraction of Stars with Planets*

So far no planets have been decisively observed outside of the Solar System. Nevertheless, the belief that planetary systems are the rule rather than the exception is not in conflict with current astronomical theory. Estimates concerning the number of stars with planets will vary according to which theory of planetary formation is adopted. The Catastrophe theory of the nineteenth and early twentieth century held that planets were the product of the debris from either collisions or the explosions of stars. Given the vast distances between stars and their durability this would suggest that planetary formation was infrequent. Catastrophe theory thus dramatically reduces the theoretical scope of f_p. However, during the past thirty to forty years the nebular theory of Descartes, Kant and Laplace has re-emerged. This theory maintains that planets are the offshoots of the large rotating clouds out of which the stars were formed. Accordingly, planets are very likely companions of stars as the theory maintains that they are products of the same creative forces. If stars and planets are thus formed together, at the rate of 10 planets for each star, there would be a minimum of 400 billion planets in the Galaxy. Nevertheless, f_p is still very much in the realm of theory; there is no compelling proof that the rotating clouds have evolved into planets, although numerous projects for the detection of planets are well established (see Cambell 1990; Scargle 1988; Vidal-Madjar 1994).

n *The Number of Planets That Could Support Life*

This, again, is an appeal to theoretical possibility. Chemical surveys suggest that there is nothing unique about Earth's material and that all the matter in the Solar System and beyond has a common origin. By extension of the theory the same building blocks which are necessary for life will be found throughout the Universe. This, however, only specifies the necessary, not the sufficient, conditions for life as other intervening factors may prevent the actual emergence of life. The essential chemicals for living structures may be in abundance but the environment might not be favourable. Extreme temperatures would inhibit life. A planet too far away from the Sun, like Mars, will freeze in a stillborn death. If it is too close, like Venus, it will suffer from a runaway greenhouse effect. Strictly speaking n_e is limited to the number of planets which are found in a habitable zone. Even on the most optimistic estimates the number assigned to n_e would be restricted to planets with an orbit similar to that of the Earth around the Sun, with a similar eco-sphere capable of sustaining life. Nevertheless, according to the authors of Project Cyclops (1971) there are probably on the order of 10^{10} potentially habitable planets in the Galaxy.

f_l *The Number of Planets That Have Actually Developed Life*

So far the factors in the Drake equation call for guesses, but appeals to fairly well-established theories provides them with a degree of plausibility. More work may be undertaken to strengthen the theory, more reliable data may be forthcoming. But from this point onwards the equation moves into the realm of conjecture. Estimates regarding the size of f_l have no empirical support and rest on a widely disputed theory of how life began. At present biologists have not reached a consensus regarding explanations as to how terrestrial life originated, which renders Earth as an unsatisfactory model for the rest of the Universe. The dominant theories refer to the evolution of organic compounds out of the oceans, when the Earth's atmosphere consisted primarily of hydrogen, nitrogen, carbon dioxide, methane and water vapour. It is argued that the action of the Sun, together with volcanic activity on Earth, was primarily causative of the emergence of organic compounds (such as the amino acids) out of inorganic molecules. If correct, this explanation provides the necessary, but not the sufficient, conditions for life. All we have is a model of the possible emergence of life; there is no guarantee that life must emerge. The theory does not show how matter came to life; this problem has yet to be resolved. As yet life has not been produced in a laboratory. Until we know much more about the origins of terrestrial life the Earth cannot be regarded as a reliable model for the emergence of life on other planets.

f_i *The Number of Planets with Intelligent Life*

Conjectures regarding the number of life-bearing planets with intelligent life are predicated on widely disputed concepts of intelligence. Philosophers, psychologists, anthropologists and sociologists have failed to produce an authoritative definition of intelligence which can be applied to life forms on Earth. There is no satisfactory definition of intelligence for either humans or other species. There are disputes whether intelligence can be attributed to non-humans, such as chimpanzees and dolphins, and whether machines can exhibit forms of intelligence behaviour. In what might be described as a 'leap of faith' SETI exobiologists maintain that life, as it evolves, is likely to evolve greater forms of intelligence; that intelligence must necessarily follow life. A study of terrestrial natural history is unlikely to support this prediction, and even if an evolution towards greater intelligence does occur, we are still left with the problem of determining the nature of intelligence. For many SETI scientists, the criterion for intelligence is bound up with the emergence of a technological civilization capable of radio-communication.

f_c *The Number of Planets Where Intelligent Life Has the Capacity to Communicate With Other Planets*

Conjectures regarding the fraction f_c rest entirely on the belief that a scientific civilization like ours is an inevitable consequence of intelligent life. This presupposes that other evolving species are likely to pass through similar stages from the Stone Age to the Nuclear Age. But if we consider the contingencies in the development of human natural and social history such a repetition appears very unlikely. Mammals might not have survived long enough to develop a human species. Natural disasters of the kind that wiped out the dinosaurs could have arrested human development allowing some entirely different dominant species to evolve. Humans might not have developed sophisticated coordination between hands and eyes, which is necessary for tool-making and hunting and eventual control over the environment; the minimum conditions for a scientific culture. Intelligent human civilizations, such as those in Ancient Greece and Ancient China, flourished and died without any likelihood of discovering Maxwell's laws of electro-magnetism; the minimum condition for radio-communication. A scientific civilization also requires a few lucky breaks from the physical environment. Despite their acknowledged intelligence and communicative skills Dolphins will never invent printing or acquire an adequate knowledge of electro-magnetism as a precursor to radio contact. But even if an ET civilization did develop an understanding of the laws of physics that would still leave open the intriguing philosophical question of whether they mean the same to them as they do for us (see Munevar 1989).

Generous estimates of the number of planets with intelligent communicative life suffered a serious set-back in 1992 following the completion of a radio search conducted by D. G. Blair at the University of Western Australia. Blair led a team of astronomers, using a Parkes radio telescope, at the 'magic' frequency of 4.462336275 gigahertz during two observations in 1991 and 1992. This frequency was selected by multiplying the frequency of hydrogen by pi, the first fundamental constant considered likely to be discovered by any intelligent civilization. The search covered the neighbourhoods of 176 stars in the F, G, and K range, within forty light years of the Earth. No signal was detected. These negative results weaken Drake's assumption that technological intelligence will inevitable emerge given enough time on an Earth-size planet near a Sun-like star (Blair 1992).

L The Length of an Advanced Technological Civilization Capable of Interstellar Communication

Estimates for this factor have neither theoretical nor empirical support. The only advanced technological society we have knowledge of is our own and it will only be after its extinction that an accurate figure for L can be supplied. But to appreciate how consideration of L might reduce the size of N (the number of civilizations currently capable of communicating) it is worth reflecting that it took about 300 million years of the cool planet period for life to evolve and a further 3.500 million years to invent the radio. And within half a century after that, intelligent life on Earth had produced nuclear weapons and began to pose an ecological threat to the planet, both of which are capable of terminating life within a very short period. Reflections on the quantification of L are bound up with a related terrestrial question: can we survive advanced technology? It may be that L is little over a hundred years. It would be a matter of cosmic irony if civilizations turn out to be capable of destroying themselves at the very moment they achieve the ability to communicate with each other. Apart from war and ecological disaster L can be further reduced by collisions with asteroids, exhaustion of natural resources, over-population, genetic degeneration and loss of interest in science, where people turn to religions like Zen Buddhism, new age cults or lapse into hedonism.

A pessimistic calculation for L was made by J. Richard Gott III (1993) who estimated an 0.2 million to 8 million years limit to longevity of our species at 95% confidence level. According to Gott the average longevity for most species is between 1 million and 11 million years, and for mammals it is about 2 million years. Gott insists that intelligence – defined in terms of the possession of self-consciousness and cognitive skills which enable abstract thought, creativity, and an ability to think about the future – confers no survival value. He points out that our early ancestor, *homo erectus*, lasted 1.4 million years and the Neanderthals lasted 200,000 years. Taking the Copernican principle – there is nothing unique or privileged about the human race – at face value, Gott argues that there is unlikely to be sufficient time for the human race to either colonize space or establish widespread contact through radio communication. Space travel, he suggests, somewhat pessimistically, has run its course. The cold war, which supported the space race, is over. Moon travel lasted a mere four years. Human activity in space is confined largely to a close to Earth orbit. With increasing pollution of radio space and an energy crisis our capacity for radio communication and space travel will soon expire. The space window may soon be closed, and will only remain open for a limited period before energy supplies fail, space debris (see Simpson 1994), or light and electromagnetic pollution (McNally 1994) close it.

With luck, perhaps, humankind will survive and remain capable of sending messages which can be picked up if there is anyone out there to listen. The only uncontestable fact in this mass of speculation is that so far the human race *has* survived.

The Drake Equation is an edifice constructed from degrees of guesswork based on a range of theories which vary in evidential support. The equation makes no claims on behalf of the immunity of any of its component theories and inferences drawn from them carry only a theoretical possibility. An unkind critic might say that the Drake Equation is nothing more than a piece of inspired guesswork; comparable to guessing the number of stray dogs in Manchester by multiplying estimates of the canine birthrate by the number of food-containing waste-bins by a guess at the average lifespan of a dog. In a more favourable light the equation might be said to provide a heuristic guide whereby new theories can be appended and various blanks filled in. Consequently the numbers which make up N will vary according to the theories currently used to underpin any of the stages in the equation and, of course, the personality of the scientist. Drake (1990), optimistically, argues that N = tens of thousands of communicating intelligences across the Galaxy.

What is the methodological status of the Drake Equation? It does not provide proof, but it gives a structure to hypothesis generation which is based on falsifiable empirical developments in major branches of science including astrophysics and biology. Drake's structure conveys plausibility, which is not destroyed by the failure of any particular search; yet the component theories and empirical statements in Drake's structure are subject to either corroboration or falsification as knowledge develops. As long as this structure remains intact it is possible to employ it when hypothesising the amount of intelligent civilizations. Yet it does not tell us where to look, and no extraterrestrial civilization can be deduced from it.

Nevertheless, as a call for exploration the Drake Equation appeals to many branches of science. R^* is the province of geophysics and astrophysics; f_p involves geophysics and atmospheric physics and n_e is to be examined in the boundary between astronomy and biology; f_l involves organic chemistry and biochemistry; f_i involves neurophysiology; f_c is bound up with anthropology, archaeology and history, whilst L is inescapably linked to politics, sociology and psychology.

3. WHERE ARE THEY? THE FERMI PARADOX

Enrico Fermi, the Italian physicist, responded to the arguments in favour of ETI with his now famous retort – where are they? This has become known as the Fermi paradox, whereby all the arguments indicate the existence of ETI but there is no evidence of their existence. Where are they, if they have had millions of years to settle here? The Earth is about 4600 million years old, with human life emerging 2 million years ago, electricity and radio less than 200 years old, and space flight less than 40 years old. But the Universe was ten billion years old when the Earth was formed, which allows plenty of opportunity for many other civilizations to have emerged. Some of them could have been in existence for millions of years before the origins of human life. This is long enough for them to have explored the entire Galaxy. They ought to have been here by now. So where are they?, asked Fermi.

The US military has deployed a worldwide space surveillance system consisting of telescopes and sophisticated radar equipment which can track spacecraft. Every object over 10 centimetres in diameter has been catalogued, of which there are over 7,000. In addition there are over 10 million fragments of debris. None of which can be attributed to an extraterrestrial source. So far no extraterrestrial artifacts have been discovered and no information regarding the communicative intentions of extraterrestrials has stood up to scrutiny. However, it might be argued that a) we would not be capable of recognising their products, and b) we might not have found them. It only takes a few generations for a whole civilization to disappear as we know of the vanished civilizations of Asia and Africa, and cities completely obliterated by the elements within a millennium or two.

The Case Against ETI; Tipler's Self-replicating Colonizing Probes

John von Neumann conceived of a program whereby robot probes could be set into motion with the purpose of colonizing the Galaxy. It would take one – out of the possible millions – of civilizations to colonize the Galaxy, given enough time. Self-replicating probes, like rabbits spreading across Australia, would eventually reach our corner of the Galaxy. David Brin (1990, p. 160) cites calculations by Eric Jones of Los Alamos laboratories which suggest that an expanding sphere of settlements could fill up the Galaxy within 60 million years. Frank J. Tipler (1980), a skeptic with regard to the possibility of contact with ETI, has argued that just one self-copying probe could generate enough probes to have one near each star in the Galaxy within 300 million years. Yet, argues Tipler, they have left no sign of their existence near or on the Earth. Even if there was only one other civilization, argued Tipler, albeit a technologically advanced one, they could have built self-replicating machines which could develop space travel and eventually colonize the planets. Even if their transport is less than the speed of light, argued Tipler, they could redevelop and colonize the Galaxy within the age of the Earth. So why haven't they contacted us already? Tipler concludes that we are alone.

Tipler's argument is familiar but, as he maintains, its force is under-appreciated. He suggests that if intelligent extraterrestrials 'did exist and possessed the technology for interstellar communication, they would already be present in our Solar System. Since they are not here, it follows that they do not exist' (Tipler 1980, p. 267). Tipler maintains that 'an intelligent species with the technology for interstellar communication would necessarily develop the technology for interstellar travel, and this would automatically lead to the exploration and/or colonization of the Galaxy in less than 300 million years' (Tipler 1980, p. 267). This assumption that an intelligent technologically communicative species will *necessarily* develop the means to explore and colonize the Galaxy is based upon questionable grounds. In fact Tipler appeals to a version of the principle of mediocrity (a belief that our own evolution is typical of life throughout the universe). Yet the thesis that intelligent life is universal is precisely what Tipler ultimately denies. It might be noted that whoever resorts to the principle of mediocrity it has a weak basis for any argument. There is no evidence other than the development of some, but not all, civilizations on Earth, on which to support the conclusion that we are typical of life forms – actual or possible – throughout the universe.

Tipler's argument is nevertheless worth exploration. He begins with the assumption, based on the predictions of certain exponents of Artificial Intelligence, that an intelligent extraterrestrial species 'will eventually develop a self-replicating universal constructor with intelligence comparable to the human level' (*ibid*: 268). This constructor, or Von Neumann machine, could then be combined with present day rocket technology and would replicate themselves out of local resources and 'eventually colonize the Galaxy in less than 300 million years' (*ibid*: 268). The cost in energy and expense would be minimal to the sending civilization who would only have to find resources for the first machine, which would be programmed to seek out construction material and replicate itself and then construct similar rockets which would transport copies to the next targets, and so on throughout the Galaxy. Improvements in design could be communicated by radio, or by means of self-learning programmes, such that future machines would be more sophisticated than their 'parents'. Tipler also envisages sophisticated machines equipped with artificial womb technology which could synthesise fertilized eggs. The probes could be programmed to construct organic beings, humans, from available raw materials. This would involve beaming human DNA information to the site at the speed of light, so that people could populate an area without having to make the journey. Whilst this suggestion, like the adaptive intelligence claimed on behalf of futuristic Von Neumann machines, lies beyond immediate scientific horizons and is indicative of those weak philosophical arguments which appeal to incredible scientific possibilities, there is a plausible basis for Tipler's transport programme. The technology required would be little more than conventional rockets designed for one way journeys to the nearest construction site. This could be accomplished by chemical fuel sources, although nuclear power could be developed for longer journeys. Nevertheless, such speculation is based on wildly optimistic forecasts about machine intelligence and instrument reliability.

According to Tipler the search for extraterrestrial intelligence would have greater empirical credentials if it were directed towards the location of Von Neumann probes. A radio search can miss its target; they might not be broadcasting at the time that we are listening, or we miss them because we are tuned to the wrong frequency. But a search for probes is more determinate; either they are there or they are not. Tipler's conclusion that we are alone is based on his argument that, given an earlier start, other extraterrestrial civilizations would have set in motion a colonization process and that by now the Galaxy should be teeming with Von Neumann probes. Because they have not been observed, he concludes that we must be unique.

How plausible is this argument? Carl Sagan (1983) drew attention to several flaws in Tipler's thesis and suggested limitations on Galactic colonization. According to Sagan, if these machines are programmed to go on replicating then they will not stop until the entire Universe has been converted into Von Neumann machines, which would then presumably cannibalise each other. It follows, argues Sagan, that these machines would be a threat to any emerging intelligent civilization and steps would have to be taken to restrict their development long before they proliferate. A prudent policy would be to prevent and disrupt their construction. If Tipler is correct the entire Universe is threatened by them and every intelligent society is in danger. If there are any other intelligent civilizations they would take steps to limit this threat and the absence of Von Neumann probes could well be due to the efforts of intelligent extraterrestrials.

Tipler's model of colonization, which has also been criticised (Sagan 1983), is based upon the mathematical theory of island colonization developed by R. H. MacArthur and E. O. Wilson (1967). This model is not typical of all Earth populations and hence scarcely typical of populations throughout the Galaxy. Sagan considers several conventional models of colonization by biological populations and he calculates that "the establishment of galactic hegemony requires a perseverance to the task for a period of a billion years" (Sagan 1983, p. 118). According to Sagan, it is therefore unlikely that any advanced extraterrestrial civilization would remain steadfastly committed to this task for even a fraction of that time. A civilization of merely 1000 years into our technological future would hardly be interested in strip mining and colonizing every planet in sight. An advanced extraterrestrial civilization would have something better to do than imitate events in recent short-term Earth history.

However, Tipler could reply that his theory of colonization by self-replicating Von Neumann machines does not require any long-term commitment. Once the first probe is launched, the process is potentially independent of the sending civilization, which may lose interest in the project or self-destruct. And it would only take one civilization at any time for the first probe to be launched.

This argument becomes less plausible once it is recognised that developing and maturing civilizations are capable of recognising past errors and of taking steps to put matters right. A civilization might come to recognise the dangerous folly of sending its probes and take steps to curtail them and even warn others about them.

Perhaps a less sophisticated, but nevertheless relevant response to Tipler's argument is to question the feasibility of a system of self-replicating probes. Such machines bear little relation to present technology or any predictable future developments. According to the arguments, these machines will approach a mineral rich environment, alone or with other such machines, and begin to extract various materials, manufacture plastics, metals and so on, without steel-mills, processing plants, and a sophisticated transport and distribution system, and then begin to construct complicated apparatus including rockets containing more probes. This is nothing short of miraculous. Using the best technology on Earth there is not even the remotest possibility of designing a machine that could go out and locate the necessary materials and manufacture a bicycle. Those who employ these kind of fictitious examples should be able to specify in considerable detail how such schemes can be practically implemented.

Hart's Case for the Uniqueness of Humans

The fact that the inhabitants of planet Earth have entered the space age and contemplate colonizing voyages throughout the Solar System and beyond has lent support to SETI's opponents. If we are but a few centuries away from interstellar voyages might it not be the case that our ET neighbours are in the same position? If we are on the threshold of colonizing the Galaxy – a mere million or two years away – then why haven't others, with an earlier start, done so already? This paradox was examined by M. H. Hart (1975) who argued that the obstacles against interstellar travel have been over-exaggerated and that the only reason the Galaxy is not teeming with intelligent life is because we are unique. Hart's attempt to demolish explanations of the Great Silence is worth elaboration. First, he rejects physical explanations which claim that extraterrestrials have never arrived because 'some physical, astronomical, biological or engineering difficulty makes space travel infeasible' (Hart 1975, p. 128). Space travel, he argues, does not present insurmountable problems. It may be the case that a one way trip to Sirius at the speed of light would take 88 years which is well beyond current technology. But Hart sees ways of overcoming such problems: youthful voyagers in states of suspended animation, using drugs to slow down their metabolism; future developments in biology which could make it possible to freeze and thaw out warm blooded animals. Hopefully they would take self-functioning instruments to do this with them as there would not be anyone waiting at the end of the journey to thaw them out. Or, as Hart speculates, extraterrestrials might not be warm blooded, thus circumventing this problem. The length of time required for the journey might not be a problem for extraterrestrials who may have life-spans lasting many thousands of years and not regard a 200 year journey as a dreary waste of one's life. Deep into the realm of speculation Hart also considers frozen zygotes sent out in spaceships piloted by robots, or the use of techniques of time dilation. A journey lasting many generations could be mounted, as long as comfort and a well-organized social structure was part of the conception. The energy requirements pose few problems for Hart, who envisages nuclear powered rockets scooping up oxygen as they speed through the Galaxy.

Hart also rejects sociological explanations of the silence, which maintain that they chose not to come because of lack of interest, motivation, organization, or suffer from a tendency to self-destruct. The weakness in these explanations, says Hart, is that they fail to consider the fact that civilizations can and do change. But these explanations, he says, may only hold for one civilization at any one time. To cite them as an explanation of silence would require showing why they hold 'for *every* race of extraterrestrials – regardless of its biological, physiological, social or political structure – and at *every* stage in their history after they achieve the ability to engage in space travel (Hart 1975, p. 132). SETI, it would seem, cannot have it both ways. If, as SETI exponents argue, life on Earth is typical then they are obligated to acknowledge the possibility that some, if not all, extraterrestrial civilization will have colonizing tendencies like ours. Of course it might be responded that we *are* typical, and that just as our colonizing tendencies evaporate with the awareness of the massive costs involved and the immorality of such conquests, so it would be with extraterrestrials.

The third category of explanations for the absence of extraterrestrials which Hart rejects is classified as 'temporal explanations'. These explanations appeal to suggestions that they have not yet had enough time to reach us. To assess the plausibility of Hart's argument one needs to know how long it could take a civilization to reach us once it has embarked on a program of space exploration. Hart considers expeditions from Earth to the nearest 100 stars as a typical example. These are all 20 light years away from our Sun. Colonies might be established near them from which further voyages could be launched. Without any pause between trips Hart estimates that the Galaxy would be traversed within 650,000 years (Hart 1975, p. 133). If we allow for a period between each wave of expansion of about the length of time it takes for each voyage this would double the total number of years. But this would still mean that an advanced civilization should have reached us by now, unless they began their colonization less than two million years ago, says Hart.

The fourth and final explanation of absence rejected by Hart is that they have visited us but that we are not aware of it. But if it is suggested that they came about 5000 years ago, says Hart, a sociological theory is required to explain why no other extraterrestrials have visited since and why they do not remain here. Moreover, he adds, if it is suggested that they came and chose not to colonize then a sociological theory is required to show why every civilization which could have colonized chose not to.

This rebuttal of explanations of absence is hard to sustain. Granted it only requires but one colonizing advanced civilization to cover the Galaxy within two or so million years but there is no reason to assume that even one civilization will emulate the behaviour exhibited by some human societies over the past few thousand years of recent terrestrial history. Moreover, there may be an infinite number of sociological reasons, or a combination of all four of the explanations of absence, which could account for the failure to colonize. Moreover, despite Hart's assertion that we have no evidence of them it is possible that the Galaxy is teeming with advanced extraterrestrial life, who are so advanced that we are unaware of them, yet being much more advanced, they would have no interest in revealing themselves to us. Hart's four arguments against explanations of absence and his solipsistic conclusions suffer from the same problems which beset Tipler's case for uniqueness. The case for colonization rests on appeals to many unknown and as yet undiscovered scientific techniques, and assumptions that practices which have existed for a small part of human history are likely to be repeated universally.

SETI's Response to the Fermi Paradox

Maybe they came, did not see anything interesting and left without trace. Maybe they have not reached us yet, or they have been and we are the descendants. Maybe we missed intelligent life: it evolved, colonized the Galaxy, and self-destructed long before humanity existed. Now we are among the very first in the second time around. Perhaps the Frank Tipler probes are really there, but are programmed not to respond until we encounter them. Maybe they came, dumped their rubbish and contaminated the Earth and out of the chemical trash emerged the seeds of prebiotic life. Thomas Gold (1960) describes this as the 'garbage hypothesis'. It is, of course, an untestable hypothesis. Perhaps they have no detectable physical structures and exist as pure states of consciousness having long ago discarded their material bodies, or having rejected their planetary homes because they regard them as dirty unhealthy places, have no wish to visit another one. Perhaps their technology evolved long ago beyond a stage where they would want anything from us. Perhaps our environment is inhospitable to the species which dominates our part of the Galaxy. Perhaps contrary to Hart and Tipler, they have no desire to colonize, no desire to migrate. Advanced ETs might not suffer from any half-crazy desire to strip-mine the Galaxy, running from planet to planet polluting and raiding its resources.

One explanation of the silence rests on a weak interpretation of the SETI programme: it concedes that the combined fractions of f_l and f_c in the Drake equation (planets with intelligent life) are much lower than previously estimated. Drake estimated a rate of one planet in every hundred. But according to the weak interpretation it can be conceded that even with the precursors of life, such as sugar and amino acids, the next steps towards life and then intelligence are more difficult than imagined. Perhaps some rare, as yet unknown, catalyst is required if life is to emerge. Perhaps some, as yet unknown, software is required if intelligence is to develop. It might also be conceded that intelligent life is more prone to natural disasters than previously thought. For example, the Cretaceous Tertiary event, plus evidence of four other mass extinctions found in the sedimentary record, destroyed most species of land animal over forty kilos in weight and prohibited any potential for intelligence in the first three billion years of Earth life. The same, even more so, might occur throughout the Galaxy.

Events in social history could reduce the fraction of L – the length of an advanced civilization. According to Drake there should be one advanced civilization per million stars. Expanding populations, driven primarily by demographic pressure, tend to wreck their environments, as the Europeans did in their colonies. If this were repeated on an interstellar scale it could be argued that the termination of an advanced civilization coincides with its capacity for interstellar communication and travel. The problem with this argument is that one cannot claim that civilizations will destruct before they colonize unless one can demonstrate that every single race will do so before it reaches the capacity for interstellar flight (Birch 1990, p. 1). For it would only require one exception for the Galaxy to be populated with ETs. There is no scientific law which prohibits colonization. It may, of course, turn out that no one wants to, but this is not supportive of a thesis which claims that no one can.

Perhaps they have no need to colonize. For example, our Sun and nearby planets can provide enough energy and materials for the inhabitants of the Earth for several millions of years. Non-aggressive, non-expansionist, civilizations consisting of pacifists might have no desire to communicate with us. Carl Sagan and Frank Drake have suggested that if a species developed life-expectancy rates approaching immortality they would be reluctant to take risks and would avoid interstellar contact out of timidity. One of Sagan's answers to Tipler is that there are two types of advanced society; aggressive and peaceful. The peaceful do not take risks, do not colonize, so we are unlikely to hear from them. But the aggressive are intrinsically unstable and are unlikely to pass through the dangerous phase that we have reached on Earth with nuclear weapons. So either they die off as the result of nuclear war and nuclear winter or they pass through the dangerous phase and become non-colonizing pacifists (Sagan 1983). We thus arrive at the following paradox: only societies that can survive the critical nuclear weapons stage are those with the potential for colonization. But in order to survive as an advanced technological society aggressive behaviour of this kind would have to disappear. Hence: 'the only societies long-lived enough to perform significant colonization of the Galaxy are precisely those least likely to engage in aggressive galactic imperialism' (*ibid*: 120).

There is an additional reason why a very advanced civilization is likely to develop pacifist tendencies. Suppose they have developed methods of combatting the ageing process or have found ways of indefinitely repairing damage caused by ageing and only encounter death through accident or murder, and throughout their long lives preserved, as Drake (1993, p. 160) suggests, "a growing set of memories of individual experience". At one level we can comprehend this, but the difference between them and us would be massive. They would have, says Drake, a fantastic obsession with safety, avoiding wars, accidents and even risky contacts with other worlds. Maybe silence would be their preferred option for survival. On the other hand they could realize that spreading their knowledge about immortality – and hence encouraging others to be as equally concerned with safety – would be an alternative strategy for survival.

Intelligent extraterrestrials might very well remain silent for the same reasons as radio-astronomers on Earth. Apart from an occasional brief signal, and inadvertent signals from radio and TV networks, which are very weak, the SETI radio searches are for listening, not transmitting. This is party because intentional broadcasts are more expensive and also because of a fear that a hostile civilization might learn too much. It is possible that everyone is listening and no one is communicating. It is also possible that our failure to observe ET communication is because they have some other means of communication than electromagnetic waves.

David Brin (1990) offers a range of explanations of the Great Silence. He points out that the 250 million year orbit around the Galaxy poses major survival problems for our Solar System. When it passes the spiral arms where new stars are formed in super explosions it is likely to be destroyed. But a very advanced civilization (for example a Dyson-Kardaschev Type II or III) may simply leave this dangerous place taking their solar system far away (Dyson 1960, 1966, 1973; Kardaschev 1964). Consequently the very advanced civilizations would be further away and less likely to make contact; only the less advanced would remain and they might not have evolved the ability to communicate over long distances. Brin also notes that there may be many unforseen disasters in the Galaxy, such as huge black holes, which could destroy potentially colonizing civilizations. There may be holocausts caused by the effects of colonization such that he colonizers leave nothing behind them. But on a more friendly note Brin suggests that it is likely that the most habitable planets – not too hot and not too cold – with ample water and oxygen, will be far less dry than ours. Hence land creatures would barely develop. In this respect we are unique. But intelligent life, such as dolphins and whales, will develop in the water, without our technology derived from the use of hand and fire, and hence an intelligence with no likelihood of reaching the stars.

A novel twist to the Fermi paradox can be derived from a consideration of what ETIs may wish to obtain from making contact with us. This approach was outlined by T. B. H. Kuiper and M. Morris (1977) who suggested that a more highly developed civilization may place a high value on forms of knowledge or experiences that we have not even learnt to appreciate, and that this resource actually grows with time. Suppose, say Kuiper and Morris, that there is a certain threshold phase to this process before which we could produce nothing of value to an advanced civilization. If they were to make contact too soon, before we reach this threshold, and expose us to the superior knowledge held by the Galactic community, "instead of enriching the Galactic store of knowledge we would merely absorb it" (Kuiper and Morris 1977, p. 620). Our capacity to produce new ideas and make a contribution to the Galactic heritage would be restrained if the best human minds were 'occupied for generations digesting the technology and cultural experiences of a society advanced far beyond our own' (*ibid*: 620). Thus silence is maintained because early contact would remove the very reason for making contact.

There is a danger in too much speculation about ET life and ET intentions: in the absence of factual evidence speculation may not only be widespread, it may take on a form of wish-fulfilment representing deeply felt psychological and even moral beliefs. This may partially explain speculations about benevolent super-civilizations who are keeping a paternalistic eye on us. But there are rules to govern speculation; it must be contained within the boundaries of scientific theory. This allows one to pursue an idea beyond the limits of so called hard-headed, all facts-and-no-nonsense-thinking, but within the plausible extensions of currently defensible theory.

The Zoo Hypothesis

Another explanation of the silence is the 'Zoo Hypothesis', proposed by J. A. Ball (1980, p. 242), who said that "the perfect zoo . . . would be one in which the fauna inside do not interact with, and are unaware of, their zookeepers". The Zoo Hypothesis suggests that benevolent super-beings look on us as a species in quarantine, at present too dirty and dangerous for contact, or that the Intergalactic Council has designated Earth as a nature reserve. Perhaps, also, if they are very advanced they do not consider us worth contacting. The Zoo Hypothesis has a number of versions, most of which are falsifiable but not confirmable. As Ball put it: "we shall never find them because they do not want to be found and they have the technological ability to ensure this. Thus this hypothesis is falsifiable, but not, in principle, confirmable by future observations" (*ibid*: 243). Intelligent ET's may have penetrated the solar system with their probes and are keeping a low profile because of an embargo on contact with any inhabitants who are still planet bound. This is to avoid lending any encouragement to plans to quit their planets prematurely, as first they must somehow demonstrate their fitness to mix with other beings. In such a case, silence would not mean absence, but an unwillingness to speak. According to M. D. Papagiannis (1988) even if no message is heard after the next million hours of search, we should not conclude that we are alone. Communicating civilizations may have an ethical rule whereby newcomers are obliged to pass an entrance test, for example, avoidance of overpopulation, disease, global war or environmental disaster. He concludes that a negative result of a long and comprehensive search need not simply mean that we are alone in the Galaxy, but that we are not ready to join a sophisticated 'Galactic Club' which has rules for membership, and the headquarters of our Galactic region has not yet issued an instruction for us to be admitted. But if we can solve our problems then we may, in time, receive an invitation. Papagiannis concludes that we need not waste efforts on expensive searches but instead just concentrate on our own problems and meanwhile try to eavesdrop on them while we wait.

The main problem with the various Zoo Hypotheses – apart from their inherent untestability – is that they require total and incredible unanimity throughout the entire Galaxy. They also fail to explain why Earth and its neighbourhood were quarantined before life appeared on it. Nevertheless, there is one version of the Zoo Hypothesis which merits further consideration, as it is a testable, if slightly implausible, response to the Fermi Paradox. It is known as the 'Peer Hypothesis' and was developed by Paul Birch (1990).

The Peer Hypothesis

Birch hypothesises a race of super ETIs with no limits on their technology, who can travel from galaxy to galaxy, and can build with planets, black holes and stars, and manipulate the topology of space-time. These super beings can manufacture their own space-times, create whole universes, and control them as we would control a small zoo. Could it be, asks Birch, that we actually live in such an engineered universe? The intelligence gap between these super ETIs and us would be immense, no less than the gap theologians pose between God, who comprehends all, and mere mortals. Assuming only one creative intelligence behind the Universe, it is likely that all intelligence made in its image will be roughly similar, argues Birch. Thus according to the Peer Hypothesis, "Many ETI have been created, but their development has been deliberately arranged and guided so as to be roughly contemporaneous with our own" (Birch 1990, p. 3). This hypothesis allows for the SF scenario of thousands of ETI civilizations throughout the Universe, but it predicts that Earth is neither advanced nor backward, but probably typical of the development of intelligence through the Universe. Birch suggests that these broad similarities would be because the creators want a universe that would be eventful and interesting, as "history is at its most diverting when many groups interact, and when their capacities and interests are neither wildly disparate nor absolutely identical" (*ibid*: 3). Of course there may be many different reasons behind the creation of such a universe; intellectual challenge, entertainment, and so on.

If the development of intelligence and technology is roughly similar throughout the entire Universe, there would be enough diversity to enable different cultures to learn from each other after first contact, but not too much so as to preclude each others cognitive thresholds. The Fermi Paradox would be resolved, as being similar to us ETIs would not yet possess the technology required for colonization. But like us they will possess it soon, and then they will come. Moreover, after first contact others will follow soon, as technological skills will not be that far apart. According to Birch there are military implications with the Peer Hypothesis. Our peers would not be so powerful so as to easily defeat us, or too weak to be insignificant. Being similar to us they would come prepared and war should be considered a probable outcome.

How should we assess the Peer Hypothesis? Is it mainly a piece of imaginative SF and at variance with conventional theories regarding the origin of the universe? It requires a race possessing super intelligence with the power to create multiple universes. However far-fetched this may seem it is not incompatible with theories concerning the origin of the Universe and other universes too. John Gribbin (1994) has advanced the view that entire universes can be created out of black holes where matter has been squeezed to the point of becoming a singularity. When the singularity begins to expand a new universe can be created which will exist in its own space-time dimensions. He suggests that there could be numerous such universes. Gribbin also considers that the squeezing operation could be an act of intention.

> Bizzare though it may seem, if you could squeeze a kilogram of butter (or anything else) hard enough to make a black hole, that black hole could be the seed of a new universe as big as, or bigger than, our own (Gribbin 1994, p. 250).

If this was possible, a super intelligence may well have created several universes along the lines suggested by the Peer Hypothesis. As a matter of fact multi-universe production may well be within our technological capability. Says Gribbin:

> The technology is not so far-fetched, and would involve a super-powerful hydrogen bomb explosion somewhere in space at a safe distance from the Earth. It is even conceivable that our Universe was manufactured deliberately in this way as part of a scientific experiment by a technologically advanced race in another universe (*ibid*: 250).

Of course the postulation of super ETI, who can create and manipulate space-time, only puts the question back: where did they arrive from? Nevertheless, the Peer Hypothesis answers the Fermi paradox, and is consistent with evidence pertaining to the existence or non-existence of intelligence throughout the Universe. But is it true? Is it likely? At least the hypothesis allows for these questions to be determined by evidential considerations and predictions. Given mankind's present ability to communicate to the stars and the likelihood of interstellar travel by 2200 AD, then assuming Earth is typical of other habitable regions in the Universe, with a similarly developing timescale, we should expect to meet our peers fairly soon.

Why Come Here When You Can Travel By Information Transfer?

Most versions of Fermi's paradox appeal to the likelihood of Galactic colonization. Given the age of the Galaxy, there has been ample opportunity for at least one advanced civilization to have colonized the Galaxy. Hart (1975), as we have seen, rests his case for uniqueness on the fact that not even one colonizing civilization has reached the Earth despite ample time to do so. Objectors have pointed to the drawbacks of colonization which include the costs of highly expensive physical interstellar travel. But as the projected number of advanced civilizations is high and as Zuckerman (1985) argues, mass migration might be the only option for long established civilizations facing the collapse of their sun, the evidence of colonization should be available. So where are they? This problem has been given a new solution by Louis K. Scheffer (1994) who sees a resolution of Fermi's paradox in an appeal to developments in machine intelligence.

Scheffer considers methods of travelling across the Galaxy alternative to expensive physical craft. He begins with the notion of teleportation which, as yet, has been confined to science fiction. The idea is that a sufficient level of data is collected and measured from particular persons, including their mental states, and then this data is transmitted to its destination via electromagnetic signals. There have been many objections to this proposal. One objection was that the very act of measuring the original object would collapse its quantum state and prevent an exact replica from being constructed at the destination. Scheffer drew attention to a theoretical counter to this objection by C. H. Bennett *et al.* (1993) who suggested that by pre-arranging the transfer of correlating but unmeasured particles, the exact quantum state could be reproduced at the destination despite collapse of the original quantum state. But this theoretical solution, says Scheffer (*ibid*: 158), does not resolve a major problem of teleportation, namely that the reproduction of any object of significant size would require sending an enormous amount of information at great expense. Thus Scheffer suggests a less expensive method involving information transfer by computer program. The questions he then poses are:

1. How does one get the receiving computer to its destination in the first place?
2. Is it possible to construct a conscious being as a computer program?

The answer to the first question is that the teleporting civilization sends out probes to various destinations which will contain robots that will build and maintain receiving computers. As far as the second problem is concerned, Scheffer considers the notion that a consciousness could be mapped and teleported by computer. Most scientists, he says, would in theory agree that the state of every sub-atomic particle of any object, including a human being, could be measured and its evolution under quantum mechanical rules could be simulated. Then, in theory, its behaviour could be predicted. This, of course, is a massive problem. A human body contains at least 10^{27} atoms and even one bit of information per atom is too much information for any foreseeable computer to handle. Moreover, as Scheffer concedes (*ibid*: 159), we have no idea how to measure a human being on a sub-atomic scale. Yet despite the impossibility of practical implementation Scheffer considers information transfer as a thought-experiment. The problem might be reduced if we scale down to the transfer of a simulated nervous system, a brain whose operations might be simulated by a computer program. This again raises major philosophical problems, which Scheffer recognizes, as it can be argued that the program would not *be* a real consciousness, but only a simulation (see Searle 1987). Nevertheless, appealing to the most optimistic predictions in Artificial Intelligence research Scheffer concedes that the reason why a complete simulation is not possible today is largely because we know so little about the human brain. Thus, for the purpose of sustaining the thought experiment, Scheffer assumes that "in some cases shorter conscious programs are possible". Despite reservations with this thesis, namely, the objection that no increase in the perfection of a simulation will transform its identity to the original object, we shall continue with Scheffer's thought experiment which is suggestive of plausible solution to Fermi's paradox.

Scheffer proposes that once consciousness is captured on a computer program then its transfer would be as follows: a computer program is started on one machine, then it is stopped and the current state of the program is sent to another machine at a different location, which then resumes the execution of the program. It would appear that the program – a conscious entity – has been teleported. To secure plausibility Scheffer cites work in today's computer networks which already allow computers to send incomplete programs which are devised in such a way to allow another machine to complete it. There are also developments in cyberspace technology where 'simulations' of conscious experience could be teleported. The fundamental philosophical objection is that if a simulation of a person is teleported the person remains behind, unless we revise our views which place a conceptual restraint on the idea of a person being in two places at the same time.

So how does Scheffer's thought experiment resolve Fermi's paradox? Information transfer eliminates the need for physical systems of transport and colonization, so they are not likely to be here in any physical sense even though their computer simulations are hurtling around the Galaxy. Once the first civilization developed the technical means of information travel, its members can go anywhere at the speed of light at a fraction of the cost of physical transport. It requires no major engineering challenge, no need for large space ships or programs for strip-mining the Galaxy. Now, one of the arguments in support of Fermi's paradox is the large numbers argument which Hart (1975) employs when he says that only one civilization out of billions will need to be colonizers. Opponents are then required to show why none, out of billions of civilizations, chose not to colonize. Scheffer's answer is that once one civilization has adopted a system of information transfer then others will have a disincentive to engage in physical travel, as it will be cheaper to join the existing Galactic teleportation travel club. Moreover, he speculates, if the club offers easy Galaxy-wide teleportation there is a likelihood that it could lead to a single culture throughout the entire Galaxy, and if this culture has decided against colonization of the Earth then the large number argument loses its force.

Scheffer's argument can also be adapted as a rebuttal of one of the arguments employed against the Zoo Hypothesis. For example, it is frequently argued that the Zoo Hypothesis requires an enforced agreement among millions of civilizations that Earth will not be colonized or visited. This, it is claimed is impossible to enforce. But an homogenous Galactic culture could enforce a code of practice in which Earth is treated as a nature reserve.

What has Scheffer established with his theory of teleportation by information transfer? In a series of steps he concluded that (1) travel by information transfer is plausible, (2) that it is cheaper than physical travel over interstellar distances, and (3) that if there is intelligent life in the Galaxy using interstellar information travel, it has very likely facilitated the emergence of one large civilization. The first step, however, encounters philosophical objections concerning the possible simulation of consciousness; the second step is clearly acceptable if practical and conceptual objections to the first can be overcome, whilst the third step is highly contingent. More communication, contact and travel of whatever form, need not lead to homogeneity: it could equally produce exaggerated diversity. Despite worldwide networks of almost instant communication, the warring nations of Europe are far from homogenous.

What observational predictions can be derived from Scheffer's thesis? We could eavesdrop on their communication network if they are transmitting at a range within our reach. But, says Scheffer, they would probably transmit at between 50–60 GHz range which would be cheaper for them, but difficult for us to detect from the ground. They may have sent ships to various places in order to repair or replace damaged computers. But this might only require one visit per stellar system, which means that we would be unlikely to find them. In any case, they could have perfected their network before the Earth existed and have no further need for further physical travel.

These considerations allow Scheffer to reverse Fermi's paradox: if humans are typical of Galactic life and such intelligent life has evolved many times in the Galaxy, and since we have not been colonized then there must be some reason why. Scheffer's answer is that there is already a large unified Galactic society, where there is no incentive for physical travel, where there is no need to trade or transport material goods.

Is Fermi's Paradox Due to the Wrong Model of Galactic Colonization?

According to David G. Stephenson (1980) Fermi's paradox is a product of a particular model of Galactic colonization. A different model, he argues, need not produce a paradox whereby the Galaxy is teeming with intelligent life of which there is no evidence within the Solar System. Most models of galactic colonization, says Stephenson (1980, p. 248), assume that interstellar colonies will have a life cycle paralleling that of a virus, where the 'virus or an interstellar vessel, travel through an alien medium with most of the information in a dormant state and being irrelevant to the surrounding conditions'. When it reaches a suitable place the information is activated and put to use in its new environment. The human equivalent of this model is nineteenth century European emigration to the US where the immigrants took with them skills and information which was of no value on the journey, but was essential in modifying the New World in order to replicate some of the conditions they had left behind in Europe. The 'virus' model presents an image of colonists travelling from planet to planet in a state of limbo but then on arrival setting up colonies similar to the home environment, or at least seeking out planets offering a similar environment to the one they have left. This model is reinforced with accounts of the vast distances in interstellar space, which suggests that the travellers live very long lives or have acquired techniques of hibernation. Once this model is adopted, then the absence of colonies can then be cited in support of claims regarding human uniqueness. But Stephenson considers an alternative model which avoids Fermi's paradox.

Instead of a passive journey where the vessel is merely a means of transporting beings with a dormant capacity to activate information on arrival, Stephenson's model depicts the vessel as an equivalent to a higher organism in a constant state of activity, with its information relevant and in use. In short, the vessel would be a travelling biosphere in which its inhabitants are completely integrated. The vessel might even grow on the journey, drawing material on the way. It would not be a mere means of transporting beings from start to destination; it would be the home of its inhabitants, adapting to conditions wherever it went. Not requiring settlements at either end of the journey, the human analogy is a nomadic tribe that has adapted to an environment that fulfils its needs as long as they are mobile. Such a community would not require planetary colonies and would be unlikely to visit us. There could be, suggests Stephenson, thousands of interstellar vehicles of this type throughout space and undetected.

Fermi's paradox arose because the 'virus' model of colonization suggested that extraterrestrials, like nineteenth century Europeans, would inevitably end up in the New World after populating all corners of the Earth-Galaxy. But a Galaxy teeming with interstellar vehicles which are independent homes would not lead to the colonization of the Earth.

Are there any observable predictions in Stephenson's thesis? Stephenson's proposal not only avoids Fermi's paradox, it also predicts likely zones for the detection of their vessels. Thus in our Solar System, he suggests (*ibid*: 248–9) 'the search should concentrate on the outermost bodies of the Solar System as being the closest to the interstellar space. It is an interesting speculation', he concludes, 'that an unusual orbit and planetary parameters of Pluto might, in part, be due to the activities of visiting interstellar craft that have reduced the body from a conventional outer planet to a cosmic slag tip of unwanted elements'.

Conclusion

So much for the Great Silence. Perhaps it is too early to draw a curtain over this debate. It is only this century that speculators about the existence of ETI have been given the technology to conduct an empirical search for it. At the present time so little of the Universe has been explored, so little (less than 1%) of the Solar System has been explored. There is no firm knowledge of how life began on Earth, no prior knowledge to rule out claims as to how it might have began elsewhere. It is too premature to speak of a paradox in the context of the Great Silence, as yet there is not enough evidence either way. "Absence of evidence is not evidence of absence". As long as the number of negative searches cannot be regarded as a statistically representative sample, all conclusions regarding the absence of ETI remain unsupported. Either side could be right. But this is not to rule out the possibility that, if negative searches were to continue, it would begin to look as if we have missed the chance of finding intelligent life in the Galaxy.

University of Birmingham, UK

REFERENCES

Ball, John A. (1980), 'The Zoo Hypothesis', in *The Quest for Extraterrestrial Life*, Donald Goldsmith (ed.) (University Science Books, Mill Valley CA), pp. 241–5.

Bennett, C. H., G. Brassard, C. Crepeau, R. Jozsa, A. Peres and W. K. Wooters (1993), 'Teleporting an Unknown Quantum State via Dual Classical and Einstein-Podolsky-Rosen Channels', *Physical Review Letters* **70**, p. 1895.

Birch, Paul (1990), 'The Peer Hypothesis', *Journal of the British Interplanetary Society*, March, pp. 1–6.

Brin, David (1990), 'Mystery of the Great Silence', in *First Contact*, Ben Bova and Byron Preiss (eds.) (Headline, London), pp. 154–83.

Cambell, Bruce (1990), 'A Place to Live: Exsolar Planets', in *First Contact*, Ben Bova and Byron Preiss (eds.) (Headline, London), pp. 129–49.

Cocconi, G. and P. Morrison (1959), 'Searching for Interstellar Communication', *Nature* **184**, pp. 844–6.

Drake, Frank D. (1990), 'The Drake Equation: A Reappraisal', in *First Contact*, B. Bova and B. Preiss (eds.), pp. 150–3.

Drake, Frank and Dava Sobel (1993) *Is Anyone Out There? The Scientific Search For Extraterrestrial Intelligence* (Souvenir Press, London).

Dyson, Freeman J. (1960), 'Search for Artificial Stellar Sources of Infrared Radiation', *Science* **131**, p. 1667 [Reprinted in Donald Goldsmith (ed.) (1989), *The Quest for Extraterrestrial Life* (University Science Books, Mill Valley CA)].

———. (1966), *Prospects of Modern Physics*, R. E. Marshak (ed.) (John Wiley, New York), p. 641.

———. (1973), 'Discussion', in *Communication With Extraterrestrial Intelligence, CETI*, Carl Sagan (ed.) (MIT Press, Cambridge Mass), pp. 188–9.

Gold, Thomas (1960), 'Cosmic Garbage', *Airforce and Space Digest*, May: 65.

Gribbin, John (1994), *In the Beginning: The Birth of the Living Universe* (Penguin, Middlesex).

Hart, Michael H. (1975), 'An Explanation for the Absence of Extraterrestrials on Earth', *Quarterly Journal of the Royal Astronomical Society* **16**, pp. 128–35.

Kardaschev, Nikolai S. (1964), 'Transmission of Information by Extraterrestrial Civilizations', *Astrononicheskii*, zh. **41**, p. 282 [English Translation in *Soviet Astronomy* A. J. **8**, p. 217].

Kuiper, T. B. H. and M. Morris (1977), 'Searching for Extra Terrestrial Civilizations', *Science* **196**, pp. 616–21.

MacArthur, R. H. and E. O. Wilson (1967), *The Theory of Island Biogeography* (Princeton University Press, Princeton).

McMullin, Ernan (1971), 'Review of R. Puccetti, *Persons: A Study of Possible Moral Agents in the Universe*', *Icarus* **14**, 291–4.

McNally, Derek (ed.) (1994), *The Vanishing Universe: Adverse Environmental Impacts on Astronomy* (CUP, Cambridge).

Munévar, Gonzalo (1989), 'Extraterrestrial and Human Science', *Explorations in Knowledge* **VI**(2), pp. 1–8.

Oliver, B. M. and J. Billingham (eds.) (1973), *Project Cyclops: A Design Study of a System for Detecting Extraterrestrial Life*, NASA Report CR-114445 (NASA/Ames Research Center, Moffet Field, Calif).

Papiagiannis, Michael D. (1988), 'Regional Jurisdiction in Our Galaxy: A Possible Explanation for the Absence of Extraterrestrial Signals', in *Bioastronomy: The Next Steps*, G. Marx (ed.) (Kluwer, Dordrecht), pp. 281–5.

Sagan, Carl (1973), *Communication with Extraterrestrial Intelligence CETI* (MIT Press, Cambridge Mass).

———. (1983), 'The Solipsistic Approach to Extraterrestrial Intelligence', *Quarterly Journal of the Royal Astronomical Society* **24**, pp. 113–21.

Scargle, J. D. (1988), 'Planetary Detection Techniques: An Overview', in *Bioastronomy: The Next Steps* G. Marx (ed.) (Kluwer, Dordrecht), pp. 79–82.

Scheffer, Louis K. (1994), 'Machine Intelligence, the Cost of Interstellar Travel and Fermi's Paradox', *Quarterly Journal of the Royal Astronomical Society* **35**, pp. 157–75.

Searle, John (1987), 'Minds and Brains Without Programs', in *Mindwaves*, C. Blackmore and S. Greenfield (eds.) (Blackwells, Oxford), pp. 208–53.

Simpson, John A. (ed.) (1994), *Preservation of Near-Earth Space for Future Generations* (CUP, Cambridge).

Stephenson, David G. (1980), 'Extraterrestrial Culture within the Solar System', in *The Quest for Extraterrestrial Life*, Donald Goldsmith (ed.) (University Science Books, Mill Valley CA), pp. 246–9.

Tipler, Frank J. (1980), 'Extraterrestrial Intelligence Beings Do Not Exist', *Quarterly Journal of the Royal Astronomical Society* **21**, pp. 267–81.

Verschuur, Gerrit L. (1980), 'A Search for Narrow Bank 21-cm Wavelength Signals from Ten Nearby Stars', in *The Quest for Extraterrestrial Life*, Donald Goldsmith (ed.) (University Science Books, Mill Valley CA), pp. 142–51.

Vidal-Madjar, Alfred (1994), 'The Extraterrestrial Life Debate', *The Cambridge Atlas of Astronomy* (CUP, Cambridge), pp. 412–7.

Zuckerman, B. (1985), 'Stellar Evolution: Motivation for Mass Interstellar Migration', *Quarterly Journal of the Royal Astronomical Society* **26**, pp. 56–9.

RAMÓN QUERALTÓ

SINCE INDETERMINACY: THE NEW PICTURE OF THE PHYSICAL WORLD AT THE END OF MODERNITY

PHILOSOPHICAL STATUS OF PHYSICAL INDETERMINISM

Indeterminism in modern physics has supposed an important change of paradigms for our scientific understanding of physical reality. In fact this change, from a philosophical point of view, is still stimulating a lot of reflections and discussions which hold a great conceptual significance, in such a way that the philosophical virtualities of the new indeterministic paradigm are very far from being exhausted. These theoretical possibilities especially refer to the ontological and epistemological levels, which, as a rule, are connected to one another a natural extent. In this work, I will mainly pay attention to some ontological consequences originating in indeterminism, that is to say, to some new philosophical categories that can be considered a legacy of science to philosophy at the end of Modernity.

The first step for a right understanding of the following lines is to state a relevant difference between indeterminism and determinism as regards the historical consequences of each one. It deals with the fact that indeterminism has not yet produced any general and well accepted picture of the physical world, whereas we possess a well defined view of physical reality derived from determinism. This means that the historical process of the conceptual development of indeterminism is now just beginning.[1] So, it is easy to understand what a deterministic conception of nature means, including its most important implications, although it is not the same case with regard to an hypothetical indeterministic picture. For instance, it seems clear that we are not able to easily determine the basic features of an indeterministic cosmological conception.[2] This situation brings about the fact that our reflections should be a theoretical proposal to discuss. So, one of our aims would be to contribute to this discussion.

At the beginning, however, a previous question arises. We should ask whether researching a general picture of the physical world is a necessary task or not. Would it not be a simple intellectual amusement? It is important to assert here that the attainment of such a picture would remarkably contribute to the internal unification of scientific knowledge as a regulative idea which could guide scientific research. Such a quest for unity is one of the natural aims of human knowledge, and in this case, of the scientific one. In this line, it is possible to argue that a picture of the physical world becomes the ground of a metaphysical research program in the Popperian sense, which would tentatively contribute to a new interpretation of the philosophical meaning of some physical theories. Therefore the investigation of a general picture of physical reality is undoubtedly a significant task.[3]

D. Ginev and R. S. Cohen (eds.), Issues and Images in the Philosophy of Science, 253–262.

Nevertheless, the contemporary scientific situation shows some important difficulties in this respect. The main one consists in the lack of unification of the basic physical theories. At first glance, it is not hard to observe that two theories dispute the supremacy: quantum mechanics and general relativity. As a matter of fact it is well known that the possible world pictures derived from each one greatly differ with regard to some relevant problems, such as the meaning of physical causality – deterministic in Einstein's theory, indeterministic in quantum mechanics.[4] It means that we do not possess the unified scientific ground, historically necessary, to obtain a coherent picture of the world, as happened in the case of Newtonian physics.

There are also other problems concerning the internal situation of quantum theory which recently originated some scientific and philosophical controversies, as for example the discussion whether quantum theory is a local theory or not. Obviously the result of such discussions could determine the consequences of the theory with regard to a philosophical conception of the physical world.

Nevertheless, it is necessary to affirm that indeterminacy is one of the new categories defining the scientific understanding of the world. It has appeared as a result of scientific knowledge and has become a specific condition of the possibility of rightly researching its objects – in the Kantian sense. In this view, the object of physics at the microphysical level is necessarily concerned with indeterminacy as long as it is identified as such object, that is to say, indeterminacy and physical object are two elements which are reciprocally involved to a necessary extent. We cannot forget this special relationship because it would otherwise be impossible to understand the way nature functions. In this manner, indeterminacy and physical reality do appear as two conditions of scientific knowledge which cannot be separated.

From this point onwards, we shall be dealing with obtaining the philosophical consequences of indeterminacy as far as a general picture of physical reality is concerned.

CAUSALITY AND INDETERMINISM

Undoubtedly the first problem to raise is that of the causal content of the new indeterministic category. This comes directly from the origin of the controversy on indeterminacy. At the beginning, the causality principle was mistakenly identified with determinism, in such a way that the fall of the latter meant for many people the breaking of causality. This brought about the fact that many physicists and philosophers rejected indeterminism since it would involve a full loss of causality, and, consequently, of the possibility of scientific knowledge. By accepting indeterminism, it was frequently argued, the order of nature was eliminated and only chaos remained. In this way, it was concluded, physical objects as possible objects of scientific knowledge disappeared. But this line of reasoning believed that the only possible way of understanding physical causality was the deterministic category. This was precisely its great mistake, because indeterminacy does not mean any sort of physical chaos nor a lack of order in reality either.[5]

The physical order established by indeterminism is probabilistic. It is true that a statistical statement cannot exactly determine the physical situation of a single case: it only refers to average values of such a situation. But this does not mean that scientific activity using statistical laws does not increase the growth of our knowledge on reality. Physical order defined by classical deterministic laws was a linear order, that is to say, there was the possibility of describing the causal line from a physical "point" to another physical "point", both being fully determined. Nevertheless, this way of understanding the causal development of phenomena is not the only possible one, and quantum physics could show that other ways were possible. All of them logically contributed to the growth of our knowledge. Moreover, this was concluded without rejecting the philosophical content of the ontological principle of causality, that is to say, by maintaining its basic meaning. Briefly, the proof of this assertion would rest on the fact of the proper historical development of quantum mechanics. This growth would have been impossible without the acknowledgement of the causality principle as a precondition of scientific knowledge.

In this manner, causality in quantum physics involves another way of understanding physical objects, which is very different from Newtonian physics, but it is always situated within the limits of the ontological principle for general causality. In a few words, microphysical indeterminism provides an explication of phenomena as long as it points out the possible reasons for the occurrence of such phenomena. These reasons refer to the average level of physical situations but not to a single case. If such reasons – derived logically from physical laws – are responsible for[6] the occurrence of phenomena, it seems clear that the conceptual nucleus of the causality principle is then maintained. Thus, in indeterministic causal processes, what is especially defined is the probabilistic frame within which the phenomenon considered has to occur, this frame and its limits being objectively determined by any scientific observer. Values of physical magnitudes referred *simultaneously* to the uncertainty parameters may be basically indeterminate, but this set of values is not indefinite; on the contrary it is ruled by mathematical laws which indicate an objective frame of probabilities. On that account, probabilistic order does not involve any chaos or full indetermination, but the existence of a physical order with respect to the parameters defining quantum systems as a physical entity.[7] Briefly, quantum phenomena are not absolutely indeterminate but can fluctuate within a definite frame of probabilistic values. This being so, we possess a probabilistic causal order and not an absolute disorder.[8]

Accordingly, by accepting indeterminism as a possible pattern of causality, that is, as a logical consequence derived from the manner by which a physical object is defined in quantum mechanics, it is possible to conclude:

- indeterminacy points out different possibilities of physical evolution of phenomena, but at the same time, they are quantitatively determined by statistical laws which define an objective frame of physical occurrences;
- this way of defining matter maintains the basic meaning of the causality principle to a full extent, such as the existence of a physical order and the effective possibility of our scientific knowledge about it.

INDETERMINACY AND DYNAMICISM

Traditionally, the physical world has been characterized, from an ontological viewpoint, as motion or change. Thus, for example, in Aristotelian philosophy – qualitative change –, and in the modern scientific outlook as well-quantitative or extensive motion. In both cases physical reality has been conceived of as a dynamic world, that is to say, dynamicism being one of its essential categories. It is now convenient to introduce an important variation in the conception of physical dynamicism, as it has been carried out by modern physics.

To begin, dynamicism in Newtonian physics occupied a secondary level in comparison with the ontological significance of matter. I mean that physical substance was ontologically identified with matter, which, in turn, was conceived of as a corpuscle. Motion (physical dynamicity in this case) was a fundamental quality but matter-substance maintained its ontological supremacy in relation to motion. This can be easily observed if we consider the concepts of matter and energy, which were understood in the former physics in a very different way from that of present-day physics. Indeed, the concept of energy, as much in general relativity as in quantum mechanics, has been situated at the first level of importance as far as the rationalization of the physical world is concerned. Furthermore, by considering every physical entity as energy, this conception has involved an essential dynamicism conceived as a mode of being. So, as a result, this category has now become the basic quality of the physical world. Likewise, as a consequence of the Einsteinian formula relating mass and energy, the role of the former is convertible into the latter. The fact then is that the old ontological duality, matter and energy, is practically eliminated. Indeed, a substantial equivalence is now established between both, the main result of this new situation being that dynamicism is no longer a property of matter but the condition of the possibility of matter from an ontological perspective. This is a decisive change. There is not any kind of priority of matter over energy, as in Newtonian physics. And in practice, energy has a greater relevance than matter. In this line, energy, that is to say, the physical expression of dynamicism as an ontological category, becomes the basic element to philosophically understand the physical world. It implies that dynamicism has to be conceived as the primary nucleus of every physical reality. Not only can dynamicism be considered as an important feature of matter: it is matter, as a physical being *qua* physical, that is, dynamicism.

Moreover, this change is also strengthened by indeterminism. In this case indeterminacy has not caused this situation but contributed to it. This is easy to understand because, according to an indeterministic view of the physical world, the ontological priority of matter as a corpuscle is especially doubtful. In fact, the corpuscular duality of physical entities, considered as their basic feature, is to a certain extent tempered because of the consequences of quantum uncertainty. Or, in other words, it deals with the ontological priority of field theories over corpuscular theories. It is true that corpuscles are yet a basic epistemological instrument of scientific research, but what we elucidate now is whether the notion of physical reality as a corpuscle maintains its old ontological supremacy. In general terms, the answer is negative. The notion of physical field has become the first ontological reference, from which corpuscular entities can be derived. Quantum field theories are an example of it. Notice that it is not asserted that corpuscular entities as such are irrelevant; this would be to misunderstand our statements. What we assert is the change of ontological relevance between both, corpuscle and field. In this sense, indeterminism contributes, in an indirect way, to the conceptual equivalence between matter and energy, and as a result, to the conception of physical dynamicism as the ontological condition of physical being *qua* physical.

Finally, it is necessary to note here that together with the emergence of indeterminism, dynamicism of physical reality acquires a philosophical relevance much greater than in Newtonian physics. Both bring about a change of paradigms. Logically, all this presents a new picture of the physical world through a change of philosophical categories: from determinism to indeterminacy, and also from dynamicism conceived as a physical quality to dynamicism as a condition of the possibility of physical being.

INDETERMINACY AND PHYSICAL INTERACTION AS A MODE OF BEING.

The transition of physical dynamicism and indeterminism to the first place of importance implies a fundamental consequence for a philosophical picture of the physical world. As a matter of fact, the mode of being of physical entities is that of interaction. So, physical reality is ontologically determined as long as it is in connection with another physical entity. Indeed, any physical being, considered in isolation, would not hold any consistence nor could it be conceived as an object of scientific knowledge. What is important to remark here is the fact that it originates from dynamicism and change as ontological modes of being. In this respect, the present situation also differs from the conception of the former physics. Physical entities were then considered self-sufficient from a substantial viewpoint, but now, on the contrary, physical being is conceived as such if it interacts with another physical reality. Thus, interaction becomes an intrinsic condition for the ontological permanency of physical entities. In the final analysis it means that physical being is to be understood as a reality which is essentially open, that is to say, wanting a constant change in order to remain as such a reality. So, to change in order to go on being would explain its peculiar ontological way of remaining.

In turn, this interactionism brings about another important quality. And that is that, as a fundamental category of philosophical understanding, the category of relation must be used, one whose ontological significance is now greater. This might well define the change of paradigms in respect of the former physics. Certainly, because physical reality is now understood through a category which is quite different from that of substance or extensiveness. A physical entity is now conceived of as something ontologically relative to another one, this last also being relative to something else, and so on. Of course it is not the case that extensiveness or other classical categories disappear but only that the category of relation goes up to the first place of philosophical reflection; and, above all, it is from this category that the picture of the physical world has to be elaborated. In this sense, it is possible to state that this situation carries on the change of philosophical paradigms, parallel to the change of the scientific ones, regarding the former physics and philosophy. Up to now this category had been considered as very secondary with respect to others such as substance or entity. In this manner, indeterminacy, interaction and relation, do constitute three philosophical concepts that are now implied in one another. Note that the case is not only that we take for granted that any physical entity, conceived as a basic interaction, necessarily refers to the category of relation; but also that this one is considered as the ground of the ontological structure of the physical world, and therefore, of its possible scientific intelligibility. In other words, in classical terms, we would assert that the category of relation would be introduced into the essential nature of physical reality. This new outlook especially differs from the ontological conception underlying the former physics, because the relation among physical entities was then imposed by an external agent – force and motion –; but now relation is to be considered as fundamentally belonging to the essence of this reality, and therefore, as responsible for its ultimate reason.

INDETERMINACY AND OPEN SYSTEMS

As a logical consequence drawn from the ideas expounded up to now, it is necessary to assert that physical systems are to be conceived as open systems. This means that a significant number of scientific variables defining a physical system have to reflect this feature. These variables are those which determine the system as such, that is to say, the variables responsible for the physical system *qua* physical. In this sense, the line of demarcation between structural and dynamical variables is practically eliminated. In fact, the first ones will be reinterpreted in the terms of the second ones. Such is, for instance, mass with regard to energy at the level of elementary particles.

Indeterminism just develops this essential form of opening in physical objects since it makes their evolution possible through physical change. Indeterminism, in allowing more than one possibility for the evolution of physical systems, also allows that the development of physical reality can provide true novelties and not only a series of predetermined traits precontained in the starting-point, as happened in Laplacean universe. In this one there was not any evolution proper, so to speak, but only a development of pre-existent features throughout the mathematical linear time. In this way, time was reduced to an extensive variable, its mathematical values being well-known. With respect to the picture of the physical world this fact led to the implicit elimination of time.[9] The evolutive feature of the physical universe, which is made possible by indeterminism, carries out a recovery of time as a flow in physical phenomena which in turn are no longer fully predetermined. So, it renders possible the emergence of new entities according to the evolution of time.

Finally, this idea leads us to underline a quality of the new possible picture of the physical world which, derived from the present-day scientific situation, may perhaps be the most important difference as regards the Laplacean picture. It is the emergent condition of a physical universe conceived under the philosophical categories analyzed up to now.

INDETERMINACY AND EMERGENCE

The emergent quality of the physical world has been repeatedly shown from many different standpoints in the last years. It appears as a reasonable consequence of the previous ideas.[10] At this stage however it is necessary to specify its most relevant features in order to avoid some misunderstandings that may originate a "metaphysical" conception of the physical world.[11]

In a first approach, physical emergence synthesizes the above qualities derived from the change of paradigms in physical theories. So, the emergent condition of the universe would involve the following structural features:

- physical change as the main mode of being of physical entities,
- the ontological necessity of evolution in every physical system, and
- the possibility of the appearance of novelties in the physical world – that is to say, new elements which were not precontained nor fully predetermined.

These features become necessary conditions of a physical emergent world, but are not sufficient conditions. It seems indispensable to add something else. This deals with the fact that this possibility of appearance of novelties must come into reality, or in other words, it has to be not only a theoretical possibility but an *actual* fact. So, the realization of novelties – not only the theoretical possibility of them – should be the sufficient condition. The reason is clear: only such a fact assures the validity of the new outlook involved by the change of scientific paradigms. It constitutes its specific testing. Emergence of new elements, in the sense of a relative dependence – not absolute – on the former physical situation, can be attested throughout cosmical evolution.

Thus, for instance, the emergence of life and the appearance of man and human language.

Nevertheless, it is convenient to remark here that an emergent universe is not a creative universe in an absolute sense. This would implyg going beyond the limits of a proposal inspired by scientific theories and their change of paradigms. In this sense, it is important to clarify that a creative universe would refer to a metaphysical level without any connection with science. On the contrary, to assert an emergent quality concerns a general category synthesizing some structural features derived from the present-day scientific situation. The notion of a creative universe would connect to a kind of knowledge quite different from the scientific one. Thus, the emergent condition maintains the connection with the scientific viewpoint to a reasonable extent. It would not even be possible to think of its creative quality in a metaphorical sense, as some authors do,[12] because in any case it leads to some contradictions as regards the strict metaphysical sense of this term. All of which would imply mistaking the emergent condition for another very different feature - for instance, the world as a whole intended as a living substance-. Emergence and creation are not the same thing.

Finally, it is convenient to show the relevance of indeterminacy for this emergent quality, indeterminism would just be the basic condition of the possibility of emergence, indeed, because without indeterminism in physical processes the emergence of novelties would be impossible. Without indeterminacy all things would be precontained and predetermined in the past, time being a mere extensive variable but not means of explaining emergence. And now it is very important to remember again that indeterminism and emergence do not involve any chaos in the physical world. Instead they establish a general order, mathematically ruled by statistical laws. In this way, emergence would be intended as the logical consequence of indeterminacy, thus becoming one of the main philosophical categories for a picture of the physical world.

CONCLUDING REMARKS

Every former philosophical category, conceived as general features of the physical world, develops an indeterministic rationality which is undoubtedly able to face the quest for a picture of physical reality. In this aspect, it is not hard to accept that indeterministic rationality has overcome the deterministic one. Indeed, because it has been possible for indeterminism to establish a way of understanding physical reality, it leads without contradictions to a reasonable picture of the physical world. So, just as determinism originated a well-defined conception of the world, indeterminism nowadays fulfills the same task by demonstrating that some other world picture is actually possible which is very different from the former one. Like determinism, this new picture is also based on scientific theories and likewise develops a proper rationality without incoherences. This means that indeterminacy, as a rational category derived from the current scientific situation, really possesses as much power of explanation as the previous ones and satisfies the heuristic necessities for rightly conceiving a new picture of the physical world. Accordingly, it is not only determinism that fulfills the demands – both scientific and philosophical — of a conception of physical reality. So does indeterminism.

All this implies the specific acknowledgement of indeterminacy as a way of understanding the physical world. The first test it had to overcome was that of causality: it was necessary to show that the indeterministic paradigm did not destroy the general principle of causality in nature, but on the contrary it did indicate a new pattern of the ontological principle of causation, which obviously maintained all its requirements. However, this was not enough, since it was also necessary to show that indeterminacy became profitable from an epistemological viewpoint. This aspect was especially important because it involved the acceptance of its possible conceptual efficacy for encompassing a picture of the physical world. Were it not so, indeterminism would have been eventually just an empty category without any significance.

Of course this new world picture, derived from indeterminism, involves a level of abstraction which is much greater than that of the Laplacean picture. This implies a loss of intuitiveness in our conception of the world. But it is the unavoidable consequence of the scientific ground it rests upon, because this very abstract process has taken place in scientific theories as well. Evidently both quantum mechanics and general relativity do overcome our ordinary intuition of the world. The Laplacean picture, on the contrary, is easily imaginable by our human intuition. On that account, it is not hard to understand that any picture of reality derived from those scientific theories brings along some categories of a higher abstractive level. It is an usual consequence of the historical evolution of scientific knowledge. But that is just a corroboration of the growth of our knowledge.

University of Sevilla, Spain

NOTES

[1] An interesting example is Popper. See, among other titles, Popper, K. R., *Quantum Theory and the Schism in Physics* (Hutchinson, London, 1982); *The Open Universe. An Argument for Indeterminism*. Idem. As is well known, Popperian philosophical thought on the physical world is based on indeterminism, which is considered the primary feature of reality.

[2] An interesting contribution to this subject is that of Kanitscheider, B., *Von der mechanistichen Welt zum kreativen Universum* (Wissenschaftliche Buchgesellschaft, Darmstadt, 1993).

[3] As a corroboration of this relevance, we could quote, for instance, Schommers, W. (ed.), *Quantum Theory and Pictures of Reality. Foundations, Interpretations, and New Aspects*. With contributions by B. d'Espagnat, P. Eberhard, W. Schommers and F. Selleri (Springer, Berlin-New York, 1989).

[4] In addition, indeterminism in quantum mechanics has received an important corroboration in the last years. See, among others, Aspect, A., P. Grangier and G. Roger, *Physical Review Letters* **47** (1981), p. 460; **49**, p. 91 (1982). More recently: Englert, B. G. and M. O. Scully, 'Quantum Optical Test of Complementarity', *Nature* **351**, **6322** (May 9, 1991), pp. 111–6.

[5] The bibliography about the causality problem in modern physics is enormous. We can only cite the following titles: Mackie, J. L., *The Cement of Universe: A Study of Causation* (Clarendon Press, Oxford, 1975). Wallace, W., *Causality and Scientific Explanation*, 2 vols. (University of Michigan Press, 1972–1974). Bunge, M., *Causality and Modern Science* (Dover, New York, 1979) (3rd. rev. ed.).
[6] Or if you prefer, "explain" phenomena.
[7] In this respect Prigogine's works are especially meaningful. In them he shows the epistemological significance of statistical statements. See for instance: *From Being to Becoming. Time and Complexity in the Physical Sciences* (Freeman, San Francisco, 1980). On the philosophical consequences of Prigogine's ideas see our work: 'El nuevo ordentiempo. En torno a la obra de Ilya Prigogine', *Thémata*, 1995, Special Issue titled: "La filosofía de los científicos".
[8] As an additional reference about this problem see my articles: 'Significación filosófica de la causalidad en la física contemporánea', *Anuario Filosófico* **X/2** (1977), pp. 145–70; 'Indeterminismo y racionalidad. En torno al problema de la causalidad en física', *Sapientia* **XLIII** (1988), pp. 69–84.
[9] See again the critique of Prigogine of this conception. For example in: I. Prigogine et I. Stengers, *La nouvelle alliance, métamorphose de la science* (Gallimard, Paris, 1979).
[10] Along this line, together with Popper's and Kanitscheider's works mentioned above, we could now add those of a very different philosopher, M. Bunge. Cf. *Treatise on Basic Philosophy*, vol. 4, *Ontology II: A World of Systems* (D. Reidel Pub., Dordrecht-Boston, 1979). Emergence seems then a well-established trait of the physical world at the end of the century.
[11] "Metaphysical" should be understood here as referred to the statements on the physical world ignoring the results of scientific investigation.
[12] This is the case of Popper. Cf. his book: *The Open Universe. An Argument for Indeterminism*, o.c.

FRIEDRICH RAPP

THE DYNAMICS OF SCIENCE

I. ANALYTIC SEPARATION

Objectively, there is only one world. But this world is subject to change and manifests itself in the most varied forms. This objective variety is perceived *subjectively* in many different ways by the epistemological subject. The point is which epochs, cultures, social groups, scientific disciplines and individuals perceive and intrepret this world, whether collectively or individually. For this reason it is sometimes said metaphorically that individual people and the different scientific disciplines – to keep to our examples – live in quite different worlds.

In this world everything depends on everything else. There really are no totally *isolated states or processes*. However one partitions the characteristics, states and dependence relationships of the world, the phenomenon or event which is being observed at any given time is related to other elements in a more or less dense network of causal and conditional connections. Thus, the most disparate events can be shown to have a causal connection, however tenuous and indirect it might be.

This world, every part of which is linked, is also *structured*. It is neither chaotic (it has observable regularities) nor amorphous (however varied the phenotypes may be). Nevertheless, it is not homogeneous (since there are generically different phenomena such as, for example, physical occurrences and historical processes). The structures of the world are intrinsically independent of mankind's epistemological exploration. This notwithstanding, they are always seen from an individual point of view, that is, they can be only perceived and grasped in a specific and distinct manner. This applies to sciences as diverse as physics and chemistry on the one hand and psychology and sociology on the other.

The problem of the epistemological process for both science as a whole and for individuals is that it has a *subjective* as well as an *objective* component. The subjective dimension consists in the spontaneity of the intellect (in other words, man's conceptual creative freedom) to choose certain questions and the hypotheses, models, paradigms and theories which constitute a certain heuristic view of the world. Without these intellectual approaches the world would remain mute. However, the object under investigation does not easily yield to the researcher's epistemological probing and exacts a price for surrendering its secrets. Every theoretical approach has to be tested empirically, it can and must be corrected and sometimes rejected on the basis of the data obtained.

Reality, the "thing-in-itself", to use Kant's terminology,[1] is characterised by two basic features: (1) It gives concrete answers to the specific questions posed

D. Ginev and R. S. Cohen (eds.), Issues and Images in the Philosophy of Science, 263–274.

by experimental research. (2) Its traits cannot be exhausted by means of a single conceptual scheme; there is always the possibility of a different understanding, of fresh approaches, and of more extensive knowledge. Without reference to this reality all scientific knowledge would be pure fantasy, it would be impossible to distinguish a scientific treatise from a science fiction story. And obliterating the distinction between the knowledge actually attained and the inherent, partly still unknown nature of the object investigated would bring history of science to a definite end and exclude any further progress in the understanding of nature. In short, we can and we do attain knowledge of reality, but this knowledge is never exhaustive.

If we restrict our analysis to the various scientific disciplines, it becomes clear that each of them concentrates on a specific aspect of reality, just as a headlight illuminates a specific area of whatever object is being examined or the way that the same landscape appears in a different perspective from different points of view. The picture changes but the object itself remains the same.[2] In order to correspond to the idea of omniscience one would have to activate all the different perspectives simultaneously – something only possible for a divine *intellectus infinitus*. Cubism was painting's attempt to activate different points of view simultaneously.

The unity of the world does not only reveal itself in the concrete, real conditional and causal connectedness between different events and phenomena.[3] Since all forms of knowledge refer to and depend on the real, concrete world, this unity must become evident on the *conceptual, theoretical* level as well. After all, each of the scientific disciplines and approaches, no matter how different they are from each other, relate to the one and same world in the final resort. Of course, there are different understandings and interpretations which determine the way in which the individual and the community understand themselves and the world. However, the latent assumption – generally not consciously articulated – of all these different ways of viewing the world is that they all relate to this one world, which has a place for every object contained in it.

This does not imply that it would be possible to conceptualise the unity and connectedness of the world in a unique and exhaustive way. The "thing-in-itself" possesses more features than we can describe by means of any given approach. To overlook this feature would amount to committing what one might call the idealistic error *esse est cognosci*, a mistake reminiscent of the empiricists' dictum *esse est percipi*. In both cases we encounter the same epistemological mistake which consists in presuming that the existence, as well as the traits, of the object perceived depend on the subjective act of sensory perception or of intellectual understanding. Yet we are not omnipotent. We are free to decide which aspect of the object investigated we will consider more closely, but we are not in control of what we discover; we simply have no choice but to accept its existence.

It is by means of categorisation, that is by conceptually distinguishing objects and the relations obtaining between them from one another and by appropriately classifying them, that we analytically structure the world in theoretical terms. However, this step is a mental act, an intellectual affair. It is a distinction made by men to optimise the pursuit of knowledge in terms of a specific intellectual interest, an interest which might, in principle, have been different.[4] The possibilities and limits of a certain cognitive perspective become clear if one takes seriously the inevitably arbitrary character of all conceptual differentiations and, moreover, bears in mind that it is only through a certain heuristic concept – in the language of the liberal sciences through a hermeneutic project – that any knowledge whatsoever can come about. At this point it becomes clear that all complex scientific investigations which go beyond a specific well-defined context cannot be viewed as true or false in an elementary sense. Rather, they must be regarded as a whole and judged on the basis of their classificatory force, their explanatory power and the heuristic benefits of the concepts applied.[5]

II. ONE SCIENCE AND MANY ASPECTS

These basic and admittedly general considerations are also valid when it comes to dealing with the comparatively special topic of the natural sciences. For the social sciences and liberal arts similar considerations can be made, although details will have to be altered.[6] The most significant factor is an obvious and apparently trivial fact, which is nevertheless of decisive importance; like other aspects of reality the natural sciences taken as a whole exhibit quite different but nevertheless intrinsically coherent features: The natural sciences are the result of a collective, social action, they represent a cultural achievement, they are an historical phenomenon, a system of knowledge, the consequence of theoretical approaches, the result of applying methodological principles, the outcome of usually tacit philosophical assumptions, and an economic factor.[7] In reality these differing and, in their own way, productive approaches have as their object the one and the same essentially indivisible phenomenon of the natural sciences. These different approaches are not independent, but they are necessarily partial in the sense that they can never encompass the whole of natural science. It is only by adherence to the ideal of taking into account or at least considering all of these approaches – an ideal that, as was pointed out, can never be completely fulfilled – that one can hope to arrive at a view of the phenomenon of natural science that is as comprehensive as possible.

Dealing with a quite specific aspect inevitably implies a simplification and a one-sidedness. But this one-sidedness is, as it were, the price which has to be paid to make possible an in-depth understanding. If one wished to grasp details, one has to look in a quite definite direction; this necessarily excludes other ways of dealing with things which in principle also could have been chosen.

If one considers the approaches (social action, system of knowledge, application of methodological rules etc.) in terms of which one can investigate the phenomenon of science more closely, it becomes evident that it always would have been possible to choose other, complementary conceptualisations and taxonomies, each offering their own specific approaches with their specific insights. In each individual case the epistemological perspective chosen isolates and partitions off certain issues and relations from the total context and thus makes their investigation possible. The epistemological perspective chosen opens up a specific avenue of understanding. But by deciding in favour of a certain point of view, and by the marginalisation of others deemed irrelevant and unimportant, one necessarily excludes other approaches which might also have been possible. Defining a thing, that is saying what it is, at least tacitly implies saying what it is not. Spinoza's "omnis determinatio est negatio" puts this in a nutshell.[8]

III. WAYS OF DIFFERENTIATION

Some examples may help to clarify the different approaches which can be used when natural sciences are being investigated:

1. An elementary, basic distinction which suggests itself is that between the *object investigated* and the *research process*, between the cognitive content and the social activity, between knowing and doing. Both aspects belong together inseparably. Like all human activities the collective research process (and by implication the activities of the individual researchers involved) are embedded in the flux of time. Every scientific endeavour – interpreted as social activity and as a process of gaining knowledge – is a historical phenomenon which takes place at a certain historical moment, has its roots in previous developments and will in its turn influence the future. As a temporally located, collective, cognitive, and social activity science nevertheless – though in a form which changes over time – refers to one and the same subject investigated and to the lawful patterns it exhibits.

2. The *temporal dimension* offers another possibility of differentiation. Let us consider the historical development of science. In this case the Cartesian co-ordinate system can serve as an illustration. If we represent the flow of time through an arrow (the x axis) and if we place intersections through this arrow by means of an orthogonal plane spanned by the y and z axes which then contain by definition everything which takes place at this moment of time, then we obtain the following two pictures. The aggregating, *synchronic* point of view includes everything which takes place at a certain instant (in the y–z plane). Adding up these individual snapshots along the x axis then ideally provides us with a continuous *diachronic* representation of the whole of the scientific process.

3. This scheme is reminiscent of distinctions between *historical* and *systematic* approaches but is not identical with them. The difference is that the systematic approach concentrates solely on the structure of the *subject matter*

in question (as considered independent of time) whereas the investigation of the development of science deals with the *historical development* of the scientific notions involved. In the case of a systematic analysis the issues are of course, too, embedded in the historical context. But they appear as they do in Hegel's speculative philosophy of history simply as instances, as specific manifestations of an all-encompassing historical process. At this point one is confronted with the difficult epistemological question of the relationship between the historically variable conceptions of natural laws on the one hand and the immutable, timelessly valid lawful patterns of the natural world "in itself" on the other.

4. H. Reichenbach developed the concept of a distinction between the subjective, psychologically explicable *context of discovery* and the systematic *context of justification.*[9] According to this view only the latter is relevant since it determines the power of a theory to explain the phenomenon investigated. In this case, the determining historic setting is reduced to social and psychological "chance" circumstances which are of an accidental nature in the process of gaining specific scientific knowledge, but have no systematic relevance. Here too there is a link to the distinction between the historical and the systematic aspects. But in Reichenbach's understanding the time dimension is only relevant for the individual epistemological process. If one were to generalise the thesis that the context of discovery is scientifically irrelevant this would mean that the history of science would be completely irrelevant for the natural sciences – a point of view which would not find wide acceptance. Such an exclusively object-related interpretation which neglects the historical dimension would be the opposite pole of a historical and social contextualisation which neglects the objective dimension.

5. In contrast to Reichenbach, Husserl assigns constitutive significance to the concrete process of gaining knowledge – or more precisely to each *cognitive act*, which by its very nature is unique and individual, but at the same time aims to reveal essential and universal traits of the subject matter considered. He sees an inseparable connection between the cognitive act directed towards a certain object and the intended cognitive content. Of course it is true, as Husserl emphasizes, that the distinction between cognitive act and its content is of a purely abstract and analytic nature. A concrete insight which I actually have is always an inseparable unity which is simultaneously the realisation of an action and the appropriation of the content targeted by the cognitive action. The object – in the case of the sciences, the regularities of the physical world – is for me only recognisable through a cognitive act and this cognitive act is always so structured that it is directed towards an object understood in the broadest sense of the term.[10]

6. By adding a further facet we obtain Popper's distinction between (1) the psychological act of thinking, (2) the content of the thought as understood by individual researchers, and (3) the *third world* of knowledge gained by

the collective process of research.[11] The introduction of a suprapersonal, reified world of knowledge that is independent of the individual scientist goes beyond Husserl's analysis. Popper regards the research results collectively gained as a reified entity independent of the subject. The concept of *mentalité* discussed by French thinkers (also from a psychological point of view) and the *objektive Geist*, which plays such an important role in Hegel's philosophy, are reminiscent of this area.

7. The distinction between an *internal immanent logic* of scientific change on the one hand and *external factors* outside the scientific process on the other hand is another dual categorisation which concentrates on the aggregated research process and neglects the historical dimension. This distinction, which originates in sociology, implies that there is an inherent logic which fosters scientific progress, a tendency which is fuelled solely by previous developments and by scientific curiosity to resolve the open questions. From this point of view those elements of scientific change which are not due to the cognitive drive of science are defined *ex negativo* and regarded as merely external.

However, this separation between the cognitive content of scientific knowledge and the motives and actions of individual and collective actors who accumulate this knowledge is by no means unambiguous. In reality there is not a strict exclusion relationship; the boundary between the cognitive content and the individual actions is permeable. This becomes clear when one realises the extent to which scientific topics are controlled by financial factors or by politically determined priorities. Such external factors do not directly influence the cognitive structure of scientific statements. Rather, their influence is of an indirect nature because specific questions and areas are given preferential treatment *in the interest of certain applications*, while other possible research projects fail to materialise because financial or political support is missing. At this point we should dispense with the illusion that *in basic research* the provision of the appropriate financial resources will lead to corresponding research results (which perhaps are proportional to the investment). In the area of applied technical research and development such assumptions may be justified. However, by its very nature, basic research eludes predictability.

If one attempts to summarise the different aspects schematically, so that the points mentioned are given their due, four distinctions emerge.

- From a *systematic* point of view one should distinguish between the research activity (the gaining of knowledge) and the research results obtained (the system of knowledge).
- From a *historical* point of view a synchronic snapshot of what is taking place at a given point has to be contrasted with the diachronic examination of the whole process.
- For the *individual person*, that is, the individual researcher, the cognitive

act comprises the activity, the process of cognition as well as its content, namely the knowledge gained. Taken as a *collective social enterprise* research is characterised by division of labour, so that the aggregated result in the form of general knowledge can be abstracted from individual cognitive acts.

- The *internal* cognitively and methodologically determined *logic of research* can be distinguished from *external influence* from the side of society (politics, finance). In terms of this analysis, the technical application and economic use of scientific research amounts to a phenomenon external to science.

The crucial insight is that all the above-mentioned distinctions have their factual justification without anyone of them exhaustively explaining the phenomenon of science taken as a whole. Clearly, a finer differentiation would produce further distinctions. One thinks of, for instance, the sociology of science (which focusses on the social roots of research activity), the politics of science (which highlights political influences on the scientific process), the history of ideas (which concentrates on the general theoretical background, especially philosophical ideas), of the publishing apparatus (encyclopaedia, periodicals, text books), which have a decisive impact on the propagation of the state of the art in the sciences.[12]

IV. PREPROGRAMMED GROWTH

As H. Plessner has put it, one of the historical laws is that mankind only discovers what it is doing from the consequences of its actions.[13] This expresses in concise form that whatever goals and expectations historical actors may have, the actual outcomes of their actions will not correspond to what they hoped for. In an altered form this statement is to a certain degree also valid for the history of science. The details of the results of basic research, which by definition are unpredictable, are a particularly striking example of this.

Nevertheless, it is undeniable that the development of natural science in modern times – notwithstanding the paradigm change theory put forward by T. S. Kuhn – is characterised by general, comprehensive tendencies. As far as we can see, these tendencies which are determined by the inherent structure of the scientific research process will continue to hold good in the future. One might even go so far as to claim that the very nature of modern natural sciences is such that, all things considered, it contradicts Plessner's thesis. In the case of the natural sciences further successes are predictable from the internal logic of the scientific enterprise. It is precisely for this reason that modern natural sciences and technology are usually seen as the guarantors of future progress.[14]

The growth potential of the natural sciences can be explained by ten factors. These factors guarantee the empirical content and the practical and technical applicability of the research results. Furthermore, and that is the most important point for the context examined here, in this way a further growth, an increase in scientific knowledge (and hence also of technical applicability) is structurally pre-programmed.

1. The empirical basis of modern scientific knowledge is ascertained by means of *experiments*. The common feature of these experiments is that they – in contrast to the unique, non-recurrent events in history – are by their very nature structured in such a way as to be verifiable and reproducible. In this way every researcher, independently of his historical, cultural and social context can test the validity of the results which his colleagues have gained under well-defined conditions and of course also makes use of this possibility. Since every scientist knows that the experimental results are critically judged by colleagues and if necessary empirically tested, he sets himself high standards from the very beginning. Negligence and fraud are to a great extent excluded. The negative sanctions for a researcher who is caught doing imprecise work or engaging in deception consist in the loss of reputation and thus the loss of the possibility of continuing to work in the same field. In this way a decent, scholarly code of conduct is imposed by the very rules of the scientific community; there is no way of profiting from unethical conduct.

2. Every experiment consists in using appropriate instruments and apparatuses in order to realise the experimental design (which has been suggested by theory) and to make accessible the observational data arrived at. In this context the distinction between science and technology is an arbitrary one. In a more general sense *every experiment* which makes use of physical, material resources *represents a technical process* and, in turn, every technical process can be interpreted as an experiment. Modern technology exploits scientific findings by following Bacon's methodological principle of dominating nature by obeying her laws. Every causal relation which is systematically investigated through scientific research and thus is known, can also be made use of technologically.

3. The *mathematical presentation* of the results gained is a *conditio sine qua non* in modern science. In the case of technical applications the precise quantification of the variables under scrutiny provides the possibility of exact prognoses and of optimisation of measurement. Summarising, one could therefore describe modern science as experimental, technical and mathematical.[15]

4. As pointed out in section I, scientific research is always coloured by the questions raised, the basic conceptions which are used and the theoretical premises made. The circumstance that in modern science questions, concepts and theoretical assumptions, whether implicit or explicit, were and are so selected that the criteria of reproducibility, of experimental/technical production and of mathematical quantification are fulfilled is decisive for the history and future of scientific research. Due to these features there is from the onset a guarantee that *future scientific research results will be technically useful* – provided they are considered worthy of economic exploitation.

5. The *sciences learn from their mistakes.* As a child of the Enlightenment science has the basic tendency to be empirical and to expect unexpected results and, if necessary, to correct previous conclusions. The rejection of the ether hypothesis and the rethinking occasioned by the introduction of the quantum and relativity theory bear witness to this. If science remains willing to learn and is faithful to the procedure of trial and error, it is, methodologically speaking, infallible in the sense that it cannot be taken by surprise. For, hower unexpected the results turn out to be, due to the procedure's flexibility science will always be able to claim them as evidence of success.

6. The extremely impressive successes of modern science with both its comprehensive theoretical conceptions and its detailed results would not be possible without a well-developed *division of labour and the specialisation* implied. In contrast to artistic achievements, which are always the work of individual geniuses, the achievements of science are the aggregated and accumulated result of a co-operative effort which draws on all the previous stages.

7. A further feature is the close, systematic connection between the different areas. In spite of the inevitable narrowing of the field of vision typical of all specialised research, the complete resources of the multiplex network of science are at the disposal of even the most specialised researcher. In principle, the individual researcher as well as a specific field can always draw on the store of all the scientific knowledge available at a given time. The *synergetic effects* thus gained make an important contribution to the progress of science.

8. International division of labour and specialisation are made productive by the free, publicly accessible *publication* of the results obtained and the world-wide *exchange of information.* Furthermore, the institutionalised mechanisms of communication and of general, public criticism make visible to all concerned the advantages and disadvantages of a certain conception, its weak points and its heuristic potential.

9. The methodological principle of starting from whatever happens to be the most advanced position ensures permanent growth. Everything which is discovered by research is based directly or indirectly on what predecessors have achieved. At any given point in time researchers are standing on the shoulders of their predecessors. Because they can start at the high level which has already been achieved, a further increase in knowledge is assured by the *principle of accumulation.*

10. The basic, implicit agenda for every scientific research project stipulates that the results obtained have to be *integrable in the state of knowledge already achieved.* If they are not integrable from the onset they will be reshaped so that they are compatible with the hitherto system of sciences – as happened to quantum theory. Thus the number of permissible questions is limited from the outset. They may – possibly through introducing aesthetic or ethical

ideas – be such that they do not fit into the conceptual and theoretical context of science.[16]

The ten points mentioned form epistemological, methodological, organisational and institutional prerequisites to which the researcher will have to comply with if he wishes to be successful in his work. Taken together, they guarantee further progress for the enterprise of modern science as a whole, a progress which, in the final resort, is limited only by the structural laws of the physical world.[17]

At first sight these prerequisites seem to be purely theoretical, methodological and thus to refer only to the cognitive sphere. But in reality the efficacy of these prerequisites depends on their activation by reciprocity, by world-wide cultural exchange and by public criticism, that is by cultural activity and by collective actions. At this juncture cognitive and social factors fuse indissolubly. Everything which happens in the cognitive sphere is, on the one hand, connected to the motives, intentions and actions of the researchers, that is, to the behaviour of the actors which is conditioned by the structure of the social system of science. On the other hand, the research process also depends on the methodological principles applied, and it has to adjust to the 'objective' features and the inherent patterns of nature.

V. CLOSED CIRCLE

From the point of view of chronology one can distinguish three stages in the relationship between the cognitive and the social dimensions of science.

1. There is the social transmission of the cultural (including the cognitive), as well as of the material, heritage. It is this heritage which constitutes the *input* of the scientific system at a given moment.
2. Within the actual research process the interrelation between the cognitive and the social sphere leads to the patterns of inherent *pre-programmed growth* pointed out in the preceeding Section.
3. The *output* of the scientific system consists in the cognitive results obtained in the research process; and these results are open to practical, technical application.

Nowadays these features are the most important social aspect of science. Society is as it were the consumer of technically applicable and economically useful research results.

The outcome of all this is the much discussed scientific-technical-industrial complex. A simplified explanatory scheme would depict a spiral movement between a state of life at a given time, the research results provided by science and the new, transformed state of life based on the new knowledge. This spiral takes as starting point an existing social form including its scientific achievements. The technical application of these achievements leads to a new state of society, the new state usually being interpreted as a more progressive one.

This general spiral movement obtains in the ideal as well as in the material sphere. As Husserl was at pains to emphasise, in the final analysis all *the ideal structures of science* grow – metaphorically speaking – out of the *life-world* (*Lebenswelt*) of our immediate experience as we live it independently of and before every theoretical, and in particular, every scientific reflection.[18] And in turn, scientific knowledge overflows in a popularised way into everyday consciousness in a frequently fragmented and possibly misunderstood form. A similar relationship also obtains in the *material sphere* with respect to the state of technology. The state of scientific research provides methods and insights which can be made use of technologically. Nowadays scientific research is frequently conducted with such aims in mind. The area of research and development is an intermediate area, in which both fields, science and technology, form a more or less indissoluble unity. The consequence of this symbiosis is reciprocal support and synergisation.

CONCLUSION

One of the effects of the mechanism described in the previous section is the ongoing development of multimedia society. The interest in science fiction and virtual reality is a sympton that the boundaries between science, technology and reality have become fluid. A new reality is being created by scientific methods and technological aids.

The attempt to influence the pre-programmed growth of the natural sciences and the technological innovations induced by it leads one to consider those factors or conditions that are in principle open to change. The most important ones in this context are on the one hand the general, social, systemic, structural conditions which channel collective behaviour, and on the other hand the motives, priorities and interests which determine the behaviour of the individual agent. Science, technology and industry now provide a cornucopia of products and potential life-styles based on them. These potentials tend to be realised in a consumer society where each individual follows the maxim, 'do whatever one can do'. Thus the potentially unlimited needs of consumers and the expansive dynamics of the economy mutually reinforce each other. Advertising positions the subject as a consumer with unsatiable desires. The scientific-technological-industrial complex provides the means to satisfy these constantly increasing needs.

If nothing is done to stop this process, this synergisation – as we all know – will in the long run lead to an ecological catastrophe. Two basic approaches are necessary to avoid this catastrophe. The first possibility consists in intervening in the normative sphere of the values, the goals, the priorities, the interests, the needs of the voters and the consumers. It is clear that in industrial countries the expectations of a civilised life-style will have to be adjusted to an ecologically acceptable standard (sustainable development). The second approach would involve intervention in the systemic sphere. The normative restrictions must be complemented by appropriate institutional and organisational regulations. In the context of the general political framework, restrictions – possibly in form of ecological and tax regulations – are needed in order to canalise the economic process in an ecologically sustainable way. In the final resort these will only be effective if they are applied on a world-wide basis. A global problem requires global solutions.

University of Dortmund, Germany

NOTES

[1] Kant, I., *Kritik der reinen Vernunft*, Weischedel (ed.) (Darmstadt, 1963), pp. 267–285 (B 293–315).
[2] This idea is pointed out by Leibniz, G. W., *Monadologie*, H. Glockner (ed.) (Stuttgart, 1975), p. 25 (§57).
[3] Whitehead, A., *Process and Reality*, D. R. Griffin and D. W. Sherburne (ed.) (New York/London, 1979), strongly underlines this point in terms of the universal categories of "nexus" and "prehension".
[4] In a striking, almost surrealist manner this is made evident in an essay of Jorge Luis Borges, in which a certain Chinese encyclopedia is described that consists of a perfectly absurd classification. Foucault, M., *Les mots et les choses* (Paris, 1966), explains in the preface that it was this very essay that inspired him to his book.
[5] In these cases, no longer the correspondence theory of truth but rather the coherence theory of truth turns out to be relevant; see Skirbekk, G. (ed.), *Wahrheitstheorien* (Frankfurt a.M., 1977).
[6] The universal character of hermeneutic interpretation in the *Geisteswissenschaften* is stressed by Gadamer, H.-G., *Wahrheit and Methode* (Tübingen, 1967). Abel, G., *Interpretationswelten* (Frankfurt a.M., 1993) explores the perspectivistic and interpretative character of every understanding of the world and of oneself.
[7] A line of reasoning akin to the one presented here was pointed out by Polikarov, A., *Methodological Problems of Science* (Sofia, Bulgaria, 1983), pp.197–222. See also Polikarov, A., *Science and Philosophy* (Sofia, Bulgaria, 1973), pp. 105–131.
[8] Letter to Jarig Jelles of 2nd June 1674, see de Spinoza, B., *Briefwechsel*, Walther (ed.) (Hamburg, 1977), p. 209f.
[9] Reichenbach, H., *An Analysis of the Foundation and Structure of Knowledge* (Chicago, 1938), §1.
[10] Husserl, E., *Ideen I, Neue Ausgabe* (The Hague, 1976), p. 202f.
[11] Popper, K. R., *Objective Knowledge* (Oxford, 1973), pp. 106–152.
[12] Concerning the latter point see Crane, D., *Invisible Colleges* (Chicago, 1972).
[13] Plessner, H., *Die Stufen des Organischen und der Mensch* (Berlin, 1928), p. 341. Even more outspoken is the French historian F. Braudel who declares that men do not make history but rather suffer it: *Pariser Gespräche* (Berlin, 1989), S. 93.
[14] N. Rescher, *Scientific Progress* (Oxford 1978) discusses possible limits to scientific growth. For the general background see Rapp, F., *Fortschritt* (Darmstadt, 1992).
[15] For more details see. Rapp, F., *Analytical Philosophy of Technology* (Dordrecht, 1981), p. 105f.
[16] A revealing example in question is the dispute between Goethe, J. W. and Newton, I. about the nature of colour, as expounded in Böhme, G., *Alternativen der Wissenschaft* (Frankfurt a.M., 1980), pp. 131–143.
[17] Limiting possibilities of technological activity (*perpetuum mobile*, velocity of light, Heisenberg's uncertainty principle, Shannon's theorem of information theory) are discussed by Tondl, L., "On the Concept of 'Technology' and 'Technological Sciences' ", in *Contributions to a Philosophy of Technology*, F. Rapp (ed.) (Dordrecht, 1974), pp. 1–18.
[18] The *locus classicus* is Husserl, E., *Die Krisis der europäischen Wissenschaften und die transzendentale Phänomenologie* (The Hague, 1962), pp. 48–54 and pp. 105–53. Recent research results are collected in *Lebenswelt und Wissenschaft*, C.F. Gethmann (ed.) (Bonn, 1991).

NICHOLAS RESCHER

THE LAW OF LOGARITHMIC RETURNS AND ITS IMPLICATIONS

Our knowledge of the real is inevitably the result of a belief-mediated grasp – or mis-grasp – of facts. Belief affords our only access to the knowledge of nature: our *theories* are the only feasible pathways to nature's *laws*. Kant was right: the "I think" is omnipresent in the cognitive domain – though in scientific contexts it generally comes into operation in the objectifying plural "we think". And this brings the characteristic mechanisms of mental operation upon the stage.

Nature consists of events and of processes. In nature things just happen. In dealing with the world's actual realities we can apply such distinctions as local/pervasive, and frequent/infrequent, terminating/nonterminating. But *facts* – even *true* facts – can take on a different set of characterizations. They can be precise or vague, well-understood or problematic, significant or insignificant. Their inherent coordination with the realm of thought equips facts with a mind-coordinated dimension of normativity and value that natural realities do not possess as such.

The particular mind-correlative appraisal category that will be at the focus of these present deliberations is the distinction: important/unimportant. We will examine its significance for the theory of knowledge and in particular endeavor both to *naturalize* it and to *quantify* it.

In epistemology and the theory of cognition certain pivotal appraisal-distinctions and classifications are commonly employed, for example, the familiar distinctions between the true and the false, and between the well evidentiated (probable) and the evidentially counter-indicated (improbable). But the portentous difference between real knowledge and mere information is generally neglected. And this is eminently unfortunate. After all, items of information were not created equal. Some claims will, even if true, be insignificant and make little or no impact upon the larger cognitive scheme of things. (What difference does my personal preference for pink over purple make to anything?) And other truths will be distinctly important – the theory of relativity for example, or Avogadro's law in chemistry. Few issues regarding cognition have as far-reaching a bearing upon matters of discovery, learning, understanding, and prediction – alike in scientific and in everyday contexts. This bearing is pivotal for our present concerns.

How can one measure the volume of information generated in a field of scientific or scholarly inquiry? Various ways suggest themselves. The size of the literature of a field (as measured by the sheer bulk of publication in it) affords one possible measure. And there are various other possibilities as well, for example: the number of investigate-hours devoted to the topics at

D. Ginev and R. S. Cohen (eds.), Issues and Images in the Philosophy of Science, 275–287.

issue, or the value of resource investment in information-engendering technology. So much for assessing information. But what about knowledge?

By "knowledge" we shall here understand *putative* knowledge that is not necessarily correct but merely represents a conscientiously contrived best estimate of what the truth of the matter actually is. But this is not all there is to it. Information is simply a collection of (putatively correct) beliefs or assertions. Knowledge, by contrast, is something more select, more deeply issue-resolving. Like all usable information it must be duly evidentiated and presumably correct. But there is also an additionally evaluative aspect, seeing that knowledge is a matter of *important* information: information that is *significantly* informative. Not every insignificant smidgen of information constitutes knowledge, and the person whose body of information consists of utter trivia really knows virtually nothing. (To reemphasize, knowledge, in the context of the present discussion is simply *putative* knowledge: it encompasses what people responsibly *think* to be correct, even when – unfortunately – they are quite wrong about it.[1])

To provide a simple illustration for this matter of significance, let us suppose a color taxonomy – for the sake of example, an over-simple one based simply on Blue, Red, and Other. Then that single item of *knowledge* represented by "knowing the color" of an object – viz. that it is red – is bound up with many different items of (correct) *information* on the subject (that it is not blue, is either red or other, etc.). As such information proliferates, we confront a situation of redundancy and diminished productiveness. Any knowable fact is always potentially surrounded by a vast penumbral cloud of relevant information – and as that information grows to be ever more extensive, those new speciable *significant* facts become more difficult to make out. Knowledge certainly increases with information, but as a far less than proportional rate.

One instructive way of measuring the volume of available information is via the opportunities for placement in a framework for describing or classifying the features of things. With n concepts you can make n^2 two-concept combinations. With m facts you can project m^2 fact-connecting juxtapositions.[2] A single step in the advancement of knowledge is here accompanied by a massive increase in the proliferation of information. Thus let us also contemplate shapes in addition to colors, again supposing only three of them: Rectangular, Circular, Other. Now when we *combine* color and shape there will be $9 = 3 \times 3$ possibilities in the resultant (cross-) classification. So, with that complex, dual-aspect piece of knowledge (color + shape), we also launch into a vastly amplified (i.e., multiplied) information spectrum over that increased classification-space. In moving cognitively from n to $n + 1$ cognitive parameters, we enlarge our knowledge addictively but expand our information field exponentially.

If we have a control mechanism that can handle three items – be it in physical or in cognitive management – then we can lift ourselves to higher levels of capacity by hierarchical layering. For we can (1) by hypothesis manage three base-level items, and then (2) by grouping three of these into a first-level complex manage nine base level items, and thereupon (3) by grouping three of these first-level complexes into a second level complex manage 27 base level items. Hierarchical grouping is clearly a pathway to enhanced managerial capacity. But in cognitive contexts of information management it is clear that a few items of high-level information (= knowledge) will correspond to a vast range of low-level information (= mere truths). Thus knowing where we stand in those three levels – that is having three pieces of *knowledge* – will position us in a vastly greater information space (viz. one of 27 compartments). And this situation is typical. A small range of knowledge (by way of high-grade information) always positions one cognitively within a vastly larger range of low-level information.

The greater a body of information, the larger the *patterns of order* that can potentially exist within them, and the greater the effort needed to bring particular orderings to light. With two-place combinations of the letters *A* and *B* (yielding the four pairs AA, AB, BA, and BB) we have only two possible patterns of order – namely "The same letter all the way through" (AA and BB), and "Alternating letters" (AB and BA). But as we add more letters, the possibilities proliferate.

It is instructive to view this idea from a different point of view. Knowledge commonly develops via distinctions (*A* vs. non-*A*) introduced with even greater elaboration to avert the problems arising with less sophisticated approaches. A situation obtains that is analogous to the "Game of 20 Questions" with an exponentially exfoliating possibility space being traced out stepwise (2, 4, 8, 16, . . . , 2^n, . . .). With n descriptors one can specify for 2^n potential descriptions that specify exactly how, over all, a given object may be characterized. When we add a new descriptor we increase by one the amount of knowledge but double the amount of available information. The *information* at hand grows with 2^n, but the knowledge acquired merely with n. The cognitive exploitation of information is a matter of dramatically diminishing returns.

The extraction of information from legibility-impaired manuscripts and papyri and inscriptions in classical paleography provides another illustration. If we can decipher 70% of the letters in such a manuscript we can reconstruct the phrase at issue. If we can make out 70% of the phrases we can figure out the sentences. If we can read 70% of its sentences, we can understand the message of the whole text. So some one-third of the letters suffice to carry the whole message. A modest body of knowledge bears the weight of a vast load of information. From the standpoint of knowledge, mere information is highly redundant.

Consider now the result of combining two ideas, to wit that

1. Knowledge is distinguished from mere information as such by its significance. In fact: *Knowledge is simply particularly significant information* – information whose significance exceeds some threshold level (say q). In principle there is room for variation here according as one sets the quality level of entry qualification and the domain higher or lower.)

2. The significance of *additional* information is determined by its impact upon *pre-existing* information. Significance in this sense is a matter of the relative (percentage-wise) increase that the new acquisitions effect upon the body of pre-existing information (I), which may be estimated (in the first instance) by the sheer bulk of the relevant body of information – the literature of the subject, as it were. Accordingly: *The significance of incrementally new information can be measured by the ration* $\Delta I/I$.

Putting the two ideas of principles (1) and (2) together, we have it that a new item of actual knowledge is one for which

$$\frac{\Delta I}{I} \geq q$$

for some threshold-indicative quantity q. Thus knowledge-constituting *significant* information is determined through the proportional extent of the change effected by a new item in the preexisting situation (independently of what that preexisting situation is).

On the basis of principles (1) and (2) it follows that the total amount of knowledge (K) encompassed in an overall body of information of size I is given by:

$$K = \int \frac{\mathrm{d}I}{I} = \log I + \text{const} = \log cI$$

where K represents the volume of actual *knowledge* that can be extracted from a body of bare *information* I,[3] with the constants at issue open to treatment as parameters of the measuring scale, so that the equation at issue can be simplified without loss to $K = \log I$.[4] In milking additional information for significant insights it is generally the *proportion* of the increase the matters: its percentage rather than its brute amount. We accordingly arrive at the *law of logarithmic returns* governing the extraction of significant *knowledge* from bodies of mere *information*.

This state of affairs clearly has the most far-reaching implications for the progress of knowledge. This is clearly illustrated in Display 1. There is an immense K/I imbalance: with ongoingly increasing information I, the corresponding increase in knowledge K, shrinks markedly.

Display 1.

I	cI (with $c = 0.1$)	$\log cI$ ($=K$)	$\Delta(K)$	K as % of I
100	10	1	–	1.0
1,000	100	2	1	0.2
10,000	1,000	3	1	0.03
100,000	10,000	4	1	0.004

The ramifications of this principle for cognitive progress are not difficult to discern. Nature imposes increasing resistance barriers to intellectual as to physical penetration. Consider the analogy of extracting air for creating a vacuum. The first 90% comes out rather easily. The next 9% is effectively as difficult to extract than all that went before. The next 0.9 is just as difficult. And so on. Each successive order-of-magnitude step involves a massive cost for lesser progress; each successive fixed-size investment of effort yields a substantially diminished return. Intellectual progress is exactly the same: when we extract *knowledge* (i.e. high-grade, nature-descriptively significant information) from mere information of the routine, common "garden variety", the same sort of quantity/quality relationship obtains. Initially a sizeable proportion of the available is high grade – but as we press further this proportion of what is cognitively significant gets ever smaller. To double knowledge we must quadruple information.

This situation is reflected in Max Planck's appraisal of the problems of scientific progress. He wrote that "*with every advance [in science] the difficulty of the task is increased: ever larger demands are made on the achievements of researchers*, and the need for a suitable division of labor becomes more pressing".[5] The Law of Logarithmic Returns at once characterizes and explains this circumstance of why substantial findings are thicker on the ground in the earlier phase of a new discipline and become ever more attenuated in the natural course of progress.

What we have here is an epistemological analogue of the Weber-Fechner law of psychophysics, which holds that a just-noticeable (i.e. minimally significant) difference in a stimulus is proportional to the magnitude of that stimulus. This law has it that inputs of geometrically increasing magnitude are required to yield *perceptual* outputs of arithmetically increasing intensity. The presently contemplated law envisions a parallelism of perception and conception in this regard. It stipulates that (informational) inputs of geometrically increasing magnitude are needed to provide for (cognitive and thus) *conceptual* outputs of arithmetically increasing magnitude. In searching for meaningful patterns, the ongoing proliferation of data points makes contributions of rapidly diminishing value.

The Law of Logarithmic Returns also connects up with – and is interestingly illustrated by – the idea of a cognitive lifespan as expounded by the sagacious Edward Gibbon in his *Memoirs of My Life*: "The proportion of a part to the whole is the only standard by which we can measure the length of our existence. At the age of twenty, one year is a tenth perhaps of the time which has elapsed within our consciousness and memory; at the age of fifty it is no more than a fortieth, and this relative value continues to decrease till the last sands are shaken out by the hand of death".[6] On this basis, knowledge development is a matter of adding a given percentage increment of what has gone before. Thus fresh experience superadds its additional increment ΔE to the preexisting total E in such a way that its effective import is measured by the proportion $\Delta E/E$. And $\int \Delta E/E = \log E$. On such an approach, an increment to one's lifetime has a *cognitive* value determined on strict analogy with Daniel Bernoulli's famous proposal to measure the *utility* value of incremental economic resources by means of a logarithmic yardstick.[7]

Gibbon's "law of learning" thus means that a body of experience that grows linearly over time yields a merely logarithmic growth in *cognitive* age. In dealing with human individuals whose experiential information also grows in a roughly linear way in the course of tried experience, this law provides for a distinction between cognitive and chronological age and engenders a rather different sort of life-expectancy forecasting. The resulting picture is set out in Display 2. There is a crucial difference between one's *chronological* life (in terms of mere years) and one's *cognitive* life (in terms of experience and learning), the latter being measured essentially in terms of the *logarithm* of the former. After all, experience is cumulative (even where its *lessons* are not), and each new instalment can only add its merely marginal increment of new lessons to the great body of what has gone before. This perspective makes for a substantial distinction between actuarial life expectancy as conceived of in chronological and in cognitive terms.

Display 2. Cognitive *vs.* chronological age.

Chronological age	log (Age)	log (Age) as % of log 80	Cognitive age (on basis of 80 year lifespan)
5	0.70	37	30
10	1.00	53	42
20	1.30	68	54
30	1.48	78	62
40	1.60	84	67
50	1.70	89	71
60	1.78	94	75
70	1.85	97	78
80	1.90	100	80

The general purport of such a Law of Logarithmic Returns as regards expanding information is clear enough. Letting $K(I)$ represent the quantity of knowledge inherent in a body of information I, we begin with our fundamental relationship: $K(I) \approx \log I$. On the basis of this fundamental relationship, the knowledge of a two-sector domain increases addictively notwithstanding a multiplicative explosion in the amount of information that is now upon the scene. For if the field (F) under consideration consists of two sub-fields (F_1 and F_2), then because of the cross-connections obtaining within the information-domains at issue the overall information complex will take on a multiplicative size:

$$I = \inf(F) = \inf(F_1) \times \inf(F_2) = I_1 \times I_2$$

With compilation, information is multiplied. But in view of the indicated logarithmic relationship, the knowledge associated with the body of compound information I will stand at:

$$K(I) = \log I = \log (I_1 \times I_2) = \log I_1 + \log I_2 = K(I_1) + K(I_2)$$

The knowledge obtained by compiling two information-domains (sub-fields) into an over-all aggregates will (as one would expect) consist simply in *adding* the two bodies of knowledge at issue.

And so, in general, when a plurality of information-domains ($I_1, I_2, \ldots I_n$) are at hand, the overall amount of knowledge that emerges from them is measurable by

$$K = \sum_i c_i \log I_i$$

where the c_i are constants appropriate to the particular information domains in view. (See footnote 3.) Where knowledge grows by mere steps, information proliferates by leaps and bounds.

Such a state of affairs clearly has substantial implications for scientific progress. With the enhancement of scientific technology, the size and complexity of this body of data inevitably grows, expanding on quantity and diversifying in kind. Technological progress constantly enlarges the window through which we look out upon nature's parametric space. In developing natural science, we use this window to scrutinize parametric space, continually augmenting our data base and then generalizing upon what we see. And what we have here is not a homogeneous lunar landscape, where once we have seen one sector we have seen it all, and where theory projects derived from lesser data generally remain in place when further data comes our way. Historical experience shows that there is every reason to expect that our ideas about nature are subject to constant radical changes as we broaden our data base. But of course the expansion of knowledge is far slower than the increase of base information.

A helpful perspective on this situation comes to view through the idea of "noise", considering that expanding bodies of information encompass so much cognition-impeding redundancy and unhelpful irrelevancy that it takes successive many-fold increases in information to effect successive fixed-size increases in actual knowledge. It is supposed that internal "noise" or variability in the information-system is such that the increment needed to effect a cognitive advance of a certain fixed size or significance is proportional to the magnitude of the starting position.

The world of knowledge and information is quantitatively stratified. It comes in layers: article/chapter, book, library, library system. Or pictographically: sign, scene (= ordered collection of signs), story (= ordered collection of scenes). In such layering, we have successive levels of complexity corresponding to successive levels of informational combinations, proportional with n, n^2, n^3, etc. The logarithm of the levels – $\log n$, $2 \log n$, $3 \log n$, etc. – reflect the all-in complexity that we encounter at each level, and thereby determines the amount of "knowledge" that is available through information that we can obtain about the state of affairs prevailing at each level. And this is something that increases only one step at a time, despite the exponential increase in information.

It is not too difficult to come by a plausible explanation for the sort of information/knowledge relationship that is represented by our Law of Logarithmic Returns ($K \approx \log I$). The principal reason for such a K/I imbalance may lie in the efficiency of intelligence in securing a view of the *modus operandi* of a relatively simple (i.e. complexity-aversive) world. For in a world whose law-structure is comparatively simple, one can learn a disproportionate amount of general fact from a modest amount of information. (Note that whenever an infinite series of 0's and 1's, as per 01010101 . . . , is generated – as this series indeed is – by a relatively *simple* law, then this circumstance can be gleaned from a comparatively short initial segment of this series.) In rational inquiry we try the simple solutions first, and only if and when they cease to work – when they are ruled out by further findings (by some further influx of coordinating information) – do we move on to the more complex. Things go along until an oversimple solution becomes destabilized by enlarged experience. We get by with the comparatively simpler options until the expanding information about the world's *modus operandi*, made possible by enhanced new means of observation and experimentation, demands otherwise. With the expansion of knowledge new accessions set ever increasing demands. At bottom, then, there are thus two closely interrelated reasons for that K/I disparity:

i. Where order exists in the world, intelligence is rather efficient in finding it.
ii. If the world were not orderly – were not amenable to the probes of intelligence – then intelligent beings would not and could not have emerged in it through evolutionary mechanisms.

Be this matter of explanatory etiology as it may, the Law of Logarithmic Returns regarding the cognitive benefit of additional information has striking implications for the *rate* of scientific progress.[8] In particular, it stands coordinate with a principle of a swift and steady decline in the comparative cognitive yield of additions to our body of mere information.

Let us look at the issue in a temporal perspective. In general our basic formula

$$K \approx \log I$$

means that

$$\frac{\mathrm{d}}{\mathrm{d}t}K = \frac{1}{I}\frac{\mathrm{d}}{\mathrm{d}t}I$$

The more information we already have, the slower will be the rate at which knowledge grows with newly acquired information. And the larger the body of information we have in hand, the smaller (by rapid decline) will be the percentage of this information that represents real knowledge. Looking to the quantity of information (I) and of knowledge (K) as a function of time, so as to address $I(t)$ and $K(t)$, respectively, it seems plausible to suppose that our body of bare *information* will increase in proportion with our resource investment in information-gathering. Then if this investment grows exponentially over time (as has historically been the case in the recent period) we shall in consequence have

$$I(t) \approx k^t \text{ and correspondingly also } \frac{\mathrm{d}}{\mathrm{d}t} I(t) = k^t$$

and accordingly

$$K(t) \approx \log I(t) \approx \log k^t = t$$

and consequently

$$\frac{\mathrm{d}}{\mathrm{d}t} K(t) = \text{constant}$$

Exponential growth in I is coordinated with linear growth in K.

On this basis, it is not all that difficult to find empirical substantiation of our law of logarithmic returns ($K \approx \log I$). The phenomenology is that while information has increased exponentially in the past (as shown by the exponential increase in journals, scientists, and outlays for the instrumentalities of research), real knowledge has expanded only linearly. Throughout modern times the number of scientists has been increasing at roughly 5% per annum.[9] Thus well over 80% of ever-existing scientists (in even the oldest specialties, e.g., mathematics, physics, and medicine) are alive and active at the present time. And scientific information has also been growing at the (reasonably constant) exponential rate over the past several centuries, so as to produce a veritable flood of scientific literature in our time. The *Physical Review* is now divided into six parts, each of which is larger than the whole journal was a decade or so ago. The total volume of scientific publication is truly staggering. It is reliably estimated that, from the start, well over 10 million scientific papers have been published and that currently some 30,000 journals published some 600,000 new papers each year. It is readily documented that the number of books, journals, journal-papers has been increasing at an exponential rate over the recent period.[10] (See Display 3.) To be precise, the printed literature of science has been increasing at some 5% *per annum* throughout the last two centuries, to yield an exponential growth-rate with a doubling time of ca. 15 years and an order-of-magnitude increase roughly every half century. And much the same story holds for the recruitment of people and the expenditures on equipment – that is for resource commitment in general.

But during this period the progress of science *per se* – the progress of knowledge as measured in the sort of first-rate discoveries of the highest level of significance has progressed at a more or less linear rate. There is much evidence to this claim. It suffices to consider the size of encyclopedias and synoptic text-

Display 3. Total number of scientific journals and abstract journals founded, as a function of date.

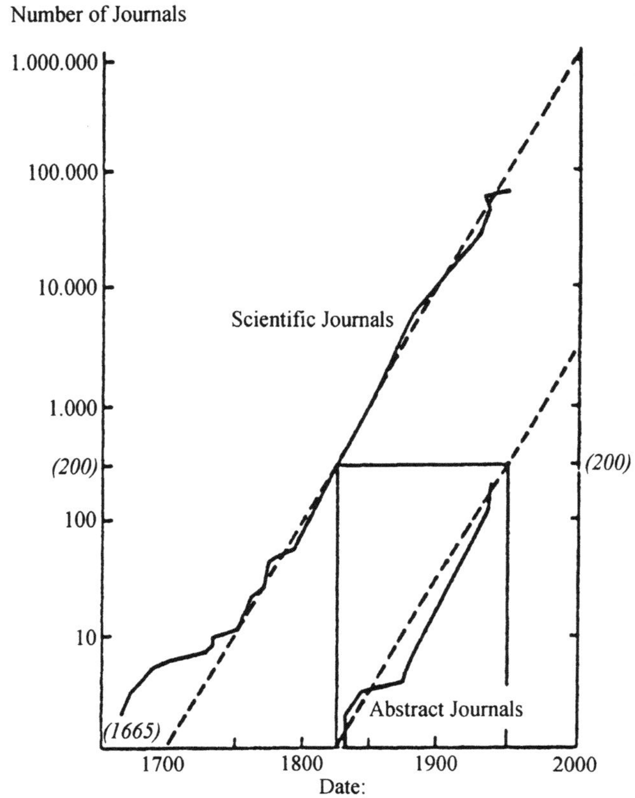

Source: Derek J. de Solla Price, *Science Since Babylon* (New Haven: Yale University Press, 1961).

books, or again, the number of awards given for "really big contributions" (Nobel Prizes, academy memberships, honorary degrees), or the expansion of the classificatory taxonomy of branches of science and problem areas of inquiry.[11] All of these measures of preeminently *cognitive* contributions indicates a comparative constancy in year-to-year progress.

And this relationship of linear progress in the face of exponential growth does the job. Clearly, if an exponential increase in information acquisition is matched by a merely linear increase in genuine knowledge, then knowledge will stand proportional with the logarithm of the information.

So much for the exponential growth past. In the future's (at best) steady state era of zero-growth resource investment, we shall have

$$I(t) \approx t \text{ and correspondingly } \frac{d}{dt} I(t) = \text{constant}$$

and accordingly

$$K(t) \approx \log I(t) \approx \log t$$

and consequently

$$\frac{\mathrm{d}}{\mathrm{d}t} K(t) \approx \frac{1}{t}$$

The constant growth of I in a steady-state era will be associated with a time-inversely proportional rate of growth of K. Where information grows linearly, knowledge too will grow – but at a swiftly decelerating rate.

To be sure, there is, on this basis, no inherent limit to the possibility of future progress in scientific knowledge. But the exploitation of this theoretical prospect gets ever more difficult, expensive, and demanding in terms of effort and ingenuity. New findings of equal levels of significance require ever greater aggregate efforts. In the ongoing course of scientific progress, the earlier investigations in the various departments of inquiry are able to skim the cream, so to speak: they take the "easy pickings", and later achievements of comparable significance require ever deeper forays into complexity and call for an ever-increasing body of information. (And it is important to realize that this cost-increase is not because latter-day workers are doing *better* science, but simply because it is harder to achieve *the same level* of science: one must dig deeper or search wider to achieve results of the same significance as before.)

The preceding deliberations do serve to indicate, however, why the question of the rate of scientific progress is somewhat tricky. For this whole issue will turn, rather delicately, on fundamentally *evaluative* considerations. Thus consider the past once more. At the crudest – but also most basic – level, where progress is measured simply by the growth of the scientific literature, there has been for centuries astonishingly swift and sure progress: and exponential growth with a doubling time of ca. 15 years. At the more sophisticated, and demanding level of high quality results of a suitably "important" character, there has also been exponential growth, but only at the pace of a far longer doubling time, perhaps 30 years, approximating the reduplication with each successive generation envisaged by Henry Adams at the turn of this century.[12] Finally, at the maximally exacting level of the really crucial insights that fundamentally enhance our picture of nature, our analysis has it that science has been maintaining a merely constant pace of progress.

It is this last consideration that is crucial for present purposes. What we have been dealing with here is an essentially seismological standard of importance. It is based on the question "If the thesis at issue were abrogated or abandoned, how large would the ramifications and implications of this be? How extensive would be the shocks and tremors felt throughout the whole range of what we (presumably) know?" And what is at issue is exactly a kind of cognitive Richter Scale based on the idea of successive orders of magnitude of impact. The crucial factor is the extent of seismic disturbance of the cognitive terrain. Would we have to abandon and/or rewrite the entire textbook, or a whole chapter, or a section, or a paragraph, or a sentence, or a mere footnote.

The plausible way to proceed here is to look at the "impact" that a given piece of work makes on the literature of its domain. This can be measured – as a reasonable approximation – by means of citations in the sense of citation indices (or perhaps by counting not citations as such, but only citations in discussions that themselves are extensively cited). In analyzing the available citation data, it emerges that the number of papers cited by at least p percent of a body of scientific literature of size N increases inversely with this size: $\Delta N/N$. Hence the total number of papers cited in (at least) p percent of the literature is proportional to $\int \Delta N/N = \log N$. On this basis increasing the total number of papers massively (exponentially) will only increase modestly (linearly) the number of citations "really important" papers. (And this circumstance is unaffected by percentage level that we fix upon as our importance-determinative threshold of citation-impact.)

And so, while the question of the rate of scientific progress does indeed involve the somewhat delicate issue of the setting of an evaluative standard, our stance here has been rough and ready – and justifiably so because the details just wash out. *Viewing science as a cognitive discipline* – one whose task is the unfolding of a rational account of the *modus operandi* of nature – the progress of SCIENCE as a cognitive product, a growing body of knowledge stands correlative with its accession of really major discoveries: the seismically significant, cartography-revising insights into nature. And this perspective lays the foundation for our present analysis – that the historical situation has been one of a *constant* progress of science as a *cognitive discipline* not-withstanding its *exponential* growth as a *productive enterprise* (as measured in terms of resources, money, manpower, publications, etc). The Law of Logarithmic Returns says it all.[13,14]

University of Pittsburgh, Pittsburgh, USA

NOTES

[1] There is thus no suggestion of cumulatively here. Clearly, knowledge, like populations, grows not just by additions (births) but also by abandonment (deaths). But with knowledge, in contrast to population, the concepts of displacement and replacement play a more crucial role. Knowledge never dies off without leaving progeny: in this domain you cannot eliminate something with nothing.

[2] Note that a self-concatenated concept or fact is still a concept or fact, even as a self-mixed color is still a color.

[3] In information theory, *entropy* is the measure of the information conveyed by a message, and is there measured by $k \log M$, where M is the number of structurally equivalent messages available via the available sorts of symbols. By extension, the $c \log I$ measure of the knowledge contained in a given body of information might accordingly be designated as the *entropy*. Either way, the concept at issue measures informative actuality in relation to informative possibility. For there are two types of informative possibilities (1) structural/syntactical as dealt

with in classical information theory, and (2) hermeneutic/semantical (i.e. genuinely meaning-oriented) as dealt with in the present theory.

[4] Some writers have suggested that the subcategory of significant information included in an overall body of crude data of size I should be measured by I^k (for some suitably adjusted value $0 < k < 1$) – for example by the "Rousseau's Law" standard of $\sqrt{I}$. (For details see Chapter VI of the author's *Scientific Progress* [Oxford: Blackwell, 1978].). Now since $\log I^k = k \log I$ which is proportional to $\log I$, the specification of *this* sort of quality level for information would again lead to a $K \approx \log I$ relationship.

[5] Max Planck, *Vorträge und Erinnerungen*, 5th ed. (S. Hirzel, Stuttgart, 1949), p. 376; italics added. Shrewd insights seldom go unanticipated, so it is not surprising that other theorists should be able to contest claims to Planck's priority here. C. S. Peirce is particularly noteworthy in this connection.

[6] Edward Gibbon, *Memoirs of My Life* (Penguin Books, Harmondworth, 1984), p. 63.

[7] This law thus suggests a parallelism between the structure of the increase in *cognitive* as compared with *material* wealth. Think here of Daniel Bernoulli's (and Laplace's) contrast between *la fortune morale* and *la fortune physique*, where the latter mode of wealth is measured not by the amount of money at stake, but merely by the logarithm thereof. See Isaac Todhunter, *A History of the Mathematical Theory of Probability* (London, and Cambridge, 1865; reptd. NY: Chelsea, 1949), pp. 213–6. On this analogy, the step from mere information to actual knowledge involves a shift reminiscent of that in moving from mere money to its effective utility, accordingly envisioning the cognitive utility of information as growing in a merely logarithmic way.

[8] It might be asked: "Why should a mere accretion in scientific 'information' – in mere belief – be taken to constitute *progress*, seeing that those later beliefs are not necessarily *true* (even as the earlier one's were not)?" The answer is that they are in any case better *substantiated* – that they are "improvements" on the earlier one's by way of the elimination of shortcomings. For a detailed consideration is the relevant issues, see the author's *Scientific Realism* (D. Reidel, Dordrecht, 1987).

[9] Derek J. deSolla Price, *Little Science, Big Science* (Columbia University Press, New York, 1963), pp. 6–8. See also Price's *Statistics of Science and Technology* (HMSO, London, 1970), p. 115.

[10] Cf. Derek J. de Solla Price, *Science Since Babylon* (Yale University Press, New Haven, 1961; 2nd ed. 1975), See Ch. 8, "Diseases of Science". Further detail is given in Price's *Little Science, Big Science* (*op. cit.*)

[11] The data here are set out in the author's *Scientific Progress* (Blackwell, Oxford. 1978).

[12] Henry Brooks Adams, *The Education of Henry Adams* (Massachusetts Historical Society, Boston, 1918).

[13] To be sure, we are caught up here in the usual cyclic pattern of all hypothetico-deductive reasoning. In addition to explaining the various phenomena we have been canvassing, that projected K/I relationship is in turn substantiated by them. This is not a vicious circularity but simply a matter of the systemic coherence that lies at the basis of inductive reasonings. Of course the crux is that there also be some predictive power, which is exactly what our discussion of deceleration is designed to exhibit.

[14] Some of the issues considered here are also addressed in the author's *Scientific Progress* (Blackwell, Oxford, 1978).

HANS JÖRG SANDKÜHLER

THE HUMAN RIGHT TO KNOW AND THE REALITY OF KNOWLEDGE

Practical-Philosophical Consequences of Epistemological Realism

Most people today, being asked the question 'is subjectivity a waning standard?', would tend to answer in the affirmative, without giving the matter a second thought. How, then, is subjectivity to provide a central perspective for philosophical thought and for living one's life – after Nietzsche's critique of humanism, Marx's analysis of the conditions under which any free association between free human beings is not possible, and under the impact of the postmodernist veto on this central category of the tradition of European Enlightenment? Without doubt the Romantic notion of *subjectivity* is no longer an issue today. And yet, I shall be defending the thesis that the age of subjectivity ushered in by Kant only now has come into existence. Modernity, in other words, has only just begun.

The subject I am dealing with is the relationship between *knowledge* and *democracy*. This is a political theme, but I shall be treating it in epistemological terms. Modern societies are *knowledge-based*; they open up space for individual action to the extent that subjectivity is *knowledge-based*.

I shall begin, under the heading 'crisis of knowledge', by briefly outlining the age we live in, an age in which nothing will be changed if people lack the *capacity to judge*. Following this, I shall make an excursion into the realm of philosophical epistemology in order to determine which concept of knowledge we need to have if we are to assert a claim for *knowledge as the precondition for democracy*. The third part of my argument will be a remembrance of the pre-history of the human right to know. Fourthly and finally, I shall plead for a world-image of individual freedom, in which democracy provides the fundamental order on which all knowledge is based.

The programme I should like to propose can be summarised as follows: Epistemologies[1] are implicates of self-images and world-images, *and* are systems on which such world-images and those universes of meanings (i.e. semantic systems for world-images) in which an individual designs *his possible world* are grounded. Anyone who reflects on the relationship between being and knowledge defines as a subject – in historically relative autonomy – a self-relationship. Epistemologies are framework theories in which decisions are made as to the choice of ontologies. This is widely acknowledged today. It is often forgotten, however, that *at the same time* they are the foundation for the historical-philosophical modes of self-understanding of human beings and give rise to practical philosophical consequences. My objective here is to show the manner in which agreement and understanding with regard to any realism are able to bring in its train a practical philosophy of knowledge and the idea of the *democracy of knowledge*.[2] I should like to substantiate

D. Ginev and R. S. Cohen (eds.), Issues and Images in the Philosophy of Science, 289–303.

the following thesis, namely that *epistemological realism is the foundation for the human right to know.* From this is derived my answer to the question posed at the outset, namely that subjectivity is not a 'waning standard'.

THE CRISIS OF KNOWLEDGE AS A GLOBAL PROBLEM AND THE FIRST REASON FOR THE HUMAN RIGHT TO KNOW

I know of no document that is more deeply impressing in its evidence for this process than H. Arendt's *Eichmann in Jerusalem: A Report on the Banality of Evil.* In the Foreword to the German translation, she renews the emphasis on her fundamental assumption that 'it is in the essence of the total apparatus of domination and perhaps in the nature of any bureaucracy that it turns human beings into functionaries, mere wheels in the administrative machine, thus dehumanizing them'. She analyses the tendency that is bound up with this, 'to make the responsibility of the agent for his deeds vanish away by explaining it in terms of the one or the other determinism'. This leads her to the core of the problem of the capacity of individuals for action, self-determination and guilt; the issue is the 'essence and the functioning of the human faculty of judgement'.[3] What interests Arendt in Eichmann is the type of human that gains personal identity through obedience within the 'banality of evil', at the same time as losing it through the incapacity to judge, to speak and to remember. In Arendt's words, 'his inability to speak was closely connected with an inability to *think*, namely, to think from the standpoint of somebody else'. Eichmann, the reader of Kant's *Critique of Practical Reason* who was capable of thinking only the thought of 'duty', explained 'that from the moment he was charged with carrying out the Final Solution he had ceased to live according to Kantian principles, that he had known it, and that he had consoled himself with the thought that he was no longer "master of his own deeds", that he was unable "to change anything".'

It is the balance of the 'terrifying normality' that Arendt draws – 'that this new type of criminal, who is in actual fact *hostis generis humani*, commits his crimes under circumstances that make it well-nigh impossible for him to know or to feel that he is doing wrong'[4] – that must be remembered in my context. Freedom is the precondition for responsibility and the ability on the part of the individual to feel guilt; freedom is dependent on a certain knowledge in which a self-image, an image of the other and a world-image coherently coexist because in total it represents the criteria for human purposes.

Our age, in which the epochal understanding of itself is bound up with the revolution in and through science and technology, could be termed the age of knowledge. When has knowledge ever been so very much a measure of possible humanity? Is the process whereby mental activity assumes priority over material labour not something that is significantly new in history? Something that was at first only a tendency within modernity, but now an objective possibility? Do the aims and objectives of humanity not derive their rationale from this point onwards from the concrete understanding of concrete knowledge? Reason, or understanding, is the human capacity to subjectively shape history according to humanist criteria. Reason is the cognitive alliance between the consciousness of historical origins and that of the possible future; it desires reality to be progress that is in accordance with its own criteria. Reason mediates between the three dimensions of knowledgeability: *memory* of the origins that continue to affect our present time, *knowledge* of reality, and *anticipation*, the anticipatory future shape of what we wish to be because we *can* be it. The great concept of *reason* in the speculative philosophy of history is nowadays considered to be antiquated. To what extent can and do individuals wish to assign to themselves a pluralistically more modest reason, an as yet constrained freedom from which a new, free historical world must arise?

The present day status of consciousness is such that this capacity to reason determines action *in reverse*: distorted memory of history transfigures the past as a better other reality than the present one, or denounces origins as a history of decline by employing a critique of rationality, enlightenment or science; *ignorance* of the relevant causes of famine, war and any social, economic and political praxis that violates human rights represents a paralysis of that possible self-determination which is the precondition for knowing these causes and the reasons for such violations; the anticipation, free of fear, of a free, equal and solidaristic human race of the future is thus subverted by a reversal to angst and the despondent fear of the uncertainty that the next day represents. One consequence of this reversal of reason is collective and individual irrationalism: politics perceives the movement for peace and against war as the destabilization of domination; individual consciousness is oriented towards enemy images, in which fear of uncomprehended political beliefs or religions is projected onto races, nations or peoples; *enemy images* are reversed self-images that constitute ideologies and in which the alien 'Other' is no longer perceived and acknowledged as the 'Other Self'.

This brief diagnosis can be summarized by saying that the rights of humans are violated not least because ignorance about precisely those legitimate claims to the universality and the indivisibility of social, economic, political and cultural human rights compels people into a defensive, angst-ridden stance as opposed to an offensive one. Where do the causes for this unconsciousness lie? Should they be assigned to subjective immatureness and the refusal to think? Do people withdraw from the rational obligation to know as an escape from responsibility? Such causes cannot be rejected in total, but it would be short-sighted to believe that they suffice as an explanation. One significant cause can be discovered in the objective process that largely determines social time, commencing with the political-economic and cultural constitution of the technologically highly developed society, radiating out over the globe with all its might.

This revolution contains within it – besides what are certainly positive aspects – the counter-revolution that leads to a *global crisis of knowledge*. And in this crisis it is necessary to redefine the field of rationality. This revolution provokes with a paradox, not despite but because it is in essence an *epistemic* revolution, a revolution of knowledge – it objectively expands the knowledge of the species – while signifying for the subjects a lack of knowledge. The notion that the species knows more is followed by the question as to the extent of individual *participation in* and *isolation from* knowledge, the question as to the freedom of knowledge. Freedom can only mean the autonomy of individuals to knowingly shape the world and themselves and to found their action in knowledge of self and the world. The crisis of the epistemic revolution does not consist of the fact that the individual cannot be a physics expert, a historian and a technologist all at the same time. It is much rather the case that *rational world images*, in which knowledge is totalised to become a unity, are suffocated in a chaotic process of increasing fragmentation in the cognitive system. This is what constitutes the crisis of knowledge.

The abyss that is opening up between the expanded potency of knowledge and subjective deficiencies is nothing new. What is new in character are the forms by which this dialectic is socialised, a socialisation in which economised technology and political domination take possession of knowledge. Of course, it is not 'the revolution' that expands any body of knowledge that the individual withdraws from; the question concerns who it is that controls this process and with this result. Information technology is accompanied – if it is not controlled according to rational criteria – by new forms of immaturity and imposed denial of self-determination; information technology is the site at which the consensus is organised between the forces of domination and those individuals who allow themselves to be oppressed without having consciousness of it. Division of labour and specialisation, disintegration of the sciences and their separation from the everyday culture, segmentation of social experience and resignation in the face of a mass of data that in itself has no significance – this all deepens alienation from knowledge, which becomes a fiction to individuals to an extent never previously known.

In what is today obviously the *deconstruction* of the epistemic-semantic whole, and separated from the self-reflexive experience of the individual, knowledge becomes an abstract possibility without a subject, a sense or an aim. Knowledge is thus no longer the subjective ability on the part of individual reason to *construct reality*. In societies with 'knowledge-based infrastructure'[5] the threat of destruction hangs over those *universes of meaning*, those semantic systems for world images that must be available to us if we are to understand ourselves as *constructors of the possible world governed by reason*.

This touches on the root of the problem and the beginning of a humanist solution to it. *We must know what we can know; only through knowledge are we able to describe ourselves; our self-descriptions, however, are the foundation of world-knowledge*. What the human being is, what hopes he may entertain and what he is permitted to do, these questions always lead us back to knowledge – Kant supplied the rationale for this. An adequate definition of humanism cannot dispense with the stating of epistemological reasons: as long as we base our assumptions on the fallacy of naive ontological (metaphysical) realism, namely that we are confronted within the process of knowledge acquisition with an external world that is somehow developing itself and which we have only to discern, then we will misconceive ourselves as helper's helpers within what are seemingly objective necessities. Or, another variation, we are only able to shape our world because we can construct it in our body of knowledge according to the laws of our mental powers; or, put another way, only through the *reality of knowledge*[6] is a world suited to humans made possible.

How is the semantics of a humanist self-description by the *citoyen* enabled, in what way is the design of a democratic civil society made possible? The humanist world image may appear for the present as being idealist on account of its epistemological foundations. But it is realistic, because it is to epistemological realism that we owe the fact that the mind and reason are the constructors of that reality in which only we are able to live as human beings. The world image and its concentric circles – *from* knowledge *to* rational praxis – in which the right and the duty to know form the centre point, arise in the epistemic crisis from oppositional thought. This is oppositional knowledge against widespread anthropological, natural philosophical and historical-philosophical determinism and fatalism. This has practical consequences. The decisive conclusion to be drawn is the extension of human rights to include the *human right to know*. This legal entitlement must be provided with a philosophical foundation, and is prior to any political claims that may be pressed.

This thesis is a controversial one among philosophers. I cannot agree with R. Rorty's plea for '*The Priority of Democracy to Philosophy*' as long as 'democracy' is defined only by means of the formal guarantees provided by liberalism. Rorty challenges Rawls's claim, that 'philosophy as the search for truth about an independent metaphysical and moral order cannot [. . .] provide a workable and shared basis for a political conception of justice in a democratic society';[7] he also conceded that 'it is not clear how to argue for the claim that human beings ought to be liberals rather than fanatic without being driven back on a theory of human nature, on philosophy'.[8] Rorty's critically distanced agreement with Rawls's dispensing with the concept of truth and replacing it with what is a qualitatively undefined 'intersubjective reflective equilibrium' of similarly undefined 'subjects' of democracy must necessarily be followed by the admission that 'so philosophy, as the explanation of the relation between such an order and human nature, is not relevant either. When the two come into conflict, democracy takes precedence over philosophy.'[9] The decisive question is not being asked, however, namely which subjects constituted in what way possess the aptitude for a democracy whose constitution is the freedom and the happiness of all?

To pose this question and provide a reasoned answer to it presumes a sufficient understanding of the relationship between 'reality' and 'knowledge'.

This means, firstly, that the guarantee promised by metaphysical, realistic ontologies – of the *representation*, through perception, of a Reality[A 10] external to consciousness – does not exist; the *adaequatio rei et intellectus* consists much rather of the fact that intellect adapts the reality about which statements can be made and in which people can made judgments and take action, as the *other-of-itself*. This principle of an epistemological, an internal realism, is the only possible alternative to a 'God's Eye' view. The latter is not only terroristic in terms of the theory of truth, but is a rationale for domination over the incapacitated.

EPISTEMOLOGICAL REALISM, NATURALISM AND RATIONALITY ACCORDING TO THE CRITERION OF HUMANITY

The epistemological realism I am talking about is that non-metaphysical and non-commonsensical realism that does not derive from an ontology of representation, but instead one which demands *ontology in the form of a philosophy of mind*. Such a realism can refer to Kant's Copernican turn – granted of course that Kant is not misunderstood as a founder of the Idealist tradition, but instead as representing the logical and critical conclusion of the rational empirical tradition initiated by Galileo and Bacon and continued by Locke and Hume (i.e. the theory of the 'twin pillars' of perception, according to which there cannot be any experience or induction that is not also interpreted by the faculty of reason). The core of this epistemological tradition comprises, firstly, the insight into the *theory innate in all experience* and in the constitution of data in the external world as *res facti*, i.e. as the transformation of sense data into mental notions in a manner corresponding to the nature of mind. N. Goodman has produced the following variation of this idea: 'Facts [. . .] are theory-laden; they are as theory-laden as we hope our theories are fact-laden. Or in other words, facts are small theories, and true theories are big facts.'[11] The second constitutive element for Kant's empirical realism must not be forgotten here either, of course, namely that the synthesis of experience and knowledge is assured by the *a priori*-ness of mind's nature.[12]

These are questions that are dealt with by philosophy today. But there is generally no awareness in the realism debate of the fact that analytic philosophy is not at all the sole descendant of the rational-empirical tradition. There is a second manner of thinking, one that has developed independently of analytic philosophy, that I wish to remind you of, because it has made a significant contribution to the philosophy of knowledge and science, I am referring to the *Epistémologie* of Gaston Bachelard.

Conscious of the radical transformations in the natural sciences, above all by relativity theory, and seeking grounds for a non-Cartesian philosophy of science, Bachelard argues – in *La Philosophie du non* (1940), for example – an express rejection of any inductivist empiricism or of any positivism of pure facticity, as well as of any evolutionist conception of a linear and cumulative history of science; his *Epistémologie* is an affirmation of the dialectic and discontinuity in the development of knowledge and science, a call to analyse spontaneous empirical obstacles to understanding and to respond to them with a rational critique. The essential core of the *Epistémologie*, however, is the assumed necessity for the dynamic and objectivity of science of a *coupure épistémologique* (an epistemological incisive split in experience) between everyday and scientific knowledge.

Again and again, Bachelard provides various grounds supporting the proposition made in *La formation de l'esprit scientifique*, that '*scientific experience* is [. . .] an experience that *contradicts habitual experience*'.[13] What appears to positivism as 'fact' is seen by epistemology first of all as an *obstacle épistémologique* (obstacle to knowledge),[14] in order to understand that 'the primary experience or, more precisely, the first observation is always an obstacle for scientific education [. . .] it is picturesque, concrete, natural, simple [. . .] we will show that there is a rift and not a continuity between observation and experience'.[15] The central thesis, that

in the formation and development of the scientific intellect the first obstacle is the first experience, the experience that exists before and above critique that of necessity forms an integrating element of the scientific mind [. . .]. For us it is important [. . .] to clearly oppose any simple philosophy that is based on a more or less open, more or less romantic sensualism, and which pretends to draw its knowledge directly from some clear, sharp, certain and constant given, that permanently presents itself to the open mind. The philosophical thesis that we advocate is this: the scientific intellect must form itself *against* nature, against the impulse and directive nature in us and external to ourselves, against the monopolisation through nature, against the brightly colouredness and multivariance of facts. The scientific mind must shape itself by reshaping itself.[16]

Le nouvel esprit scientifique is a response to the phenomenon of the 'new' that is constructively created by the modern natural sciences. In taking what is 'new' into consideration, scientific experience can no longer be understood to be a representation of what is naturally and materially given. The epistemological revolution mediates a rational empiricism and principles from Kant's transcendental philosophy. In *Le rationalisme appliqué* (1949) it is stated that 'in the labour of making scientifically precise, elements of a Copernican turn in objectivity can be recognised. It is not the object that defines precision, but the method.'[17]

This is precisely one of the relevant points at which analytic philosophical thought and theory of science is brought to bear, not without contradictions of course. The toughest veto is that of naturalism. What is now acknowledged as the classical opponent of epistemological realism and philosophy of mind has been well-known at least since Quine's *Word and Object*; here, the problems of ontological relativity and the indeterminacy of translation have moved to the centre of a strategy that naturalises epistemology:

epistemology [. . .] falls into place as a chapter of psychology and hence of natural science. It studies a natural phenomenon, viz. a physical human subject. This human subject is accorded a certain experimentally controlled input [. . .] and in the fullness of time the subject delivers as output a description of the three-dimensional external world and its history.[18]

Here, just as in the deterministic '*eliminative*' *scientific materialism*, there is no longer any epistemological foundation for a human self-description that proceeds from the (historically relative) autonomy of the cognitive constitution of reality, allowing practical-philosophical demands for freedom and the human rights to knowledge. Naturalism has fortunately stumbled on the contradiction inherent in Putnam's work, for example.

The aporia of hard realism comes to light in the forms of naturalised cognitive theory especially, since the Reality^{A} that it strives to define through naturalisation is itself, *as* stated reality, nothing other than *theory*. Merit must be accorded to H. Putnam for the turn he was already making in the mid-70s – which became explicit in *Reason, Truth and History* – towards a transcendentally oriented 'internal realism', in that he reopened the philosophical dimensions of the theory of knowledge by coming out against any widely held naturalist[19] features in it. He strives 'to break the stranglehold which a number of dichotomies appear to have on the thinking of both philosophers and laymen. Chief among these is the dichotomy between objective and subjective views of truth and reason'; he argues first and foremost against any 'copy' theory of truth.[20] He asserts that 'the only criterion for what is a fact is what is *rational* to accept' and that, correspondingly, '*value facts*' must be admissible. Putnam, by agreeing with the 'subjectivist philosophers' 'that our conceptions of reason evolve in history', and opposing any acceptance of the 'positivist' notion 'that the scientific world is in some way constructed out of "sense data"', but also by arguing that 'the mind *makes up* the world',[21] takes an opposing stance with respect to 'metaphysical realism' and for an '*internalist* perspective, because it is characteristic of this view to hold that "*what objects does the world consist of*?" is a question that it only makes sense to ask *within* a theory of description'.[22] While there are no 'inputs *which are not themselves to some extent shaped by our concepts*', Putnam challenges the idea – as in P. Feyerabend's *Anything goes* – that this could justify any 'facile relativism'. 'Our conceptions of coherence and acceptability [. . .] define a kind of objectivity, *objectivity for us*, even if it is not the metaphysical objectivity of the God's Eye view, objectivity and rationality, humanly speaking, are what we have; they are better than nothing.'[23]

Paths lead from here to the epistemological, practical-philosophically expressed foundation for the human right to know; the concept of knowledge is centred around that concept of *reality*^{B} (by which is meant reality in the sense of the phenomenal world) which we can have as constructors of the world 'according to human criteria'.

HISTORICAL TRACES OF THE HUMAN RIGHT TO KNOW

As early as Comenius, for example, the humanist demand was 'to teach all people all things'. Little now remains of such optimism. Public life in bourgeois society has sought to do justice to this postulate in the codification of education within the social contract, or, in the case of Condorcet, in the idea of the democratisation of knowledge. Kant, in his cosmopolitan and republican answer to the question *What is Enlightenment?*, expressed a similar aim – the idea of the duty to know. 'A person can choose to defer his own enlightenment with regard to that which he has a duty to know, although only for a limited time; but to dispense with this duty, whether individually or for his descendants, is tantamount to violating and trampling under foot the sacred rights of Man.'[24] Fichte's concept of culture as self-activity,[25] or Hegel's risky and speculative elevation of knowledge in the *Phenomenology of Spirit* or *The Science of Logic* to a moment in the self-determination and self-manifestation of the Absolute in a manner that violates the rights of the subject, or, finally, Humboldt's educational reform, signify that philosophical Idealism and New Humanism in Germany went further towards founding reality in knowledge: 'The structure of the "general education of humanity" results from the problem as to how scientific knowledge can become real as subjectivity, or as the general in the individual.'[26]

Such ideas were soon to be contradicted by the sobering and disillusioning materialist critique. Taking up the insights of French materialists such as Helvétius, that if the human being is formed by circumstances, then the precondition for any progress towards reason is the humanisation of those same circumstances, Marx posited the great adversative – social being is prior to social consciousness. The cultural revolution can only follow in the train of socioeconomic revolution. In turn, the liberal reformist answer to the 'social question' under capitalism was to regard 'education' as the precondition for any fundamental transformation. The scientification of knowledge (with the accent placed on the natural sciences) and the democratization of knowledge were the 'signature of the age' during the bourgeois revolutions of 1848. Diesterweg, the educational reformer, was the first to seek a wider audience for the demand that knowledge must become a *general human right*.

Supranational codifications of human rights now manifest a tendency towards recognition of the human right to know. The *General Declaration* of December 10, 1984 takes up the classical Enlightenment topos of the relationship between 'human dignity' and 'reason', extending freedom of opinion to the right 'to seek, obtain and disseminate information and ideas with all available means for communication, without regard for national borders' (Article 19). Article 27 is of particular importance for extending human rights to encompass the right to knowledge – here, cultural human rights are specified for the first time in such a way that 'every human being has the right to freely participate in the cultural life of the community, to enjoy the arts and to take part in scientific progress and the benefits it produces'.

The memory of historical developments to date thus do not lead into no-man's-land. But how can the diagnosis of a global crisis of knowledge – and therefore of behaviour – be reconciled with a non-Utopian hope for an *actual* right to knowledge? Understanding that there is no simple cause-effect relationship between social relations and knowledge provides an initial answer to this question; the humanisation of relations is no less dependent on the materialisation of the right to know than is this right dependent on the humanness of the respective social relations.

Where are there signs of a path towards human beings gaining a world-image and self-image, the measure and norms for which are determined by reason? *Pluralism*,[27] legitimated through the differences between cultures, world-images and life goals and indispensible for any free constitution of life, contains its dialectic reverse: ' "public life" is reduced to the space in which disparate truths coexist indifferently, the space occupied by understanding and its manipulation by commercial or political interests, and not the sphere of conscious shaping of social and cultural relations.'[28] The pluralist uncertainty of world-images and forms of praxis is one form taken by the crisis of knowledge, one that the following arguments seek to oppose.

EPISTEMOLOGY AND THE DEMOCRACY OF KNOWLEDGE

Knowledge and ignorance determine increasingly whether or not liberation for self-determination in the reality of knowledge is possible. This is the foundation for an offensive to achieve the human right to know, and reason enough for the demand that 'a constitution of scientific and technological progress can and must enable the moral reflection of progress'. But I see only pragmatic grounds, and no categoric reasons for the conclusion that such a constitution 'cannot, however, refer to any universal principles that give rise to the prospect of consensual judgement in society. *The moment that the moral ambivalence of progress becomes manifest, human rights, too, will lose their normative clarity*'.[29]

Philosophy represents *one* path for finding a *critique* of knowledge and arriving at a holistic self-description[30] – philosophy is a form of participation in the integral of knowledge. Its function as a rational theory of the political, of human rights and democracy shall only be mentioned indirectly here. I am concerned rather to analyse the dimensions of understanding and knowledge in which democracy – epistemically and politically – appears to be the precondition and the result.

Knowledge is a result of individual behaviour, that of the *construction of reality*. Our concept of knowledge must therefore be so designed as to already contain the idea of the autonomy of the subject within the determination of the conditions and criteria for *truth*.[31] Only on the basis of the subjectively constituted truth, for which responsibility can be rationally assumed, is a rational praxis possible, since knowledge is a mode of freedom, depending on its possibility. Knowledge is a result of the self-realisation of individuals, and thus of successful individuation. Subjectivity is successful epistemic and moral individuation. The subject must be aware of his function, namely to constitute the data obtained from perception of the world as objects of knowledge within the framework of a semantical world-image system, by assigning his own meanings to them.

Knowledge is thus *translation*. As such it is relative with respect to world-images, self-images and theoretical frameworks. Possible worlds are determined within constellations of knowledge, although not arbitrarily, of course. Even N. Goodman, who opts for a 'radical relativism under rigorous constraints' in *Ways of Worldmaking*,[32] protests against the assumption 'that right versions can be arrived at casually, or that worlds are built from scratch:'[33] 'Willingness to accept countless alternative true or right world-versions does not mean that everything goes [. . .], that truths are no longer distinguished from falsehoods, but only that truth must be otherwise conceived than as correspondence with a ready-made world.'[34]

Being is *intentionally* translated into that subjectively acquired and construed *Reality*^{B}. As *reality of knowledge* this is not simply some dependent variable of material Reality^A. Knowledge constitutes phenomenal reality not simply *in* the subject, but *through* the subject, in whose activity *experience* and *construction* are united in such a way that translations arise within the framework of universes of meanings. Only through this union of world-experience and world-creation does knowledge lead to norms for praxis. The precondition for these norms to be complied with is the *individual, free* possession of knowledge and the competency to translate. This is the epistemological basis for the ethical dimension of the right to know.

D. Henrich, in his important essay 'On some preconditions for the understandability of human rights', asserted that reasons for the authority and acceptability of human rights are not dependent solely on historical-social contexts.[35] Once again, it is from knowledge, theories, and world-images that human rights gain their *meaning* and *significance*. Ideal conditions must be met in addition to the practical, if norms 'are to be accepted and become effective in actual action'. D. Henrich distinguishes terminologically between 'the three conditions [. . .] which are those of relevance, applicability, and acceptance of norms'. Human rights can only be considered relevant if they are understood within the context of 'thoughts on the correct order for our relations to the world'. The conditions for their applicability depend strongly on the 'variability of self-descriptions on the part of human beings'; 'norm types' and 'self-description types' form a single unit, with differences in self-description on the part of the agent arising depending on whether he

(a) views the world as the intentional content of his active behaviour, or
(b) sees in it only an ordering framework for his action, or
(c) perceives the world only as the area that is to be structured according to norms, and within which norms are to be asserted.

The third form of relationship to the world is constitutive for the idea of the rights of Man.[36]

Only world-image (c), and this form of self-description, offer appropriate conditions for the acceptance of norms motivating behaviour. This raises the question as to the 'world orientation of the person presuming to comprehend human rights as a fundamental norm'. This is the context within which epistemological reasons become central to establishing the grounds for the human right to know.

These grounds are made up of three components. Firstly, a world-image[37] constituted in such a way that reality is not conceived of as a world 'ready-made' and given for subjectivity, independent of consciousness, a world-image that opposes any form of fatalism in the philosophy of history; this is a world-image of Man that defines him as the *constructor of reality*. Secondly, a world-image that contains a concept of knowledge in which the subjectivity of knowledge is linked to transcendental conditions for the equality of subjects,[38] thus requiring *a priori* conditions enabling knowledge; *internal realism* as a reasonable philosophy of mind, with its concept of the epistemic-semantic constitution of 'reality' and by means of general and necessary conditions, thus guards rational understanding and the knowledge arising through it from any relativism of knowledge and norms. Thirdly, a subjective self-description of the human being, one that attributes to itself a competency for the epistemic construction of reality and therefore the human right to know as a *precondition* for the realisation of human rights in their totality.

There is a need today to grasp philosophically that, in the age following the scientific revolutions of the 17th century and the Enlightenment, as well as in the epoch of the scientific technological revolution, relationships of knowledge in which freedom arises from having access to the entire body of knowledge and in which democracy arises from self-determined knowledge – i.e. a *logic of intellectual anticipation of the new* – is able to supercede the old logic comprising the mere repetitive representation of the existing.[39] In this sense it is true to say that the age of subjectivity, i.e. Modernity, has only just begun. In the normative concept of self-determined rationality there can be a reunification of the moments of *knowledge* and *responsibility*[40] that were separated to the detriment of any rationality of cognitive behaviour. Within the scope of the epistemology and ethics of knowledge, the centre of which is the subject liberated for knowledge of the whole, knowledge of the historical possibilities in the future enters an alliance with the moral '*judgement*' of that for which responsibility can be taken. This is precisely the issue that Hannah Arendt was addressing. I am not prevented from taking up this same issue, despite the fact that my epistemological, transcendental argumentation takes a different route than the one she chose to take.

The origin on which the right to know can be universally based is the *equal nature of mind*, the possibility of reason in accordance with human criteria. Inherent in the world-image of the autonomy and the epistemic-semantic construction of reality is the idea of the difference of possible historical-cultural worlds in accordance with which conditions for relevance, applicability and acceptance are determined. At the same time, it also contains the idea of the possible transcultural identity of the world of the human race, a concept of *humanity*. Individuals who are denied autonomy in the epistemic constitution of reality – which in turn requires participation in the entire body of knowledge within a semantical system of world-image – cannot be historical subjects nor associate themselves with humanity. An epistemology that grasps this fact will be able to contribute to the establishing of the grounds for *democracy as a rational order of knowledge.*

The freedom and the variety of knowledge, in addition to the right to all knowledge, are constituents of the very essence of human knowledge. Any knowledge is relative, depending on the ontic and epistemic constitution of subjectivity, to that *third world of objective knowledge*[41] in which the objective epistemic foundations of the world-image and human self-description exist within universes of meaning. Individuals can only comply with practical norms of truthfulness to the extent that they have access to coherent semantic worlds. In our reality, such access leads increasingly to *theoretically formulated worlds*, mainly, but not only, those of the sciences. Living in the reality of knowledge, human beings are also living in possible worlds in which the coherence or non-coherence of the self-description of knowledge and the description of the world determine the extent to which that world complies with the criteria of reason.

Human dignity is violated as long as the subject does not know in what sense it is a *constructor* of reality, and that it must assert, as a creator of human reality, its right to freedom and to the democracy of knowledge. This is a rarely found perspective in the realism debate and in analytical philosophy too. It is within this perspective that epistemological realism gains its practical philosophical significance: Knowledge must be interpreted in order to be changed. There is a widespread fear of change; the courage to interpret has lost its potency in the postmodern world-image. It would surely be illusory to claim that self-determination is only possible on the basis of thought and knowledge. But nothing at all would be possible if self-determination of subjectivity is not founded on knowledge.

University of Bremen, Germany

NOTES

1 The term 'epistemology' is used here to refer to the systemic unity of the philosophy and the history of knowledge and science. See Sandkühler 1990a, on the historical development.

2 See my book with the same title, Sandkühler 1991b.

3 Arendt 1990, pp. 59, 65.

4 Arendt 1963, pp. 44, 121, 253.

5 See the theoretical deliberations on constitutional aspects in Preuss 1990, p. 80, as well as pp. 73–89.

6 See my book with the same title, Sandkühler 1990.

7 Rawls 1985, p. 225, quoted in Rorty 1990, p. 284.

8 Rorty 1990, p. 290.

9 *Ibid.*, p. 291.

10 'Reality^A': Tokens of most current observable common-sense types objectively exist independently of the mental. 'Reality^B': Only through the theory-laden reality of knowledge is a world suited to humans made possible.

[11] Goodman 1978, pp. 96–7.
[12] Putnam (1990, p. 117) argues against this presumption of a 'transcendental nature'.
[13] Bachelard 1984, p. 44.
[14] *Ibid.*, p. 50ff.
[15] *Ibid.*, p. 51.
[16] *Ibid.*, p. 59.
[17] *Ibid.*, p. 138.
[18] Quine 1971, p. 97.
[19] See Putnam 1981 on the mind-body problematic and his anti-naturalist argumentation.
[20] *Ibid.*, p. IX.
[21] *Ibid.*, pp. X–XI.
[22] *Ibid.*, p. 49.
[23] *Ibid.*, pp. 54–5.
[24] I. Kant, in answer to the question, 'What is Enlightenment', in *Kant's Werke, Akademie Textausgabe*, Vol. VIII, p. 39.
[25] This concept of culture is valuable for solving the aporia that is apparently given by the universality of human rights, namely that European thought is forced upon other historical cultures, i.e. that universality and plurality form a contradiction.
[26] Bracht *et al.* 1990, p. 925.
[27] If pluralism of socio-cultural identities and political democracy is defined as indispensible, this does not signify a recognition of a pluralism of cognitive theories and relativism of truth, such as that of W. James.
[28] Bracht *et al.* 1990, p. 935.
[29] Preuss 1990, p. 86.
[30] This holistic concept, referring back to semantic universals, takes up Kant's concept of the synthesis of all multifariousness within thought.
[31] Here, 'Truth' assumes the concept of rational acceptibility, but not the strongly metaphysical-realistic assumption of agreement between statement and facts that are independent of consciousness, not the extremely opposing radical constructivist replacement of 'truth' by the evolutionary-pragmatic concept of 'viability'.
[32] Goodman 1978, p. X.
[33] *Ibid.*, p. 97.
[34] *Ibid.*, p. 94.
[35] Cf. Henrich 1990, p. 277.
[36] *Ibid.*, p. 282ff.
[37] See Whitehead 1925, on the significance of world images for the sciences in modernity.
[38] Equality of subjects in Kant's sense is a much more powerful concept than that of mere discursive or hermeneutic intersubjectivity; it is founded in a concept of the 'equal nature of mind'.
[39] This logic of the repetitive corresponds to a theoretical model of representation as reproduction, as is frequently encountered in Marxist theory under the name 'reflection'.
[40] Cf. Sève 1985.
[41] Cf. Lektorski 1985.

REFERENCES

EE = *Europäische Enzyklopädie zu Philosophie und Wissenschaften*. H. J. Sandkühler (ed.), 4 Bde (Hamburg, 1990).

Arendt, H. (1963), *Eichmann in Jerusalem. A Report on the Banality of Evil* (New York).

———. (1990), 'Vorwort', *Eichmann in Jerusalem. Ein Bericht von der Banalität des Bösen* (Leipzig).

Bachelard, G. (1974), *Epistemologie. Ausgew. Texte.* (Frankfurt a.M., Berlin, Wien).

———. (1978), *Le nouvel esprit scientifique* (Paris).
———. (1980), *Le matérialisme rationnel* (Paris).
———. (1984), *Die Bildung des wissenschaftlichen Geistes. Beitrag zu einer Psychoanalyse der objektiven Erkenntnis* (Frankfurt/M).
Bracht, U., B. Fichtner, Th. Mies and G. Rückriem (1990), 'Art, Erziehung und Bildung', *FE*, Bd. 1 (Hamburg).
Franzen, W. (1982), 'Die Bedeutung von "wahr" und "Wahrheit" ', *Analysen zum Wahrheitsbegriff und zu einigen neuen Wahrheitstheorien* (Freiburg/München).
———. (1985), 'Vernunft nach Menschenmass': Hilary Putnams neue Philosophie als mittlerer Weg zwischen Absolutheitsdenken und Relativismus', *Philos. Rundschau* **32**, H. 3/4.
Goodman, N. (1978), *Ways of Worldmaking* (The Harvester Press, Sussex).
Heidelmeyer, W. (ed.) (1982), '*Die Menschenrechte*, Erklärungen, Verfassungsartikel', Internationale Abkommen (Paderborn/München/Wien/Zürich).
Henrich, D. (1990), 'Über einige Voraussetzungen der Verstehbarkeit von Rechten der Menschen', *Ethik zum nuklearen Frieden* (Frankfurt/M).
Holz, H. H., L. Lambrecht and G. Stuby (Hg.) (1987), *Die Rechte der Menschen* (= *Dialektik* 12) (Köln).
Lektorskij, V. A. (1985), *Subjekt, Objekt, Erkenntnis. Grundlegung einer Theorie des Wissens* (Frankfurt a.M., Bern, New York) [tr. from Russian].
Preuss, U. K. (1990), *Revolution, Fortschritt und Verfassung. Zu einem neuen Verfassungsverständnis* (Berlin).
Putnam, H. (1981), *Reason, Truth and History* (Cambridge, London, New York).
Quine, W. V. O. (1971), 'Epistemology Naturalized', in Aktes des XIV. *Internationalen Kongresses für Philosophie*, Wien 2–9. September 1968, Vol. Vi (Wien).
———. (1968), 'Ontological Relativity', *The Journal of Philosophy* **XVI**(7), April 4.
Rawls, J. (1985), 'Justice as Fairness: Political not Metaphysical', *Philosophy and Public Affairs* **14**.
Rorty, R. (1990), 'The Priority of Democracy to Philosophy', in *Reading Rorty*. Critical Responses to *Philosophy and the Mirror of Nature*, A. R. Malachowski (ed.) (London).
Rosenbaum, A. S. (1980), *The Philosophy of Human Rights. International Perspectives* (Westport Conn).
Sandkühler, H. J. (1987), Rechte der Menschen und Moralität in der Wissenschaft – Humanismus, Wahrheit, Fortschritt. Zur Einführung. In *Dialektik* 14 "Humanität, Vernunft und Moral in der Wissenschaften" (Köln).
———. (1988), 'Dialektik, Krise des Wissens, Enzyklopädie und Emanzipation', *Dialektik* **16** 'Enzyklopädie und Emanzipation. Das Ganze wissen" (Köln).
———. (1990), 'Das Recht der Menschen auf Wahrheit, Handeln und Hoffen', in *Zur Architektonik der Vernunft*, I. Berthold (ed.) (Berlin).
———. (1990a), Art. 'Erkenntnis, Erkenntnistheorie', *EE* Bd. **1**, 772–904 (Hamburg).
———. (1991a), *Die Wirklichkeit des Wissens. Geschichtliche Einführung in die Epistemologie und Theorie der Erkenntnis* (Frankfurt/M).
———. (1991b), *Demokratie des Wissens. Aufklärung, Rationalität, Menschenrechte und die Notwendigkeit des Möglichen* (Hamburg).
Seve, L. (1985), 'Wissen und Verantwortung', in *Philosophie in weltbürgerlicher Absicht und wissenschaftlicher Sozialismus*, M. Buhr and H. J. Sandkühler (eds.) (Köln).
Sing, H. (1986), 'Universalität der Menschenrechte – Pluralität kultureller Identitäten', *Concordia* **9** (Frankfurt/M).
Whitehead, A. N. (1925), *Science and the Modern World* (MacMillan Publishing Co., London).

HOWARD SANKEY

KUHN'S ONTOLOGICAL RELATIVISM

In *The Structure of Scientific Revolutions*, Thomas S. Kuhn presented a model of scientific development on which science is characterized by periods of unified research intermittently disrupted by revolutionary change of paradigm. Ever since Kuhn first proposed this model of scientific theory change, relativism, in one form or another, has been associated with his work. There has, for example, been widespread discussion of Kuhn's suggestion that scientific rationality varies relative to the changing rules and standards employed by different paradigms. There has also been much discussion of his account of conceptual change in science by philosophers who saw in it an extreme conceptual relativism of radically incommensurable conceptual schemes. Yet in recent years Kuhn has retreated from many of the claims which were responsible for these earlier reactions to his position. In his later work, Kuhn presents instead an ontological form of relativism, which involves an anti-realist denial of objective natural kinds.

According to the new Kuhnian position which has lately begun to emerge, scientific theories are the source of alternative sets of taxonomic categories which are imposed by theories on the world. A set of such categories constitutes a localized complex of interconnected concepts, such that terms for such categories are unable to be translated from one set of categories into another such set. Rather than reflecting reality, these categories constitute, at most, ways of ordering experience; such categories do not reflect reality because it is not possible to do so. Given that there is no objectively right way to represent the world, and that the sets of categories imposed on the world vary with theory, there is a sense in which, as theories change, the world changes with them.

The aim of this paper is to document Kuhn's move away from conceptual relativism and rationality-relativism, and to provide an analysis of his present ontological form of relativism. I will start by discussing Kuhn's shift away from a relativistic stance about rationality and conceptual schemes. I will then turn to matters of ontology by considering Kuhn's earlier idealist-sounding talk of world-change and his later idea of changes in the taxonomic categories which theories impose on the world.

RELATIVISM ABOUT RATIONALITY

In *The Structure of Scientific Revolutions*, Kuhn made a number of claims about methodological standards, which suggested that the rationality of scientists' epistemic choices is relative to operative paradigm. He claimed that paradigms

D. Ginev and R. S. Cohen (eds.), Issues and Images in the Philosophy of Science, 305–320.

"are the source of the methods, problem-field, and standards of solution accepted by any mature scientific community at any given time" (1970a, p. 103). Because of the paradigm-dependence of methodology, "when paradigms change, there are usually significant shifts in the criteria determining the legitimacy of problems and of proposed solutions" (1970a, p. 109). Such criteria and standards cannot, however, be applied to the choice between paradigms, since "the choice is not and cannot be determined merely by the evaluative procedures characteristic of normal science" (1970a, p. 94). Yet there are no extra-paradigmatic standards to govern this choice, since, "as in political revolutions, so in paradigm choice – there is no standard higher than the assent of the relevant community" (1970a, p. 94).

Critics were quick to object to this combination of the paradigm-dependence of methodological standards and the absence of extra-paradigmatic standards. Popper saw Kuhn as an advocate of *the myth of the framework*, according to which "the rationality of science presupposes a common framework", so that rational choice and communication break down in the absence of a shared framework provided by a paradigm (1970, p. 56). On Lakatos's interpretation of Kuhn, "each paradigm contains its own standards" and "there are no super-paradigmatic standards", hence "*scientific revolution is irrational, a matter for mob psychology*" (1970, p. 178). While for Siegel, Kuhn's "irrationalist portrayal of theory choice makes scientific knowledge relative as well, since judgments of factual and theoretical adequacy are on this picture relative to the incompatible criteria of evaluation fostered by rival paradigms" (1987, p. 54).

The key relativist tendency in Kuhn's position detected by these critics centers upon the combination of the claim of paradigm-dependent evaluative criteria with the denial of higher-order criteria. For without any possible appeal to paradigm-independent criteria of theory-choice by means of which to decide between paradigms, there may be no objective, rational basis for the decision to accept one paradigm over another. Thus, if there is any sense in which scientific practice and theory-acceptance may be rational, it can at most be dependent on the operative standards of normal science, which vary with and are internal to paradigms. As a result, rationality in science is relative to an accepted paradigm, while the decision between paradigms cannot be made on rational grounds.

The relativist tendency of Kuhn's original position is so pronounced that some of those sympathetic to Kuhn have attempted to defend him by presenting a more defensible version of relativism. Gerald Doppelt, for example, criticizes the interpretation of Kuhn as a conceptual relativist, only to provide a novel interpretation of "Kuhn's epistemological relativism" (1982). Doppelt objects to the undue emphasis placed on Kuhn's meaning variance thesis in the conceptual relativist interpretation, and draws attention instead to the extent to which the problems dealt with by paradigms are incorporated into the evaluative standards employed by scientists. According to Doppelt's interpretation of Kuhn, the incommensurability of paradigms is due to variation in their problem-solving agendas, and rationality is relative to the paradigm because scientists' standards of evaluation depend on these agendas.

By the early 1970s, however, an apparent change of stance can be found in Kuhn's writings. In several publications dating from about 1970, Kuhn insists on the existence of generally applicable methodological criteria, allows an active yet limited role for rational argument in scientific theory choice, and rejects a mechanical or algorithmic view of such choice (1970a, pp. 184–5, 199–200; 1970b, pp. 259–66). This modified position is developed at greatest length in 'Objectivity, Value Judgment and Theory Choice', where Kuhn claims that there is a partially shifting, though broadly invariant set of methodological criteria, which function as values rather than as rules, and which serve to guide or influence scientists in their choices of theory (1977, pp. 322–5, 335). The set of values he describes (e.g., accuracy, consistency, simplicity) does not, however, unequivocally determine choice between theories, since the values may conflict in application and are not preferentially ordered. Moreover, Kuhn claims, particular values may be subject to variant interpretation, and so do not even themselves yield unambiguous choice of theory.

In the years after publication of *The Structure of Scientific Revolutions*, then, Kuhn has progressively moved away from the relativism about scientific rationality which characterized his original position. The position he later developed is one according to which rational factors play an important role in choice between scientific theories, though there are limitations on what rational argument can achieve in the course of such decisions. These limitations are in part due to the intrinsic inability of the various applicable methodological criteria to unambiguously determine choice in favour of one theory as opposed to an alternative. While there is, on this later view, scope for rational disagreement between advocates of rival paradigms, the position avoids a radical relativization of scientific rationality to variant methodological standards which are entirely dependent on paradigm.

CONCEPTUAL RELATIVISM

A second form of relativism often attributed to Kuhn is the doctrine of *conceptual relativism*. In relation to Kuhn's model of scientific theory change, this doctrine is usually associated with the ideas of meaning variance and semantic incommensurability. Kuhn holds that, in a revolutionary transition between paradigms, there is "a need to change the meaning of established and familiar concepts", which leads to a "displacement of the conceptual network through which scientists view the world" (1970a, p. 102). A number of different consequences have been held to flow from such meaning variance, such as the inability to translate or communicate between theories, absence of overlap between the consequences of theories and incomparability of theoretical content.

The doctrine of conceptual relativism may be formulated in a variety of different ways. Davidson, for example, presents it as the thesis that there may be totally untranslatable languages, to which reality and truth are relative. However, a version of conceptual relativism appropriate to Kuhn's model requires a close connection between paradigms and the conceptual apparatus which they employ. For, on Kuhn's model, significant conceptual variation occurs in the transition between paradigms, with the result that rival paradigms are the source of divergent conceptual schemes. In light of Kuhn's frequent remarks to the effect that "when paradigms change, the world itself changes with them" (1970a, p. 111) and that in the transition between paradigms a "whole conceptual web" had to be "shifted and laid down again on nature whole" (1970a, p. 149), it is tempting to interpret the conceptual variation involved in paradigm change as a profound change resulting in replacement of an entire conceptual scheme.

If paradigm change is taken to involve wholesale displacement of conceptual scheme, semantic incommensurability may be interpreted as radical incomparability of paradigms due to conceptual disparity. On such an interpretation, there is translation failure between the languages employed by rival paradigms, as well as communication failure between the adherents of such paradigms. As a result of translation failure, incommensurable paradigms are incomparable for content, since no consequence of one paradigm may be matched against an identical consequence of a rival paradigm or the negation of such a consequence. Moreover, the conflict between paradigms which are incomparable for content may not be resolved by means of empirical test, since such paradigms share no observational consequences in common. Indeed, given that observation is itself thoroughly impregnated by theoretical assumptions originating from background paradigm, the very possibility of objective empirical evidence for or against a theory is thrown into serious doubt. Ultimately, the ideas of objective truth and reality also come under threat. For without the possibility of an objective test or comparative evaluation of paradigms, the prospects of obtaining an accurate reflection of theory-transcendent reality seem poor.

It is doubtful that Kuhn ever meant to endorse such a radical conceptual relativism. Nevertheless, a number of Kuhn's philosophical commentators have taken such relativism to be a central feature of his work, and have objected to it accordingly. Dudley Shapere, for example, traces the relativism he attributes to Kuhn to the incomparability of paradigms due to meaning variance, and objects that such incomparability makes it inexplicable how incommensurable paradigms are able to constitute genuine rivals (1984, pp. 45–6, 82–3). Moreover, in his all-out attack on conceptual relativism and the dualism of conceptual scheme and empirical content on which it depends, Donald Davidson places Kuhn among a group of thinkers who are in the clutches of the conceptual scheme idea (Davidson 1984). In the course of his attack, Davidson raises a number of objections to conceptual relativism, the main thrust of which is to seriously challenge the idea that we may coherently conceive of the possibility of a totally untranslatable language.

Kuhn himself shed little light on the issue of conceptual relativism until the early 1980s. In his paper 'Commensurability, Comparability, Communicability', Kuhn explicitly addresses objections of incoherence raised against the incommensurability thesis by authors such as Davidson and Shapere. Instead of a relativism of radically incommensurable conceptual schemes, Kuhn there endorses a thesis of *local incommensurability*. According to this thesis, there may be localized failure of exact translation, within the context of an inclusive natural language, between the special languages employed by theories. Such languages contain complexes of terms, which are holistically interdefined, and which are unable to be translated in piecemeal fashion into another complex of terms in which the relevant semantic relations do not obtain.

The restricted untranslatability thesis enables Kuhn to meet Shapere's rivalry objection, since language peripheral to non-intertranslatable complexes of terms provides sufficient common ground for partial comparison of the content of theories. It also enables Kuhn to meet a key objection of Davidson's that the argument for translation failure typically proceeds within the very language into which translation allegedly fails. For one may argue, within some fragment of a background natural language taken as metalanguage, that a pair of alternative theoretical languages fails to be intertranslatable (see Sankey 1994, Chapter 4).

While the local version of the incommensurability thesis permits Kuhn to avoid radical conceptual relativism and various associated objections, the account he offers of the reasons for translation failure contains the seeds of his ontological relativism. For Kuhn claims that translation fails due to variation in the taxonomic structures which theories impose on the world. Before turning to that topic, however, I will discuss Kuhn's idealist-sounding talk of world-changes in his earlier work.

THE WORLD-CHANGE IMAGE

The Structure of Scientific Revolutions contains numerous suggestions that the world itself changes in the transition between competing paradigms. Kuhn remarks, for instance, that a historian considering past science might be inclined to say that "when paradigms change, the world itself changes with them", for "it is rather as if the professional community had been suddenly transported to another planet" (1970a, p. 111). Remarks such as these are accompanied by talk of new entities coming into existence and scientists seeing different things when they observe the world. For example, Kuhn says that "pendulums were brought into existence by something very like a paradigm-induced gestalt switch" (1970a, p. 120), and "Lavoisier . . . saw oxygen where Priestley had seen dephlogisticated air" (1970a, p. 118).

Although Kuhn's use of the world-change image is usually qualified, philosophical critics nevertheless detected a strong idealistic tendency in his views. However, this was not entirely due to Kuhn's use of the world-change image. Kuhn endorsed a strong version of the thesis of theory-dependence of observation, and denied that empirical factors determine choice of theory. This created the impression that reality does little to constrain theory on his model of science. In addition, the apparent conceptual relativism of Kuhn's original model portrayed scientists as if they were cut off from reality and isolated within radically variant conceptual schemes. Thus, rather than the world-change image by itself, it is Kuhn's use of the image conjoined with the anti-empirical, conceptual relativist flavour of his model, which suggests idealism. For they present a picture of science on which a drastically reduced role is played by an independent reality external to human thought and experience.

Such a denial of a role to external reality is consistent with two forms of idealism. The first form of idealism is a mentalistic doctrine which denies altogether the existence of an independent reality beyond thought and experience. There are, however, strong grounds against attributing this form of idealism to Kuhn, since, as has been argued by a number of authors, Kuhn assumes the existence of an independent reality throughout his work (e.g., Devitt 1984; Mandelbaum 1982). The assumption of such a reality is consistent with a second, *constructivist* form of idealism, which admits an independent reality but denies the possibility of epistemic access to it. The latter doctrine is a broadly Kantian position, according to which, despite the impinging of external reality on us in sense perception, the world inhabited by human cognizers is at least partly constituted by our own conceptual contribution.

On such a constructivist reading of Kuhn's metaphysical stance, different 'phenomenal worlds' are constituted by the conceptual schemes of alternative paradigms.[1] Thus, in the transition between paradigms, the phenomenal world of one paradigm is exchanged for the phenomenal world of another. While the phenomenal world of a paradigm is not reality itself, since reality is inaccessible, the phenomenal world with which a scientist is epistemically engaged depends on the paradigm accepted by the scientist. Such a constructivist reading of Kuhn, therefore, yields a sense in which the world is relative to an operative paradigm.

TAXONOMIC CHANGE AND TRANSLATION FAILURE

The third, ontological, strand of relativism has been a persistent theme throughout Kuhn's work. As we have just seen, the idea that the world is somehow relative to a paradigm was already present in his idealistic handling of the world-change image in *The Structure of Scientific Revolutions*. However, in Kuhn's later work the idea has taken on a novel form as Kuhn has developed the idea that scientific revolutions involve changes of taxonomic categories.

Since the early 1970s, Kuhn has repeatedly stressed that scientific revolutions produce changes in the systems of classification employed by scientists. Here I quote an early statement of his view, though numerous similar passages might be cited from his more recent work:

One aspect of every revolution is, then, that some of the similarity relations change. Objects which were grouped in the same set before are grouped in different sets afterwards and *vice versa*. Think of the sun, moon, Mars, and earth before and after Copernicus; of free fall, pendular, and planetary motion before and after Galileo; or of salts, alloys, and a sulphur-iron filing mix before and after Dalton. Since most objects within even the altered sets continue to be grouped together, the names of the sets are generally preserved. (1970b, p. 275)

Thus, a scientific revolution is not merely a transition between theories which make conflicting claims about entities which they classify in the same way. Rather, entities which are classified as belonging to one category by one theory may be classified as belonging to a different category by another theory. This is because the explanatory purpose of a theory may be best-served by classifying the entities in its domain of application differently from previous theories, as, for example, classifying the Earth as a planet served the explanatory purposes of Copernican astronomy.

A number of important features of Kuhn's view of categorial change may be gleaned from the above quotation. First, the categorial change at issue is not a wholesale displacement of classificatory framework. Rather, change in membership is restricted to only some categories within a classificational system. Second, change of category-membership is not restricted to redistribution of individual objects among different classes. Rather, sets of objects may also be assigned to new categories, as, for example, the alloys were shifted from the class of compounds to the class of mixtures (Kuhn 1970b, p. 269). Third, it is possible to identify at least some of the objects and sets of objects as the same things across classificatory schemes. Thus, there is a common, or at least a broadly overlapping, domain of objects and sets of objects, which is shared between alternative theoretical systems of classification.

Kuhn's views about categorial change have important semantic consequences for the kind of terms involved in such change. To the extent that there is retention of terminology across classificatory change, there may be extensional, as well as intensional, variation affecting such terminology. As Kuhn comments,

. . . the distinctive character of revolutionary change in language is that it alters not only the criteria by which terms attach to nature but also, massively, the set of objects or situations to which those terms attach. (1987, p. 19)

Because such semantic change involves membership redistribution among interconnected categories, such change is not isolated, but has a holistic effect:

What characterizes revolutions is, thus, change in several of the taxonomic categories prerequisite to scientific descriptions and generalizations. That change, furthermore, is an adjustment not only of criteria relevant to categorization, but also of the way in which given objects and situations are distributed among preexisting categories. Since such redistribution always involves more than one category and since those categories are interdefined, this sort of alteration is necessarily holistic. (1987, p. 20)

The holistic nature of the changes brought about by categorial change is, according to Kuhn, directly responsible for failure to translate from the language of one theory into the language of another.

Where I have spoken of the language of a theory, Kuhn now tends to speak of a lexicon. A lexicon is a "structured vocabulary" (1990, p. 300), which incorporates a taxonomic structure that is employed in describing the world. Such a taxonomy, which Kuhn sometimes calls a "lexical structure", is what provides the "invariants of translation" (1983, pp. 682–3). For, in order to translate a word from the lexicon of one theory into the lexicon of another, there must be a "homology of lexical structure" (1983, p. 683).[2] Because items are redistributed among categories in revolutionary transition between theories, the categories of one theory are unable to be mapped onto the categories of another. Translation fails because the meaning of a name for a given category depends upon terms which refer to other categories within the taxonomy. Because of the holistic way in which such terms are interdefined, they are unable to be translated into a lexical structure which employs a variant categorial system.

In philosophical discussions of semantic incommensurability, the claim of meaning variance has met with less resistance than has the claim of referential variance. Thus, a philosopher sympathetic to the claim that terms may shift their meaning in the transition between theories, might nevertheless object to Kuhn's claim that the names of taxonomic categories change their reference in scientific revolutions. For, as has been argued by advocates of the causal theory of reference, the reference of natural kind terms may be fixed, independently of theoretical descriptions of the kinds to which they refer, by means of direct causal relations with members of such kinds. Thus, it might be thought that Kuhn's thesis of translation failure between theories is objectionable because it mistakenly rests on a thesis of the referential variance of natural kind terms.

However, Kuhn's thesis of categorial change is not on as shaky ground as this may suggest. The application to science of the idea of non-descriptive reference-fixing at initial naming-ceremonies has proven deeply problematic in the context of theoretical terminology. Rather than reference being fixed once and for all at initial naming-ceremonies, the reference of terms used in science is subject to variation, and there tends instead to be a shift in the pattern of groundings by which terms are applied to their referents (cf. Devitt 1979). Moreover, it is necessary to incorporate into the causal theory a role for descriptions in securing reference to unobservable entities, which creates the potential for variation in the reference of theoretical terms with significant variation in descriptive content (cf. Kroon 1985; Nola 1980). Given the need to allow reference change subsequent to original term-introduction and to grant a reference-determining role to descriptions, the causal theory does not provide a basis on which to reject Kuhn's thesis of referential variance in the course of scientific revolution.

Yet, while there may be reasons internal to the theory of reference for thinking Kuhn's reference change thesis is defensible, the significance of such change to Kuhn's philosophical position is not confined to merely semantic issues. In particular, his thesis of change of taxonomic structure plays a major role in his ontological relativist position, according to which the existence of natural kinds or categories is relative to the phenomenal world of a theory. In preparation for that topic, I will now discuss Hacking's suggestion that Kuhn's position amounts to a new form of nominalism.

REVOLUTIONARY TRANSCENDENTAL NOMINALISM

Ian Hacking has suggested that Kuhn's views on the nature of scientific categorization amount to a form of nominalism, which he calls *revolutionary transcendental nominalism.* On such an interpretation, Kuhn is not to be read as an idealist who denies that there is a reality existing independently of human thought. Rather, Kuhn denies that the kinds to which individual things belong have any existence prior to thought.

The common thread running through all versions of nominalism is the thesis that all that exists are individual objects. There are neither Platonic forms existing over and above individual objects, nor do the kinds or categories to which such objects belong have any existence independently of human classificatory activity. As such, nominalism is a distinct doctrine from idealism or constructivism. For, rather than deny the mind-independent existence of reality or of objects, nominalism denies only that the classification of objects into kinds may represent kinds which exist independently of the mental.

As we saw previously, Kuhn holds that the changes of classificatory scheme which take place in scientific revolution are partial rather than total. Hacking's nominalist rendering of Kuhn preserves this aspect of Kuhn's position:

> Kuhn like some other contemporaries might be called an empirical realist and transcendental nominalist. That is, a great many of our commonplace sortings are a given fact of the interactions of any human group and the world in which it lives. That is the empirical realism . . . [According to] transcendental nominalism, there is not some uniquely right conceptualization of the world, nor is the world of itself constituted by more than merely superficial "kinds of things". The "kinds" that enter our theoretical speculations are man-made . . . (Hacking 1979, p. 230)

Thus, according to Hacking, Kuhn is an "empirical realist" because he grants the existence of "commonplace" kinds:

> many of our prescientific categories *are* natural kinds: people and grass, flesh and horseflesh. The world simply does have horses and grass in it, no matter what we think, and any conceptual scheme will acknowledge that. (1983, p. 110)

However, at a level which transcends such ordinary empirical groupings of things, the world is not itself divided up into kinds of things; at the trans-empirical level, kinds depend on human classificatory activity. Such a combination of realism and nominalism fits well with Kuhn's example of the alloys; they constitute an empirical kind which survives alteration in higher-level categories of compound and mixture.

A second feature of Hacking's interpretation which accords well with Kuhn involves the instability of trans-empirical kinds. For Hacking argues that, unlike the classical nominalist, Kuhn holds that human-imposed categorial schemes are subject to revision in the course of scientific revolution. This is why Kuhn's is a *revolutionary* form of transcendental nominalism.

> The old-fashioned nominalist of times gone by held that our systems of classification are products of the human mind. But he did not suppose that they could be radically altered. Kuhn has changed all that. The categories have been altered and may be altered again. (Hacking 1983, p. 110)

Thus, on the overall picture which emerges from Hacking's reading of Kuhn, while there are empirical kinds, trans-empirical kinds depend on human classificatory activity, and are subject to variation with change of theory.

Hacking's nominalist rendering of Kuhn permits a novel reading of Kuhn's world-change image. For while the world itself may not change, the world of kinds may do so:

> The world does not change, but we work in a new world. The world that does not change is a world of individuals. The world in and with which we work is a world of kinds. The latter changes; the former does not. After a scientific revolution, the scientist works in a world of new kinds. (1993, p. 306)

Since the world of individual objects is unaltered by change of theory, there is a robust sense in which the world is stable. Yet since we must think and interact with the world in terms of categories supplied by us, the world of kinds which we inhabit is a world in flux.

Hacking's interpretation of Kuhn as a kind of nominalist fits well with Kuhn's suggestion that there may be taxonomic change with change of theory, and it makes plausible sense of the world-change image. Yet there remains in Kuhn's work a strong tendency toward the mind-dependence of objects. Recently, for instance, Kuhn has suggested that the individuation of things as objects depends on our application of sortal concepts which enable the identification of particular objects (1991b, pp. 20–1). And he has explicitly responded to Hacking that the latter's

> nominalist version of my position – that there are real individuals out there, and we divide them into kinds at will – does not quite face my problems – I need a notion of 'kinds' . . . that will populate the world as well as divide up a preexisting population. (1993, p. 316)

It therefore appears that Kuhn's position differs from Hacking's nominalist interpretation of it by denying that individual objects are to be conceived as existing entirely independently of human conceptual activity.

Nevertheless, Kuhn's apparent commitment to the view that there are both individual objects (e.g., the sun, moon and Earth) and kinds (e.g., alloys, salts), which survive variation of higher-order category (e.g., planet, compound), suggests an intermediate view. While ultimately objects and kinds depend for their individuation upon classification, lower-level empirical objects and kinds tend, on the whole, to survive changes in higher-order, theoretical classification. Thus, Kuhn's transcendental nominalism is combined with a mitigated empirical realism, according to which low-level objects and kinds, though by and large resistant to change, are classification-dependent.

ONTOLOGICAL RELATIVISM

On the interpretation of Kuhn's ontological relativism which I propose, Hacking's transcendental nominalism provides a key element of Kuhn's position. According to transcendental nominalism, beyond the level of commonplace empirical groupings, the world does not itself contain divisions between naturally occurring kinds of things. Rather, classification of the trans-empirical world into taxonomic kinds depends entirely on human conceptual contribution. Such classificational systems are developed in the course of scientific theorizing, and they are subject to revision in the transition between theories.

However, in addition to transcendental nominalism, I suggest that Kuhn's metaphysical stance be interpreted as a Kantian one of the kind earlier described as constructivism. On such a view, there is indeed a reality independent of all human mental activity. But such a reality is, Kuhn says, "ineffable, undescribable, undiscussible" (1991a, p. 12). Presumably, it is also largely, if not entirely, unknowable (cf. 1979, p. 418). Instead of such a thoroughly mind-independent reality, the world experienced by humans is a phenomenal world that is a joint product of sensory input, deriving ultimately from reality itself, and of our human conceptual contribution. Such a phenomenal world is a constructed world which contains the kinds of entities which are described by the categorial scheme of the operative theory.

It must be emphasized that Kuhn's view is not that the phenomenal world experienced by the scientist is entirely produced by the categorial scheme of a theory. Rather, the taxonomic categories of the scheme provide a structure for possible experience:

> Insofar as the structure of the world can be experienced and the experience communicated, it is constrained by the structure of the lexicon of the community which inhabits it. (1991a, p. 10)

The idea that the lexicon provides a structure which constrains experience is, as Kuhn notes, heavily Kantian: "like the Kantian categories, the lexicon supplies preconditions of possible experience" (1991a, p. 12). And again,

> Both [lexical structures and Kant's *a priori* categories] are constitutive of *possible experience* of the world, but neither dictates what that experience must be. Rather they are constitutive of the infinite range of possible experiences that might conceivably occur in the actual world to which they give access. (1993, p. 331)

Thus, Kuhn's position is one on which the manner in which incoming sensory input is experienced is determined by the categorial scheme, and so the phenomenal world of the scientist varies relative to the operative categorial scheme.

Such constructivist variation of the phenomenal world with the categorial scheme, combined with the transcendental nominalist rejection of mind-independent trans-empirical kinds, provides the basis for my reading of Kuhn's ontological relativism. This interpretation of Kuhn takes over from transcendental nominalism the thesis that there are no higher-level trans-empirical natural kinds for the categorial schemes of theories to reflect accurately or inaccurately. And it conjoins with such nominalism the constructivist thesis that the phenomenal world experienced by the scientist depends on the categorial scheme of the theory employed by the scientist.

On the metaphysical picture yielded by this combination of nominalism and constructivism, the taxonomic structure of the phenomenal world of a theory depends on the categorial scheme employed by the theory. As a result, the phenomenal worlds of scientific theories associated with different categorial schemes contain divergent systems of natural kinds. Thus, the set of natural kinds constitutive of the taxonomic structure of the phenomenal world of a theory depends on the categorial scheme of the theory. Given that such phenomenal worlds vary relative to the categorial scheme of operative theory, the existence of a set of natural kinds which populates the phenomenal world of the scientist is therefore a form of existence which is relative to prior choice of scientific theory.

KUHN'S VIEW OF TRUTH

As further evidence that Kuhn's ontological relativism is a position of the kind I have just outlined, I wish now to discuss Kuhn's views on the nature of truth. Kuhn has been a long-standing critic of the application of the correspondence theory of truth to the relation between scientific theories and reality (cf. 1970a, p. 206). In his recent work, Kuhn continues to oppose the correspondence theory, and has also sketched his position about the nature of truth in the context of the idea of variant lexical structures.

According to Kuhn's present views on the subject of truth, the correspondence theory of truth must be rejected, though there remains a necessary role to be played by a weaker conception of truth. The required weaker notion of truth must have an application that is internal to lexical frameworks. For, while a claim may properly be said to be true or false within the context of a given lexicon, the categorial system embedded in the lexicon is not itself capable of being true or false.

In rejecting the correspondence theory of truth, Kuhn wishes to reject the idea that the categorial structure of a theory might accurately reflect the way the world is independently of theory. That such structures cannot themselves be correspondence-true is suggested by the following passage, in which Kuhn claims that the form of life associated with a given lexicon cannot itself be true or false.

> Experience and description are possible only with the described and describer separated, and the lexical structure which marks that separation can do so in different ways, each resulting in a different, though never wholly different, form of life. Some ways are better suited to some purposes, some to others. But none is to be accepted as true or rejected as false; none gives privileged access to a real, as against an invented, world. The ways of being-in-the-world which a lexicon provides are not candidates for true/false. (1991a, p. 12)

Such a denial that the taxonomic structures of theoretical lexicons may even constitute possible candidates for truth or falsity accords well with the present reading of Kuhn's ontological relativism. For on such a view, the world itself has no natural kind structure for categorial schemes to correspond with, and taxonomic structures only come into play once one has entered a given phenomenal world.

While Kuhn rejects application of the correspondence theory to the relation between categorial systems and reality, he holds that a weaker notion of truth is required, which may be applied internally to the lexical structures of theories:

> . . . lexicons are not . . . the sorts of things that can be true or false. A lexicon or lexical structure is the long-term product of tribal experience in the natural and social worlds, but its logical status, like that of word-meanings in general, is that of convention. Each lexicon makes possible a corresponding form of life within which the truth or falsity of propositions may be both claimed and rationally justified, but the justification of lexicons or of lexical change can only be pragmatic. With the Aristotelian lexicon in place it does make sense to speak of the truth or falsity of Aristotelian assertions in which terms like 'force' or 'void' play an essential role, but the truth values arrived at need have no bearing on the truth or falsity of apparently similar assertions made with the Newtonian lexicon. (1993, pp. 330–1)

Kuhn thus allows that there is a notion of truth which has a valid use within the context of a given lexicon; the notion he has in mind might, he says, be provided by "something like a redundancy theory of truth" (1991a, p. 8).

Since Kuhn makes application of the concept of truth internal to lexicon, it might appear that he adopts a relativistic view of truth. However, Kuhn does not make the truth of scientific claims relative to operative theory. It is rather the case that a claim which may be true within the lexical framework of one theory fails to correspond to any comparable claim asserted or denied by an alternative theory. This point is closely connected with the incommensurability of such theories:

> Within the world of each practice, true laws must be universal, but some of the laws governing one of these worlds cannot even be stated in the conceptual vocabulary deployed in, and partially constitutive of, another. The same no-overlap principle that necessitates the universality of true laws bars the practitioners resident in one world from importing certain of the laws that govern another. The point is not that laws true in one world may be false in another but that they may be ineffable, unavailable for conceptual or observational scrutiny. It is effability, not truth, that my view relativizes to worlds and practices. (1993, p. 336)

Thus, rather than a relativistic view on which the truth of shared claims about the world varies with theory, Kuhn's view is one on which claims about the world may fail to be shared across such theories. Hence, true claims made by one theory are unable even to be expressed within the context of another theory.

Kuhn's remarks about truth comport well with my interpretation of his ontological relativism. For one thing, consider Kuhn's rejection of the correspondence theory of truth. Kuhn denies that a categorial scheme may accurately reflect reality in the sense of the correspondence theory of truth. This accords with the transcendental nominalist denial that reality is itself divided up into natural kinds independently of human conceptual intervention. For another thing, Kuhn's notion of truth internal to a lexicon sits well with the constructivist thesis that the phenomenal world of the scientist depends on the categorial scheme of accepted theory. For, given that scientists occupy a particular phenomenal world, they will be able to decide on questions of truth and falsity arising within such a world. Yet, due to differences in the categorial structure of theories, questions of the truth-value of a particular claim made by a theory need not arise within the context of a theory with which it is incommensurable.

CONCLUSION

I wish to conclude by re-stating some of the central themes I have developed here. One of my central claims has been a historical one about the development of the relativistic position which characterizes Kuhn's philosophy of science. As originally elaborated in *The Structure of Scientific Revolutions*, Kuhn's position appeared to contain both a relativistic stance towards matters of scientific rationality and a radical conceptual relativism of incommensurable conceptual schemes. However, both of these claims were moderated, as Kuhn admitted the existence of extra-paradigmatic methodological factors informing rational theory-choice, and reduced the scope of conceptual variation between theories with his thesis of local incommensurability.

However, there continues to be a strong tendency towards relativism in Kuhn's work. This tendency centers on his denial of the existence of a reality which has an inbuilt natural kind structure independent of human conceptual intervention. This aspect of Kuhn's relativism places his views in sharp contrast with those scientific realists who hold that there is a mind-independent reality, replete with objective natural kinds, the existence and constitution of which are completely independent of human mental activity. A second key feature of Kuhn's ontological relativism is his commitment to the Kantian view that the world phenomenally presented to the scientist is in large part determined by the taxonomic structure which theories impose on the world. This aspect of his position places Kuhn in close proximity to those idealist or idealistically-inclined philosophers who have insisted on the impossibility of extricating ourselves from our conceptual frameworks to compare our thoughts and concepts directly with reality. Finally, Kuhn's rejection of a correspondence theory of truth in favour of truth being internal to a lexicon represents both a rejection of standard forms of scientific realism, as well as an attempt to present a relativistic position which avoids familiar objections to relativism about truth. In short, while Kuhn's ontological relativism sets him in conflict with classic forms of scientific realism, his work shares many common themes with anti-realist, internalist and pragmatist approaches currently much in vogue.[3]

University of Melbourne, Australia

NOTES

[1] I take the expression 'phenomenal world' from Paul Hoyningen-Huene, who draws an explicit parallel between Kuhn and Kant: "For both Kant and Kuhn, epistemic *subjects* are constitutive of [the phenomenal world]". Drawing an analogy with Kant's idea of a thing-in-itself, Hoyningen-Huene contrasts the phenomenal world of a scientist with "the world-in-itself", which is both invariant and unknowable. See Hoyningen-Huene (1993, pp. 32–5).

[2] It is important to note that the requirement of sameness of taxonomic structure across lexicons is meant by Kuhn to be stronger than a merely extensionalist requirement that the taxonomic categories of different classificatory schemes have the same items in their extensions. The extensions of such categories must not only be specified as objects belonging in the extension; they must also be represented in some way as constituting a natural kind (1983, p. 676). Presumably, this requires that there must be some minimal retention of sortal or categorial vocabulary across taxonomic systems. However, Kuhn appears to hold that the same kinds may be picked out within different systems of classification, even though no criteria of categorization are shared across classificatory system (1983, p. 681–3). For discussion of this issue, see my (1994, pp. 95–100).

[3] This paper was written while I held a Visiting Fellowship at the Center for Philosophy of Science at the University of Pittsburgh. I am grateful to the Center for hospitality and for providing a stimulating environment in which to pursue research. I also thank Dimitri Ginev for the invitation to contribute to this volume.

REFERENCES

Davidson, Donald (1984), 'On the Very Idea of a Conceptual Scheme', in *Inquiries into Truth and Interpretation* (Oxford University Press, Oxford).

Devitt, Michael (1979), 'Against Incommensurability', *Australasian Journal of Philosophy* **57**, pp. 29–50.

Devitt, Michael (1984), *Realism and Truth* (Blackwell, Oxford).

Doppelt, Gerald (1982), 'Kuhn's Epistemological Relativism: An Interpretation and Defense', in *Relativism: Cognitive and Moral*, M. Krausz and J. W. Meiland (eds.) (University of Notre Dame Press, Notre Dame).

Hacking, Ian (1979), 'Review of *The Essential Tension*', *History and Theory* **18**, pp. 223–236.

———. (1983), *Representing and Intervening* (Cambridge University Press, Cambridge).

———. (1993), 'Working in a New World: The Taxonomic Solution', in *World Changes: Thomas Kuhn and the Nature of Science*, P. Horwich (ed.) (MIT Press, Cambridge).

Hoyningen-Huene, Paul (1993), *Reconstructing Scientific Revolutions: Thomas S. Kuhn's Philosophy of Science* (University of Chicago Press, Chicago).

Kroon, Frederick W. (1985), 'Theoretical Terms and the Causal View of Reference', *Australasian Journal of Philosophy* **63**, pp. 143–66.

Kuhn, Thomas S. (1970a), *The Structure of Scientific Revolutions*, second edition, (University of Chicago Press, Chicago).

———. (1970b), 'Reflections on my Critics', in *Criticism and the Growth of Knowledge*, I. Lakatos and A. E. Musgrave (eds.) (Cambridge University Press, Cambridge).

———. (1977), 'Objectivity, Value Judgment and Theory Choice', in *The Essential Tension* (University of Chicago Press, Chicago).

———. (1979), 'Metaphor in Science', in *Metaphor and Thought*, A. Ortony (ed.) (Cambridge University Press, Cambridge).

———. (1983), 'Commensurability, Comparability, Communicability', in *PSA 1982*, Volume 2, P. D. Asquith and T. Nickles (eds.), Philosophy of Science Association (East Lansing Michigan), pp. 669–88.

———. (1987), 'What Are Scientific Revolutions?', in *The Probabilistic Revolution*, L. Krüger, L.J. Daston and M. Heidelberger (eds.) (MIT Press, Cambridge).

———. (1990), 'Dubbing and Redubbing: The Vulnerability of Rigid Designation', in *Scientific Theories: Minnesota Studies*, Volume 14, C. W. Savage (eds.) (University of Minnesota Press, Minneapolis).
———. (1991a), 'The Road Since Structure', in *PSA 1990*, Volume 2, A. Fine, M. Forbes and L. Wessels (eds.), Philosophy of Science Association (East Lansing, Michigan), pp. 2–13.
———. (1991b), 'The Natural and the Human Sciences', in *The Interpretative Turn: Philosophy, Science, Culture*, D. R. Hiley, J. F. Bohman and R. Shusterman (eds.) (Cornell University Press, Ithaca).
———. (1993), 'Afterwords', in *World Changes: Thomas Kuhn and the Nature of Science*, P. Horwich (ed.) (MIT Press, Cambridge).
Lakatos, Imre (1970), 'Falsification and the Methodology of Scientific Research Programmes', in *Criticism and the Growth of Knowledge*, I. Lakatos and A. E. Musgrave (eds.) (Cambridge University Press, Cambridge).
Mandelbaum, Maurice (1982), 'Subjective, Objective and Conceptual Relativisms', in *Relativism: Cognitive and Moral*, J. W. Meiland and M. Krausz (eds.) (University of Notre Dame Press, Notre Dame).
Nola, Robert (1980), 'Fixing the Reference of Theoretical Terms', *Philosophy of Science* **47**, pp. 505–31.
Popper, Karl (1970), 'Normal Science and its Dangers', in *Criticism and the Growth of Knowledge*, I. Lakatos and A. E. Musgrave (eds.) (Cambridge University Press, Cambridge).
Putnam, Hilary (1981), *Reason, Truth and History* (Cambridge University Press, Cambridge).
Sankey, Howard (1994), *The Incommensurability Thesis* (Kluwer, Dordrecht).
Shapere, Dudley (1984), *Reason and the Search for Knowledge* (Reidel, Dordrecht).
Siegel, Harvey (1987), *Relativism Refuted* (Reidel, Dordrecht).

ERHARD SCHEIBE

THE PROBLEM OF REDUCTION IN THE THEORY OF SPECIAL RELATIVITY

INTRODUCTION

Recently I suggested a new theory of reductions in physics (Scheibe 1993, 1995, 1997). Beginning this paper I want to point out two innovations of that theory. First, the concept of reduction was not obtained by the usual analytic method of explication, i.e. no attempt was made to start out from some definite, if only vaguely understood explicandum, next to put up conditions of material adequacy in order to end up with a formally precise explicatum. Instead a *synthetic method* was suggested according to which we begin with a variety of rather special kinds of reductions and obtain the more general cases by using a natural *product operation* by concatenation. Second, in the new approach it is not required that the reduction of a theory has to include its fundamental assumptions. Rather a concept of *partial reduction* was developed in which we have to satisfy ourselves by reducing only this or that *consequence* of the theory to a corresponding consequence of the reducing theory. Both novelties were not the result of a deliberate act but were dictated by the very situation that we encounter in physics. It seems that widely differing theory reductions, in particular approximative ones, are accepted by physicists (although they might not be aware of this) and that the uselessness of the reduction theories hitherto developed goes back to the circumstance that each of them is characterized by one single, very general concept of reduction totally inadequate to cope with the complexity of the situation. Moreover, there are notoriously recalcitrant cases of theories as, for instance, classical mechanics *vis-à-vis* quantum mechanics that defy all attempts of their total reduction whereas we do know of many partial reductions in the sense indicated. Both phenomena, the internal complexity of the world of accepted reductions and the widespread use of partial reductions in the difficult cases should be considered in a general theory of reductions.

We will give an account of the first phenomenon by considering a concrete case. Let us suppose we wished to reduce Kepler's 3^d law to Newton's theory of two gravitating bodies, say, the sun and one planet. Kepler's law says that

$$\text{(1a)} \quad \frac{a^3}{T^2} = \frac{1}{4}\pi^2 \cdot \mu$$

where a is half the length of the major axis of the elliptic orbit of the planet, T its sidereal period and μ a constant being the same for all planets. A first explanation concerns this μ – the Kepler constant. The assumption that μ equals

D. Ginev and R. S. Cohen (eds.), Issues and Images in the Philosophy of Science, 321–342.

the mass M of the sun allows for the first trivial reduction

$$\text{(1b)}\qquad \frac{a^3}{T^2} = \frac{1}{4}\pi^2 \cdot M \wedge \mu = M \vdash \frac{a^3}{T^2} = \frac{1}{4}\pi^2 \cdot \mu$$

Of course, we know about this assumption only by knowing that we can successfully reduce the new Kepler law in (1b) and, consequently, that we can do the same with the original law *if* only we make that additional assumption. But this is no argument against the *fact* of the reduction (1b) which, in the sense of our earlier classification, is a (trivial) *refinement* (cf. Scheibe 1993). The meaning of this concept will become clear below.

The new Kepler law (as well as the old one) is empirically false: The right-hand side of (1a) is not independent of the planet, it depends on its mass m. Already here we meet with the situation that reductions comes to an end if we did not admit reductions by approximations. Such is our next step

$$\text{(1c)}\qquad \frac{a^3}{T^2} = \frac{1}{4}\pi^2 \cdot \mathrm{M}\left(1 + \frac{m}{M}\right)^{-2} \wedge m \ll M \vdash \frac{a^3}{T^2} = \frac{1}{4}\pi^2 \cdot M$$

where on the left-hand side the Newtonian correction of Kepler's law occurs. This *asymptotic reduction* teaches us that, as seen from the Newtonian correction the original Kepler law is still valid with high accuracy if only m is small with respect to M. Thus, by the corrected law we come to know the range of validity of its predecessor. However, we have to go ahead with our inquiry and ask: Whence comes the new m-dependent law?

Our third step, answering this question, again is a refinement. The reduction is to Newton's gravitational equations (gravitational constant = 1)

$$\text{(1d)}\qquad \begin{cases} \mathrm{d}^2X/\mathrm{d}t^2 = -m(X - x)\,|X - x|^{-3} \\ \mathrm{d}^2x/\mathrm{d}t^2 = -M(x - X)\,|x - X|^{-3} \end{cases}$$

with the help of the two additional assumptions

(1e) the centre of gravity of the system is at rest
the motions X and x are periodic.

It is only these assumptions that now make possible the definitions

(1f) T = period of the motion x
a = half the length of the major axis of the path of x

From (1de) together with (1f) the Kepler law to be reduced (on the left-hand side of (1c)) follows. This reduction is a refinement in the following sense: The description of the system by $\langle M, X; m, x\rangle$ is more refined than that given by $\langle M; m, T, a\rangle$ because, whereas exactly one description of the latter kind is associated with any of the former, in the reverse direction the mapping is one-to-many. We gain information, therefore, in this direction and loose it in the opposite.

The reduction chain to be established can now be completed by a direct *gen-*

eralization: We dismiss in (1de) the additional premises (1e) and are left with (1d) as a generalization of (1de). This step, too, is a reduction. So on the whole we have obtained a reduction chain among whose four members we find three reductions of pairwise different kinds. Let us now envisage the following question: How can these reductions, different in kind as they are, *and* their various combinations of still other kinds all be subsumed under one non-trivial concept of reduction that would not be accomplished in the recursive manner suggested here? The examples given and many others available make it appear hopeless to give a generally satisfactory answer, and even if an answer were at hand it would almost *per definitionem* iron out the variety of reductions that we were at pains to display. Now as before we would have to elaborate this aspect of the matter.

We can now come to a closer look at the second phenomenon mentioned at the beginning: the partial reductions. For the reduction of (1a) to (1d) we can take also another route than the one we have chosen. This makes it possible to view, say, (1c) as a partial reduction that, just as by reductions we explain theories, can in turn be explained as a reduction. For it is possible to reduce Kepler's 3^{d} law at first not as in (1c) to its Newtonian correction, but to the Newtonian 1-body theory

(1g) $$\frac{d^2x}{dt^2} = -Mx|x|^{-3} \text{ and } x \text{ periodic}$$

This can be done by a refinement which in the 1-body case is the precise analogue of the refinement in the third step of the previous reduction chain. (1c) then appears as a partial reduction of (1g) to (1de) and would be explained by a corresponding total (approximate) reduction of (1g) to (1de) which in this case provides no difficulties (cf. Scheibe 1997, pp. 178ff). This procedure can be visualized by the closed *reduction square*

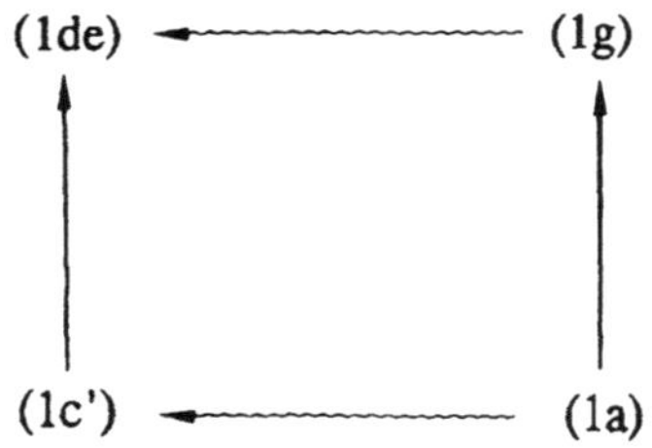

Figure 1.

where (1c′) stands for the corrected Kepler law on the left-hand side of (1c). In fact, one is inclined to view (1c′) as being *the* correction of (1a) in (1de). Moreover, we here have an explanation of the approximate reduction below by the approximate reduction above, because (1c′) is obtained from (1de) by the *same* derivation, in a sense, as the one by which (1a) is derived from (1g).

There are, however, cases, and very prominent cases at that, in which the reduction square remains *open*:

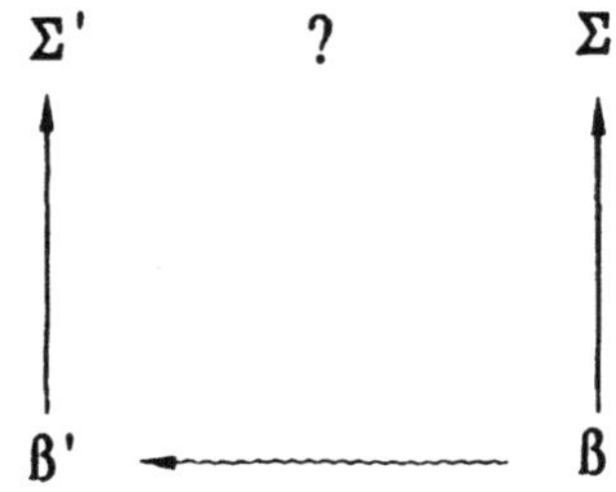

Figure 2.

A notoriously difficult case in point is the reduction of classical (statistical) mechanics (for Σ) to quantum mechanics (for Σ'). 'Consequences' β' and β can easily be found such at 1) β is approximately reducible to β' and 2) β' is in a sense *the* correction of β in Σ'. To give but one example, β may be the classical formula

(2a) $\quad E = kT$

for the mean energy E of an harmonic oscillator in thermal equilibrium at temperature T and β' the corresponding quantum mechanical formula

(2b) $$E = \frac{1}{2}\hbar\omega + \hbar\omega\left\{\exp\left(\frac{\hbar\omega}{kT}\right) - 1\right\}^{-1}.$$

With the help of

(2c) $\quad \hbar\omega \ll kT$

(2a) is (asymptotically) reducible to (2b). Moreover, there can be no doubt that (2b) says in terms of quantum mechanics what (2a) expresses classically. This can be argued from the way in which (2a) is inferred from classical statistical mechanics and (2b) from quantum mechanics (cf. Scheibe 1997, pp. 212ff). Although in the present case there are some obscurities in this respect, the serious problem is the completion of (Figure 2) in the upper row: There is generally no procedure of reducing classical (statistical) mechanics to quantum mechanics *in general*. Therefore, in order not to stand empty-handed it is better to admit partial reductions in a case like this.

In this paper we attack the problem in which sense, if at all, the geometry of the Galileo group can be reduced to the geometry of the theory of special relativity (STR) or Minkowski geometry. This problem already belongs to the more difficult cases and has become controversial among philosophers of science and physicists. The possibility of such a reduction and, moreover, the reduction of Newtonian mechanics to relativistic mechanics has been denied by some (e.g. Feyerabend 1965; Kuhn 1970, Ch. IX) while others have undauntedly asserted it (e.g. Havas 1964; Ehlers 1983 and 1986). The alleged reason for the impossibility to establish reductions in this field was the so-called incommensurability of the theories involved. But in spite of the vast literature concerned with the matter it has remained unclear what incommensurability means in general or even in the case before us. In the following the treatment is confined to the geometries in point. For it is mainly their reduction that makes the trouble. Two possibilities are described. A limiting case reduction takes seriously the formula $c \to \infty$ often used in text books on the matter. The reducing theory here is *not* ordinary Minkowski geometry but a slight generalization of it which, however, leads to a partially metaphysical status of the theory. The reduction to ordinary Minkowski geometry, coming with the formula $v \ll c$ and the like, is by an asymptotic procedure. The synthetic method is used; also partial reductions are admitted. Although, therefore, the reader's readiness for tolerance in the sense of this introduction is claimed it is hoped that also some insight into the matter is provided.

II. ASSIMILATION OF THE FORMALISMS

Among the difficulties mentioned in the introduction there are already those conceptual inhomogeneities between two spacetime theories that can eventually be removed by the proof of their *equivalence*. Equivalences of theories are borderline cases of reductions (cf. Scheibe 1997, Ch. IV.2). In our present context they are important not for their own sake but in their auxiliary function to assimilate two theories that are *not* equivalent by passing over to equivalent versions of each that are better adapted to the task of the former's reduction. It is true, to hold the view that the situation of comparison can be improved by conceptual assimilation is to admit that the possibilities of reduction are not independent of the accepted basic concepts: They are not the same *modulo* equivalence. Rather one has to view also the equivalences as genuine reductions and to invoke our synthetic principle emphasizing the necessity of combining reductions. Indeed we are concerned here with an analysis of the form

$$\text{(CA)} \qquad T_1 \approx T < T' \approx T_1'$$

where T_1 and T_1' are given and then replaced by their equivalents T and T' respectively for which finally the reduction proper $<$ has to be performed. There are therefore up to three steps necessary to complete the reduction.

In our context T_1' could be the spacetime theory that Einstein had introduced in his original paper on STR (1905), and T_1 could be Weyl's formulation of Galileo spacetime geometry (1923, §20). If we take up Einstein's theory in the 4-dimensional version that was given to it by Minkowski (1908) we have the following two theories before us for comparison: *Galileo Geometry* $\Sigma^{\circ}_{\text{gal}}$ is about a spacetime

$$\text{(3a)} \qquad \langle M; F, t, h \rangle$$

where F is a complete set of affine coordinate systems on M, t a covariant affine vector $\neq 0$ and h a two-contravariant symmetric affine tensor of signature $(0 + + +)$ with $h^{ab} \cdot t_b = 0$. Roughly put, t stands for time and h for space. By contrast, *Minkowski Geometry* $\Sigma^{\circ}_{\text{min}}$ deals with a spacetime

(3b) $\langle M; F, G\rangle$

where $\langle M; F\rangle$ is as before and G is a two-covariant symmetric affine tensor of signature $(+ - - -)$. It determines the new metric situation of spacetime, and in doing this *alone* it becomes already clear that space and time are closer connected in Minkowski than in Galileo geometry where *two* quantities had to be used for their representation. Desirable as it may be in a different context to emphasize this contrast, for purposes of comparison as we pursue them this inhomogeneity is unfavourable. In terms of (CA) we have to look for theories T and T' equivalent to T_1 and T_1' respectively and showing greater homogeneity with respect to their basic concepts. The following solution suggests itself (cf. Havas 1964).

In the Galileo case we choose a spacetime

(4a) $\langle M; F, g, h\rangle$

and in the Minkowski case accordingly

(4b) $\langle M; F, G, H\rangle$

where g and G are two-covariant symmetric affine tensors of inertial index 1 and h and H are two-contravariant symmetric affine tensors of inertial index 3. Moreover in the Galileo case

(4c) $g_{ab} \cdot h^{bc} = 0$

whereas in the Minkowski case

(4d) $G_{ab} \cdot H^{bc} = -\delta^c_a$

is required. From these theories Σ_{gal} and Σ_{min} the existence of a distinguished coordinate system follows in which the metric tensors have the normal forms

(4e) $g = \text{diag}\{1, 0, 0, 0\}, \quad h = \text{diag}\{0, 1, 1, 1\}$

and

(4f) $G = \text{diag}\{1, -1, -1, -1\}, \quad H = \text{diag}\{-1, 1, 1, 1\}$,

respectively. Σ_{gal} is equivalent to $\Sigma^{\circ}_{\text{gal}}$ and Σ_{min} to $\Sigma^{\circ}_{\text{min}}$. The relation between t and g is $g_{ab} = t_a \cdot t_b$, h remaining the same, while in the Minkowski case H occurring in (4b) is simply defined by $H = -G^{-1}$, in this case G remaining the same.

The step from (3) to (4) is no doubt a step of conceptual assimilation in the sense we wanted it for our later comparison. Before embarking on this we will go one step further by using still another process of assimilation (cf. Ehlers 1986). It is meant to remove the following difference. The metric tensors occurring in (4) are determined only up to a positive factor corresponding to an arbitrariness in the choice of units. Assuming that g and G have the dimension [time]^2, h and H the dimension [length]$^{-2}$ we are led to the following scale transformations: if we multiply *in* Σ_{gal} a time unit by α and a length unit by β we get as transformations of the tensors

(5a) $\quad g' = \alpha^{-2} \cdot g, \quad h' = \beta^{-2} \cdot h,$

and these transformations leave invariant, as it should be, the requirements of Σ_{gal} including (4c). In Σ_{gal}, therefore, we can choose time and length units *independently* of each other. It is different *in* Σ_{min}. There H is uniquely coupled to G (and *vice versa*), and therefore (4d) is only invariant for $\alpha = \beta$. This is a consequence of the unification of space and time in this theory: Only one unit at a time can be chosen arbitrarily. If we choose 1 second as time unit, we get as a natural length unit the length of the distance travelled by light during 1 second. If instead we choose 1 meter for length, we obtain as natural time unit the time light needs to cover 1 meter. We are left, therefore, with the transformations

(5b) $\quad G' = \alpha^{-2} \cdot G, \quad H' = \alpha^2 \cdot H$

only.

If we want to remove this difference between Σ_{gal} and Σ_{min} in favour of an independent choice of units also in the latter (which again is a bit against the spirit of this theory), we have to generalize it (cf. Ehlers 1986 and Lottermoser 1988). We obtain a *common generalization* $\Sigma_{\geq 0}$ of Σ_{gal} and Σ_{min} by introducing an additional real parameter $\lambda \geq 0$ of dimension [speed]$^{-2}$. $\Sigma_{\geq 0}$ is about a spacetime

(6a) $\quad \langle M; F, G, H, \lambda \rangle$

submitting it to the same assumptions (common to Σ_{gal} and Σ_{min}) as before except for (4cd) that are replaced by

(6b) $\quad G_{ab} \cdot H^{bc} = -\lambda \cdot \delta_a^c.$

The scale transformations now become

(6c) $\quad G' = \alpha^{-2} G, \quad H' = \beta^2 H, \quad \lambda' = \beta^2 \alpha^{-2} \lambda,$

and (6b) is now invariant under these transformations. The new normalforms, replacing (4ef), are

(6d) $\quad G = \mathrm{diag}\{1, -\lambda, -\lambda, -\lambda\}, \quad H = \mathrm{diag}\{-\lambda, 1, 1, 1\}.$

The theories Σ_{gal} and Σ_{min} are special cases of $\Sigma_{\geq 0}$ for $\lambda = 0$ and $\lambda = 1$ respectively. But we also have a new rival for Σ_{gal} in form of the *generalized Minkowski geometry* $\Sigma_{>0}$ by restricting $\Sigma_{\geq 0}$ to $\lambda > 0$.

It is necessary to protect this theory against a notorious misunderstanding. It consists in the objection that λ as the reciprocal square of the maximal speed on the light cone is a constant subject to dimension that can be given *any* numerical value by appropriate scale or coordinate transformations. The variable λ, therefore, nothing but feigns that we have here really different cases and a genuine generalization before us. Now it is true that we have the following procedure: if we apply the transformations (6c) to the normalforms (6d) they get lost, and we obtain

$$G' = \mathrm{diag}\{\alpha^{-2}, -\beta^{-2}\lambda', -\beta^{-2}\lambda', -\beta^{-2}\lambda'\}$$

$$H' = \mathrm{diag}\{-\alpha^{-2}\lambda', \beta^{2}, \beta^{2}, \beta^{2}\}.$$

By means of a subsequent coordinate transformation

$$y^{\circ} = \alpha^{-1}\, x^{\circ}, \quad y^{i} = \beta^{-1} x^{i}$$

we regain the normal forms (6d) with λ replaced by λ'. In the literature one often finds the special case $\alpha = c^{-1}$, $\beta = 1$, $\lambda = c^{-2}$ leading to $\lambda' = 1$ and being a transition from an uncoupled scale system, e.g. meter/second, with c as the velocity of light to a coupled system, e.g. meter/light meter, with the velocity of light 1.

It is therefore evident and immediately to be admitted that we can give the speed of light – read off from (6d) as $\lambda^{-1/2}$ – any value we want *without thereby changing the metric* – solely by scale and coordinate transformations. It is, however, an entirely different matter if we, *by fixing the scale and coordinates*, go from one spacetime (6a) with G, H, λ to another one with G_1, H_1 $\lambda_1 \neq \lambda$. From the existence of the first mentioned procedure there does not follow the non-existence of the second. Rather the various instances of (6d) represent objectively different, not scale equivalent and even non-isomorphic possible spacetimes. The situation is similar to that in ordinary geometry where Euclidean spaces were generalized to spaces with non-zero constant curvature with this scalar curvature as the new parameter. The only disputable objection that can be brought forward against the generalized Minkowski geometry is that in fact we meet with one empirical instance of (6a) only and that, moreover, we *can* meet with no more. The assumption of the variety of cases (6a) corresponding to different λ as representing so many physically possible spacetimes is, therefore, a non-empirical and even metaphysical assumption that can be made only in the subjunctive mood. And yet, for the sake of the argument, we will allow ourselves the introduction of this theory which seems to be the only one allowing for a theory of reduction worth the name within the scope of STR.

Besides the tensor formulation of the geometries introduced there are also essentially *group theoretical formulations*. In his Erlangen programme of 1872 Felix Klein had suggested to treat the geometries known at that time under the unifying aspect of their classification by their automorphism groups. In our paper, the affine geometry, underlying all other geometries introduced, is already formulated in this way. It has been defined by a complete class F of affine coordinate systems involving the group of all affine transformations.

(7a) $\quad x^{\mu'} = A^{\mu}_{\lambda} \cdot x^{\lambda} + a^{\mu}$

of $\mathbb{R}^4$ as the group of coordinate transformations belonging to F. We now submit also the metric to a similar treatment. We envisage structures

(7b) $\langle M; F, f, \lambda\rangle$

where $f \subseteq F$ and there exists a coordinate system $\varphi \in F$ such that $f = \Gamma_\lambda\varphi$. Given $\lambda \geq 0$, Γ_λ is the group of transformations

(7c) $$\begin{cases} t' = \gamma\{t - \lambda(\mathbf{v} \cdot A\mathbf{x})\} + a) \\ x' = (A\mathbf{x} - \mathbf{v}t) + (\gamma - 1)\mathbf{v}/v\{\mathbf{v}/v \cdot (A\mathbf{x} - \mathbf{v}t)\} + \mathbf{a} \end{cases}$$

of $\mathbb{R}^4$. Here A is orthogonal, $v = |\mathbf{v}|$, $v < \lambda^{-1/2}$ and

(7d) $\gamma = (1 - \lambda v^2)^{-1/2}$.

Each of these transformations, parametrized by $\langle A, \mathbf{v}, \mathbf{a}, a\rangle$, is a product of
1. a 'spatial' rotation A,
2. an inertial motion with speed $\mathbf{v}$,
3. a translation a in 'time' and $\mathbf{a}$ in 'space'
   in this order (cf. Fock 1960, §10; Møller 1972, §2.4).

The geometry $\Sigma^*_{\geq 0}$ defined in this way is equivalent to $\Sigma_{\geq 0}$, and this equivalence induces equivalences of the special cases Σ^*_{gal} ($\lambda = 0$), Σ^*_{min} ($\lambda = 1$) and $\Sigma^*_{> 0}$ ($\lambda > 0$) to the corresponding special cases Σ_{gal}, Σ_{min} and $\Sigma_{> 0}$ respectively. In particular we obtain as the corresponding transformation groups the Galileo group Γ_{gal}, the Poincaré group Γ_{min} and the family $\{\Gamma_\lambda\}_{\lambda > 0}$ of generalized Poincaré groups. The equivalence in question is easily established by the fact that we have introduced the groups Γ_λ only by their normal forms (7c) corresponding to the normal forms (6d) of the tensors: Given the spacetime (6a), f is the set of coordinate systems distinguished by G, i.e. where G has the normal form (6d), and given a spacetime (7b), G is the tensor having its normal form (6d) in the coordinate systems of f.

In view of the visibly increasing number of geometries entering our investigation it should be emphasized that the number of *inequivalent* geometries among them is only four. But we have given each of them several equivalent formulations. The following synopsis will clarify the situation:

Galileo geometry ($\lambda = 0$): $\Sigma^\circ_{\text{gal}}$, Σ_{gal}, Σ^*_{gal}
Minkowski geometry ($\lambda = 1$): $\Sigma^\circ_{\text{min}}$, Σ_{min}, Σ^*_{min}
Generalized Minkowski geometry ($\lambda > 0$): $\Sigma_{> 0}$, $\Sigma^*_{> 0}$
λ-geometry ($\lambda \geq 0$): $\Sigma_{\geq 0}$, $\Sigma^*_{\geq 0}$

Our problem of reduction is the problem of reducing Galileo geometry (GG) to either Minkowski geometry (MG) or generalized Minkowski geometry (GMG) or both. The importance of λ-geometry (λG) for our comparison will become clear in the sequel.

III. ASSIMILATION OF MODELS

Up to this point our considerations were concentrated on the *formalisms* of our geometries: The point was to assimilate them to each other and to unify them. Something similar has now to be done with their *models*. Roughly speaking, models of a formalism are the objects that satisfy it. For our geometries they were indicated already by the formulas (3ab), (4ab), (6a) and (7b). Our first question concerning models is which ones can be distinguished as physically possible models. Secondly, it is to be asked how the models thus distinguished for each geometry must be related to each other if these relations shall reflect the situation of competition between the geometries as well as their partial coincidences. Since all our geometries are but special cases of one of them, i.e. λ-geometry, the second question will be answered once the first one is settled for λ-geometry. However, for the sake of clarity we will begin with the more familiar cases GG and GMG.

From the model theoretical point of view these theories are categorical, i.e. any two of their models are isomorphic. This means that everything that can be done to make them as strong as possible *has* been done. And yet there could be countless pairwise isomorphic models if no further restrictions were imposed. This contrasts with the age-old, more or less tacit assumption that space and time or at least spacetime in the sense of STR are *unique*. Kant has even denied the existence of any alternative to this view. "We can" says he, "imagine only one single space, and whenever we speak of many spaces we mean to speak only of parts of one and the same unique space" and correspondingly "different times are only parts of one and the same time" (*Critique of Pure Reason* B 39 and B 47 resp.). For Kant this view was a consequence of his fargoing thesis that space and time are not concepts at all but rather pure forms of intuition. Nevertheless his quoted words still hold for the spacetime of STR without its being a pure form of intuition: The ordinary formulations of relativistic theories do not provide for several possible spacetimes as they *do* provide for various possible particles, fields etc. Therefore, we have to imagine that GG and MG each as firmly connected with a model, e.g. (4ab) in the tensor versions, as the subject about which it makes assertions. It is only the pair, consisting of a formalism *and* a model, that forms a theory in this case.

However, with λ-geometry and GMG we have introduced geometries also containing *variables*: first the parameter λ, but together with it also the tensors G and H (and f as a set of distinguished coordinate systems in the group theoretical version). In this way the question arises in what multiplicity models of these theories are admitted as physically possible. The answer must be compatible with the decision already made concerning GG and MG, these being but special cases of λ-geometry. Now that decision (trivially) satisfies the principle to admit exactly one model from every class of isomorphic models. Since all special geometries corresponding to the possible values of λ are categorical, the same principle suggests itself also in general. By further considering the fact that any two models belonging to different values λ are *not* isomorphic, we are led to admit the following models of $\Sigma_{\geq 0}$ (and correspondingly of $\Sigma^*_{\geq 0}$): One affine structure $\langle M°; F° \rangle$ is fixed, one coordinate system $\varphi° \in F°$ chosen, and then for every $\lambda \geq 0$ the G_λ and H_λ given by the normalforms (6d) in $\varphi°$ (and correspondingly $f_\lambda = \Gamma_\lambda \cdot \varphi°$) complete the model. In the sense of this selection $\Sigma_{\geq 0}$ (and $\Sigma^*_{\geq 0}$ respectively) has got a 1-parametric family of *standard models*

(8a) $\langle M°; F°, G_\lambda, H_\lambda, \lambda \rangle$ (for $\Sigma_{\geq 0}$)

and

(8b) $\langle M°; F°, f_\lambda, \lambda \rangle$ (For $\Sigma^*_{\geq 0}$),

respectively. In particular all standard models have one distinguished coordinate system in common.

Having done with our first problem, we now see that we already have taken the decisive step in answering our second question: We have got the foundation for the comparison of GG with MG and with GMG. Let us argue for this in the following way. Intuitively our geometries are *rivals* of each other: They say incompatible things about the same subject, and if we assume MG or GMG we have to reject GG. To bring this situation to closure it seems necessary to describe our subject independently of the geometries. It may be that this is possible with the help of a rudimentary physical language. However, we do not want to follow up this route. A second option is to assume that one of the rival geometries is correct and, accordingly, base the comparison on a model of this geometry. Let us take as such a standard model μ of GMG. In μ we could interpret also the competing GG: we need only to give the symbols in (4a), one by one, the same interpretation as was given the symbols in (8a) in μ (for some $\lambda > 0$). In doing this we make full use of the conceptual assimilation achieved in the previous section. The competition between GG and GMG would then show up in that Σ_{gal} now makes certain *false* statements about μ. We would get, for instance, that any two events, simultaneous in one coordinate system, are simultaneous in every other coordinate system. In μ, however, as being a model of $\Sigma_{\geq 0}$, two different events, simultaneous in one inertial system, are not simultaneous in any other inertial system. We could produce any number of such inconsistencies and in this manner bring to bear the rivalry in question. For a genuine comparison, however, this would be too little even if, besides these, we also pointed out as many statements of GG that are valid in μ. What is missing is a positive idea of what *would* be the case *throughout* if GG were true.

With such an idea we can be furnished only by means of a model also of GG. And here we face our main question: How are we to choose this model, say, μ_{gal} relative to μ? The answer is given by our suggestion which models have to be counted as physically possible ones. μ being a standard model for $\lambda > 0$, μ_{gal} is naturally chosen as the (only) standard model for $\lambda = 0$. By this choice μ_{gal} becomes as similar to μ as is allowed by the incompatibility of GG and GMG. For one thing, the affine part of GG has never become controversial within STR. Accordingly, all standard models of λG have the affine space $\langle M°; F°\rangle$ in common. It is different with the metric and the coordinate systems distinguished by it. On the one hand, the requirement of one such coordinate system common to all standard models is absolutely essential, and without it no substantial comparison could be made. Moreover, it *can be* fulfilled. On the other hand, once it is fulfilled the metric is uniquely determined for all λ. There is no further choice open to us. The relative position of the standard models to each other cannot be submitted to further postulates that would not also concern their absolute position. And this, i.e. the choice of $\langle M°; F°\rangle$ and of the common coordinate system $\varphi° \in F°$, should be left untouched.

IV. LIMITING CASE REDUCTION

There remains the question whether our geometries thus conceived can justifiably be said to stand in a relation of *reduction*, i.e. whether GG (including its standard model) appears to be reduced to either MG (including its standard model) or to GMG (together with its range of standard models). The answer is in the affirmative if, in the spirit of the Introduction, one is ready to accept partial reductions and different kinds of approximate reductions. In fact, the relation between GG and MG can be understood to be a kind of *asymptotic* reduction whereas GG is related to GMG as a *limiting case* for $\lambda \to 0$ (cf. Scheibe 1993 and 1997, Ch. V). In both cases there are reasons to speak of partial reductions only. Independently of that distinction of two kinds of reduction, each of them can be applied in order to reduce concepts of a theory as well as its consequences.

Our tentative way of speaking of reductions of theories to other theories (or – for that matter – of geometries to other geometries) will first be substantiated by the idea of *concept reduction*, i.e. of reductions of concepts of a theory to concepts of another. In fact, it was mainly concept reduction that the critics of orthodox philosophy of science had in mind during the discussion about the meaning of scientific progress and related processes. Concepts were said to change their meaning in the course of history, and the problem was in what relation the older meanings stand to the newer and whether some idea of reduction could be used for this relation. Now, formally the concepts of a theory are defined by certain terms out of the basic concepts. Materially they have a well-defined extension in every model of the theory: the set of instances falling under the concept in question. Both parts – form and matter – play an important role in concept reduction. The form serves to relate concepts intertheoretically as having the *same* form and, therefore, (possibly) being about the same subject. This in turn makes them *candidates* for reduction. Their extensions, usually being different, will then actually be related by *approximations*.

In the limiting case reduction we are in fact confronted with the relation not of one concept to another concept but to a family of concepts, parametrized by a real variable which in the case at hand is our λ for $\lambda \geq 0$. The first question is which pairs, consisting of one concept of GG and a family of concepts of GMG, we have to choose as being pairs where the GG concept is the limit of the λ-family of GMG concepts for $\lambda \geq 0$. A simple method of producing this situation is to take any concept of λG (!) and restrict it to $\lambda = 0$ on the one hand and $\lambda > 0$ on the other. There is, of course, no guarantee that the limit relation holds in all these cases, and there are certainly the trivial cases where the λ-concept chosen does not even implicitly depend on λ and the pair becomes a pair of twins. But in many cases the method does provide us with exactly the kind of reductions that we are looking for.

In the *tensor version* $\Sigma_{\geq 0}$ of λG we would start out with a term $B(G_\lambda, H_\lambda)$ where B is explicitly defined for all $\lambda \geq 0$ but only implicitly dependent on λ. The restrictions mentioned provide us with a concept $B(G_0, H_0)$ of $\Sigma_0 = \Sigma_{\text{gal}}$ and the family $\{B(G_\lambda, H_\lambda)\}_{\lambda \geq 0}$ of concepts of $\Sigma_{> 0}$. The limiting case situation then means the following. We have for $\lambda \to 0$

(9a) $\quad \lim G_\lambda = G_0, \quad \lim H_\lambda = H_0$

for the basic concepts and

(9b) $\quad \lim B(G_\lambda, H_\lambda) = B(G_0, H_0)$

for the concept under discussion. It follows

(9c) $\quad \lim B(G_\lambda, H_\lambda) = B(\lim G_\lambda, \lim H_\lambda),$

and we have the closed reduction square

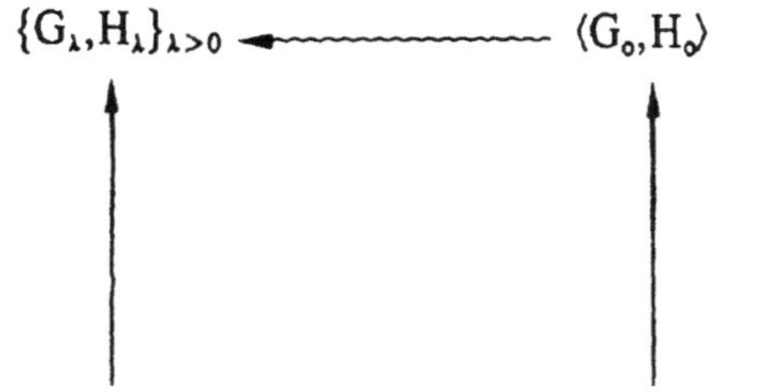

Figure 3.

Here the reduction in the lower row is explained by the upper one. We understand that $B(G_\lambda, H_\lambda)$ can be replaced by $B(G_0, H_0)$ for small λ by the corresponding behaviour of $\langle G_\lambda, H_\lambda \rangle$ with respect to $\langle G_0, H_0 \rangle$.

To clarify (9a) – the basis of it all – we recall that G_λ and H_λ (for given λ) as affine tensors are functions on F. For prooftechnical reasons we replace F by the affine group Γ. Because of $\Gamma\varphi = F$ for any $\varphi \in F$ this is no essential change. The tensors then become functions

(10a) $G_\lambda, H_\lambda\colon \mathrm{GL}(n) \times \mathbb{R}^n \longrightarrow \mathrm{Sy}(n)$

where $\mathrm{GL}(n)$ is the full linear group in n (real) dimensions and Sy(n) the set of symmetric n-dimensional matrics. With respect to a common distinguished coordinate system (one system for all λ!) and by using the normal forms (6d) we obtain the functions

(10b) $$\begin{cases} G_{ik}(\lambda) = (A^{-1})^1_i(A^{-1})^1_k - \lambda \cdot \sum_{j\geq 2} (A^{-1})^j_i(A^{-1})^j_k \\ H^{ik}(\lambda) = -\lambda \cdot A^i_1A^k_1 + \sum_{j\geq 2} A^i_jA^k_j \end{cases}$$

from (7a). It then follows immediately that

(10c) $$\begin{cases} G_{ik}(\lambda) = G_{ik}(0) - \lambda \cdot \sum_{j\geq 2} (A^{-1})^j_i(A^{-1})^j_k \\ H^{ik}(\lambda) = H^{ik}(0) - \lambda \cdot A^i_1A^k_1 \end{cases}$$

We now turn the set of functions to which G_λ and H_λ belong according to (10a) into a uniform space by providing it with the uniform structure of *pointwise convergence* with respect to the ordinary uniform structure on $\mathrm{Sy}(n)$ (cf. Querenburg 1976, Ch. 14A). (9a) then follows in the sense of pointwise convergence. However, the convergence is not uniform, and this will be the prevailing situation in limiting case reduction.

From the position of satisfaction in partial reduction, the validity of (9a) must be viewed as an unusual favour, so much the more since it is not confined to this axiomatisation. We find the corresponding result

(11a) $\lim f_\lambda = f_0$

for the group theoretic version $\Sigma^*_{\geq 0}$. To see this we turn (11a) into the equivalent formula

(11b) $\lim \Gamma_\lambda = \Gamma_0$

for the groups themselves. As we know from (7c) the general element of Γ_λ is a product of a (3-dimensional) rotation, an inertial motion and a translation. λ occurs in the second factor only. Its decomposition

(11c) $$\begin{pmatrix} 1 & 0 \\ -\mathbf{v} & 1 \end{pmatrix} + \begin{pmatrix} \gamma - 1 & -\gamma\lambda\mathbf{v} \\ -(\gamma - 1)\mathbf{v} & (\gamma - 1)v_iv_k/v^2 \end{pmatrix}$$

with the Galileo contribution as the first member shows immediately: If we view the matrices as functions of $\mathbf{v}$ we have again pointwise but not uniform convergence for $\lambda \to 0$ in the appropriate function space topology. (For a general approach, the so-called group contraction, see Inönü/Wigner 1953 and Saletan 1961.)

The limiting case reductions of Σ_{gal} to $\Sigma^0_{>0}$ and Σ^*_{gal} to $\Sigma^*_{>0}$ *explain* each other if we take the mutual definitions of their basic concepts into consideration. We have here a double closed reduction square before us that, for the sake of simplicity, we draw with double arrows:

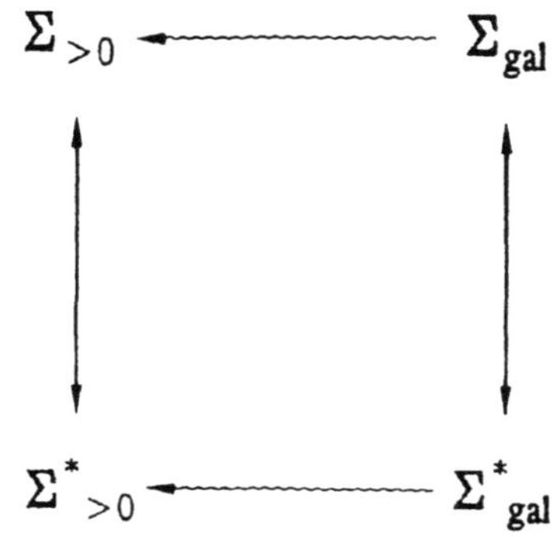

Figure 4.

G_λ and H_λ approach G_0 and H_0 for small λ because the f_λ *corresponding* to the former approach f_0 *corresponding* to the latter. This could also be formulated more precisely by using the mutual defining terms B as in (9). If, moreover, we look at Figure 4 as a compound reduction of Σ^*_{gal} to $\Sigma_{>0}$ (or of Σ_{gal} to $\Sigma^*_{>0}$) the role of $\Sigma^*_{>0}(\Sigma_{>0})$ as being *the* correction of $\Sigma^*_{\mathrm{gal}}(\Sigma_{\mathrm{gal}})$ in $\Sigma_{>0}$ ($\Sigma^*_{>0}$) can be understood only because we have used (formally) *identical* definitions (in λG) on both sides of Figure 4.

Limiting case reductions concerning basic concepts such as (10a) and (11a) are still not worth much as long as there is no general proof that the reductions can be extended to this or that class of derived concepts in the sense of (9c). The mere circumstance that a reduction is successful for the basic concepts may not seduce us to believe that such an extension is possible. Single results such as Figure 4 may, it is true, encourage us in our conviction that a total reduction is possible. They are, however, no proof thereof. What general consequences cases like (10a) and (11a) have seems to be unknown at present, and one has to be quite clear about the difficulties connected with the extension and its proof. In particular, it has to be noticed that a limit statement to make sense has to be related to a topology. Now the sets $B(G_\lambda, H_\lambda)$ are confined to scale sets over M and $\mathbb{R}$. Therefore it is possible in principle to embed the $B(G_\lambda, H_\lambda)$ (as elements or subsets) in a topological (or uniform) space that is independent of G_λ and H_λ. By means of this topology the limit statements could then be formulated and proved. It is, however, still quite unclear how such an assignment of topologies to concepts can be established *in general*. The demand that the resulting limiting cases be physically plausible makes a general and adequate selection rather questionable.

If on these grounds concept reduction remains partial from within, it remains partial also from outside: There are independent aspects of reduction, and one of them is the reduction of *consequences*. Similar to the situation in concept reduction we can have formally identical consequences in λG that split into consequence in GG and GMG having different extensions in any given standard model. The general situation is described by the formula

(12) $\Sigma_{\geq 0}(M \ldots \lambda) \wedge A(\mathrm{M} \ldots \lambda, E)) \wedge E' = D(\mathrm{M} \ldots \lambda, E) \vdash \beta(\lambda, E')$.

Here A are conditions for any given geometrical objects E under which new objects E' can be defined by means of D. β is then a consequence relating the E' independently of their origin. $\beta(0, E)$ is in general different from any of the $\beta(\lambda, E)$ for $\lambda > 0$ but in important cases it can be seen to be a limiting case of them as $\lambda \to 0$ with respect to a uniform space S such that $\{E' | \beta(\lambda, E')\} \subseteq S$ for all $\lambda \geq 0$.

Transformation laws for the representations of a geometrical object in different coordinate systems are typical consequences β in the sense of (12). They may depend on λ or not. The coordinates of an event p, for instance, are transformed by means of arbitrary affine transformations (7a). If we look at (7a) as being a 3-termed relation between x^i, $x^{i\prime}$ and $\langle A, a\rangle$ then we have an excellent illustration of a consequence β which, in this case, is independent of λ. The same is true for the tensors G_λ and H_λ although they themselves do depend on λ. But there are also λ-dependent transformation laws. An example is the transformation law for the momentary *3-velocities* of a particle. Let C be the timelike curve representing the motion of the particle, and let p_0 be an event from C. We now choose a distinguished (!) coordinate system $\varphi \in f_\lambda$. In φ as distinguished, C has a unique representation $\mathbf{x}(t)$ where t is the time coordinate in φ. With t_0 as the time of p_0, $(\mathrm{d}\mathbf{x}/\mathrm{d}t)_0$ is the 3-velocity of the particle at t_0. There is nothing peculiar to the transformations of 3-velocities by spatial rotations or translations. But for inertial motions with relative velocity $\mathbf{v}$ the transformation law for $\mathbf{u}$ into $\mathbf{u}'$ is

$$\text{(13a)} \quad \begin{cases} \mathbf{u}' = a_\lambda(\mathbf{u}, \mathbf{v}) \cdot \mathbf{u} + b_\lambda(\mathbf{u}, \mathbf{v}) \cdot \mathbf{v} \\ a_\lambda(\mathbf{u}, \mathbf{v}) = \gamma(\lambda, \mathbf{v})^{-1}(1 - \lambda(\mathbf{u} \cdot \mathbf{v}))^{-1} \\ b_\lambda(\mathbf{u}, \mathbf{v}) = \{(1 - \gamma(\lambda, \mathbf{v})^{-1})\, (\mathbf{u} \cdot \mathbf{v})/v^2 - 1\}\, (1 - \lambda(u \cdot \mathbf{v}))^{-1} \end{cases}$$

(cf. Møller 1972, ch. 2.7). For $\lambda = 0$ we obtain the well known Galileo formula

$$\text{(13b)} \quad \mathbf{u}' = \mathbf{u} - \mathbf{v},$$

and for parallel $\mathbf{u}$ and $\mathbf{v}$ the famous Einstein formula

$$\text{(13c)} \quad u' = \frac{(u - v)}{(1 - \lambda uv)}$$

follows. (13b) is a function mapping the 6-dimensional space of points $\langle \mathbf{u}, \mathbf{v}\rangle$ into the 3-dimensional space of points $\mathbf{u}$. (13a) is a λ-family of such functions. Again taking the uniform structure of pointwise convergence with respect to the usual uniform structure of the space of the $\langle \mathbf{u}, \mathbf{v}\rangle$ (cf. Querenburg 1976, Ch. 14A), the Galileo formula becomes the limiting case of the said family. And again we do not obtain uniform convergence as can readily be seen by the special case (13c) *vis-à-vis* the Galileo formula

$$\text{(13d)} \quad u' = u - v.$$

V. ASYMPTOTIC REDUCTION AND OUTLOOK

The limiting case reduction presented in the previous section is essentially a reduction to the *generalized* Minkowski geometry (GMG). It is therefore open to the objection of being a reduction to an empirically inaccessible theory. We would escape this objection if a reduction of GG to the *ordinary* Minkowski geometry (MG) were at hand. Unfortunately this is in a bad way. It is true enough that the method of comparison in this case is very simple and even more symmetric than in the limiting case reduction: Whether the reduction of concepts or of consequences is at stake, their intensional identity, if it holds, is guaranteed by the interchangeability of g with G and h with H according to (4ab) in the definitions and consequences of the two theories. The ensuing investigation of the corresponding extensions has always a definite result if it is based on the standard models (8) for $\lambda = 0$ and $\lambda = 1$ respectively. This result may even be positive, i.e. the two extensions may turn out to be identical, too. All standard models have the affine space in common after all. But as soon as the metrical concepts come in, the situation changes drastically.

I am anxious to emphasize that already the mere diversity of the extensions in question means an inconsistency in so far as one of the geometries asserts physical possibilities that are denied by the other. But it may even happen that the two extensions have no element in common. Such is the case for the fundamental tensors g and G as well as h and H. Their diversity in one coordinate system according to (4ef) is immediately transferred to any other coordinate system by means of (7a). It is true that we meet with a similar situation also in λG for any two λ's. However, whereas here we have at least an approximation of g and h for $\lambda \to 0$, nothing of the sort is in sight for GG and MG – neither for the fundamental tensors nor for many of the derived concepts. If we take, for instance, the timelikeness of two events, it follows immediately that any two events, timelike according to MG, are also timelike in GG, whereas the reverse, of course, does not hold. So we have even an inclusion of one of the two concepts by the other. But there is no question of an approximation of one by means of the other.

The situation is similar if we compare the Galileo transformation

(14a) $\quad t' = t, \quad \mathbf{x}' = \mathbf{x} - \mathbf{v}t$

with the proper Lorentz transformation

(14b) $\quad \begin{cases} t' = \gamma(t - (\mathbf{v} \cdot \mathbf{x})) \\ \mathbf{x}' = (\mathbf{x} - \mathbf{v}t) + (\gamma - 1)\mathbf{v}/v\{\mathbf{v}/v(\mathbf{x} - \mathbf{v}t)\} \end{cases}$

(cf. (7c)). The common parametrization of all groups Γ_λ according to (7c) gives us the possibility of comparing pairwise the *single transformations* for any *given* $\mathbf{v}$ as being consequences β in the sense of (12) for $\lambda = 0$ and $\lambda = 1$. Obviously there are no common solutions $\langle\langle t, \mathbf{x}\rangle, \langle t', \mathbf{x}'\rangle\rangle$ of (14a) and (14b). There is, therefore, again a particularly strong competition here which, moreover, is of empirical relevance. The various (metrically) distinguished coordinate systems can be constructed, at least in principle, by a uniform method with the help of measuring rods and clocks. It is then an empirical question how one coordinate system is related to any other. In so far as the transformations relating them can be subjected to an empirical test, the Galileo transformations would come out wrong – and appreciably wrong for large $\mathbf{v}$. As long, however, as $\mathbf{v}$ is fixed, no approximate relations between the transformations are to be seen also in this case.

The somewhat negative balance for the GG/MG case up to this point invites us to look to some of the critics of reductionism as they were already mentioned in the Introduction. The criticism includes the possibility of a reduction of Newtonian mechanics to relativistic mechanics. But it is not confined to it. Other revolutionary developments as they are typically exemplified by the two major achievements of 20th century physics – the general theory of relativity and quantum mechanics – were also suspected of lying outside the usual understanding of advances and progress in science. The very successfully presented doubts and objections of Feyerabend and Kuhn in the sixties were soon followed by vital philosophical debates which, however, in spite of their vastness hardly included any careful *systematic* case studies as the present one pretends to be. Lakatos/Musgrave 1970 is a book that mirrors the early phase of the controversy, and the recent collection Horwich 1993 shows in which direction the discussion has turned in the meantime. In neither book do systematic case studies play a role, and thoroughgoing investigations by physicists like Ludwig 1990 have gone almost unnoticed by philosophy of science. Accordingly, the question who was the winner of that debate in the relevant respect cannot be answered. The following concluding remarks are not meant to be an answer. I am fully aware of the fact that most of the problems of Kuhn and Feyerabend are different from problems as they typically arise in attempts of a rational reconstruction of closed developments in physics. But there are points of contact, and two of them may still be touched upon.

The magic word dominating the criticism of Kuhn and Feyerabend is *incommensurability*. Quite recently Kuhn has buttressed this emphasis when he said: "My own encounter with incommensurability was the first step on the road to *Structure* [= Kuhn ²1970], and the notion still seems to me the central innovation introduced by the book" (Kuhn 1993, pp. 314f). As to the question of the meaning of incommensurability Kuhn very soon took a semantical point of view. Incommensurability was to be a semantical concept expressing somehow the mutual untranslatability of two languages or parts thereof (cf. Kuhn 1983; see also Hoyningen-Huene 1989, pp. 209ff). Now untranslatability has a trivial aspect as long as the languages in question do not have a certain universality. A language that is just sufficient to describe the furniture of my room certainly is untranslatable into a language fit for the description of an ascent of Mont Blanc. But this is a trivial case of untranslatability due to the *disparity* of the 'worlds' to be described. For non-universal languages as we certainly have in geometry non-trivial, untranslatability and with it incommensurability can only mean a kind of incompatibility occurring in cases where the two languages are meant to describe the *same* state of affairs. In this sense the languages of Σ_{gal} and Σ_{min} are structured in a way that strongly suggests a formal mutual translatability of a large number of concepts on the basis of their intensional identity *without* yielding also the same extensions throughout. As we have seen from the examples given at the beginning of this section it is these pairs of concepts, arising from each other by the exchange of g with G and h with H, and usually being called by the same name, that are meant to refer to the same objects but in fact do not do it and therefore are the most likely candidates for pairs of incommensurable concepts. Timelikeness in the sense of GG is incommensurable with timelikeness in the sense of MG. Simultaneity in the sense of GG is incommensurable with simultaneity in MG. The Galilean transformation law for coordinates of an event is incommensurable with the corresponding law in MG etc. With respect to these occurrences the position of Kuhn and Feyerabend seems to be fairly strong, if in addition no approximate reductions are in sight.

In this situation the reductionist has one of two moves. He can point out *other* pairs of incommensurable concepts or consequences for GG and MG where an asymptotic reduction *is* possible or fall back upon the limiting case reductions of the previous section where we have seen reductions even for the fundamental tensors. The anti-reductionist strategy must then be directed against approximate reductions of any kind, claiming that they do not achieve what they pretend to achieve. In fact, referring to an alleged derivation of Newtonian mechanics from relativistic mechanics Kuhn argues (1970, p. 102):

> Unless we change the definitions of the variables in the [derived statements they] are not Newtonian. If we do change them, we cannot properly be said to have *derived* Newton's laws, at least not in any sense of "derived" now generally recognized. . . . Our argument has, of course [sic!], explained why Newton's laws ever seemed to work. . . . [But . . . it] has not . . . shown Newton's laws to be a limiting case of Einstein's. For in the passage to the limit it is not only the forms of the laws that have changed. Simultaneously we have had to alter the fundamental structural elements of which the universe to which they apply is composed.

Similarly, we hear from Feyerabend (1970, p. 221):

> It is of course true that the relativistic scheme very often gives us *numbers* which are practically identical with the numbers we get from CM [= Newtonian mechanics] – but this does not make the *concepts* more similar. Even the case $c \to \infty$ (or $v \to 0$) which gives strictly identical predictions cannot be used as an argument for showing that the concepts must coincide at least in this case.

What Feyerabend here says is entirely true. But it seems that he wants to infer from this Kuhn's statement that, in STR, the Newtonian formulas cannot come out literally, not even approximately, because they are based on Newtonian and not on relativistic concepts. And this would be false.

It has, of course, to be admitted that in a derivation in the proper sense, i.e. in a strictly logical deduction (including definitions), the conclusion as such cannot contain concepts in any essential way that were not already contained in the premises. There is no meaning change in an ordinary derivation. And since the basic concepts of GG cannot be defined from concepts of MG there can be no question of an ordinary derivation of GG from MG. But this situation already occurs in cases where we seem to have no problems with incommensurability whatsoever. It does occur, for instance, in the Kepler/Newton case mentioned in the Introduction. Although T and a of (1a) have exactly the same meaning in both theories and μ can be defined by M, the Kepler laws cannot be derived from Newton's equations. Only an asymptotic reduction according to (1c) is possible. And precisely in this wider sense an approximate derivation is possible also in cases of incommensurability.

Let us take *length contraction* as an example. Here we obtain as a *consequence* of the respective geometries GG and MG that a rod of length $1°$, at rest in an inertial system $I°$, has the same length $1 = 1°$, according to GG, in an inertial system I moving with velocity $v > 0$ in the longitudinal direction of the rod, while, according to MG, it has the different length $1 = 1°\gamma^{-1}(v)$ in I. Thus in the 3-dimensional space of $\langle v, 1°, 1\rangle$, with $v < 1$ and $1, 1° > 0$ we have the hypersurface

(15a) $\quad 1 = 1°$

according to GG and the hypersurface

(15b) $\quad 1 = 1°(1 - v^2)^{1/2} \qquad [= 1°\gamma^{-1}]$

according to MG as representing the relation between the three quantities. The following statement of *asymptotic reduction* is an immediate consequence of (15ab):

(15c) $\left\{\right.$ For any given $\varepsilon > 0$ there exists $\delta > 0$ such that, if v, $1°$ and 1 are related by (15b) and $v < \delta$ then there exists v', $1°'$, $1'$ such that (15a) holds for them and $|v' - v|/v'$, $|1°' - 1°|/1°'$, $|1' - 1|/1' < \varepsilon$.

This means that arbitrarily close to any given solutions of (15b) *there are* solutions of (15a) if only v is small enough.

Now I am pretty sure that Kuhn and Feyerabend hold that the concept of the length of a moving rod in GG is incommensurable to the concept of the length of a moving rod in MG. And indeed they *are* incommensurable in the well-defined sense developed above. But the 'inference' from a solution of (15b) *as a consequence of MG* to the existence of a close-by solution of (15a) is an inference to this solution *as a consequence of GG*. It is true that (15c) can be proved with respect to (15ab) without knowing anything about the conceptual origin of v, $1°$ and 1. But once the meaning we have given to (15b) as a consequence of MG is taken into account the meaning of (15a) as a consequence of GG is also fixed: it is the *same* rod with respect to inertial system $I°$ and I to which we refer in both cases. This can be made explicit if we ask for the derivations of (15b) and (15a) in MG and GG respectively.

If 1° and 1 in (15b) pertain to a rod whose end points have the world lines $x_i(t) = x_i^\circ$ in I° then *in MG*

$$1^\circ = |x_2(t) - x_1(t)| = |x_2^\circ - x_1^\circ|,$$

the representatives of the same world lines in I are

$$x_i'(t') = \gamma^{-1}x_i^\circ - vt'$$

and finally

$$1 = |x_2'(t') - x_1'(t')| = \gamma^{-1}|x_2^\circ - x_1^\circ| = \gamma^{-1}1^\circ.$$

If now it is asked what is meant by the v', 1°′ and 1′ whose existence is stated in (15c) for given v, 1° and 1 the answer simply is: they refer to the very same scenario as it is described *in GG*. In the present case the choice of the v', 1°′, 1′ can even be made more specific than is demanded by (15c). We again begin with

$$1^{\circ\prime} = |x_2(t) - x_1(t)| = |x_2^\circ - x_1^\circ|$$

in I° and setting $v' = v$ continue with

$$x_i'(t') = x_i^\circ - vt$$

to end up with

$$1' = |x_2'(t') - x_1'(t')| = |x_2^\circ - x_1^\circ|.$$

In this way we obtain (15a), i.e. $1' = 1^{\circ\prime}$, and the approximation of $\langle v, 1^\circ, 1\rangle$ as demanded if we choose $\delta^2 = 1 - (1 - \varepsilon)^2$ for given ε. Therefore, as far as a derivation of the formula (15a) can be said to have been given in MG at all, the meaning of this formula is precisely the meaning that is given to it according to GG. And the same is true in other cases, e.g. the case (13) of the *3-velocities* – now with $\lambda = 1$ in (13a): For small v (13a) can be replaced by (13b) with any prescribed accuracy, and the meaning in which (13b) is 'derived' in the approximate sense from (13a) is the Galilean meaning as soon as (13a) is given its Minkowskian meaning. This is in need of proof and can be proved. The result is again a kind of closed reduction square concerning *additional* geometrical objects as, for instance, a timelike curve on the upper row of the square and its 3-velocities on the lower.

Hamburg, Germany

REFERENCES

Ehlers, J. (1983), 'Relations Between the Galilei-Invariant and the Lorentz-Invariant Theories of Collision', in *Space, Time, and Mechanics. Basic Structures of a Physical Theory*, D. Mayr and G. Süssman (eds.) (Reidel, Dordrecht), pp. 21–37.

Ehlers, J. (1986), 'On Limit Relations Between, and Approximate Explanations of, Physical Theories', in *Logic, Methodology and Philosophy of Science*, Volume 7, B. B. Marcus *et al.* (eds.) (North-Holland, Amsterdam), pp. 387–404.

Einstein, A. (1905), 'Zur Elektrodynamik bewegter Körper', *Ann. der Physik*, XVII, pp. 891–921. (Also in Lorentz *et al.* (1922), pp. 26–50).
Feyerabend, P. (1965), 'On the "Meaning" of Scientific Terms', *Journal of Phil.* **62**, pp. 266–74.
———. (1970), 'Consolations for the Specialist', in Lakatos/Musgrave, pp. 197–230.
Fock, V. (1960), *Theorie von Raum, Zeit und Gravitation* (Akademie-Verlag, Berlin).
Havas, P. (1964), 'Four-Dimensional Formulations of Newtonian Mechanics and their Relation to the Special and General Theory of Relativity', *Rev. Mod. Phys.* **36**, pp. 938–65.
Horwich, P. (ed.) (1993), *World Changes* (MIT Press, Cambridge, Mass.).
Hoyningen-Huene, P. (1989), *Die Wissenschaftsphilosophie Thomas S. Kuhns* (Vieweg, Braunschweig).
Inönü, E. and E. P. Wigner (1953), 'On the Contraction of Groups and their Representations', *Proc. Nat. Acad. Sci. (U.S.)* **39**, pp. 510–24.
Kuhn, Th. S. (1970), *The Structure of Scientific Revolutions* (University of Chicago Press, Chicago). Second Edition.
———. (1983), 'Commensurability, Comparability, Communicability', in *PSA 1982*, Volume 2, P. D. Asquith and T. Nickles (eds.) (East Lansing, Mich.), pp. 669–88.
———. (1993), 'Afterwords', in Horwich, pp. 311–41.
Lakatos, I. and A. Musgrave (1970), *Criticism and the Growth of Knowledge* (Cambridge University Press, Cambridge).
Lottermoser, M. (1988), Über den Newtonschen Grenzwert der Allgemeinen Relativitätstheorie und die relativistische Erweiterung Newtonscher Anfangsdaten. Ph.D. Thesis: München.
Ludwig, G. (1990), *Die Grundstrukturen einer physikalischen Theorie* (Springer, Berlin).
Lorentz, H. A., A. Einstein and H. Minkowski (1922), *Das Relativitätsprinzip* (Teubner, Leipzig).
Minkowski, H. (1922), 'Raum und Zeit', in Lorentz *et al.*, pp. 54–56.
Møller, C. (1972), *The Theory of Relativity* (Clarendon Press, Oxford).
Querenburg, B. V. (1976), *Mengentheoretische Topologie* (Springer, Berlin).
Saletan, E. J. (1961), 'Contraction of Lie Groups', *Journal Math. Phys.* **2**, pp. 1–21.
Scheibe, E. (1993), 'A New Theory of Reduction in Physics', in *Philosophical Problems of the Internal and External Worlds: Essays on the Philosophy of A. Grünbaum*, J. Earman *et al.* (eds.) (University of Pittsburgh Press, Pittsburgh), pp. 249–71.
———. (1995), 'The Rationality of Reductionism', in R. Zwilling (ed.), *Natural Sciences and Human Thought* (Springer, Berlin).
———. (1997), *Die Reduktion physikalischer Theorien* Teil I: Grundlagen und elementare Theorie (Springer, Berlin).
Weyl, H. (1923), *Raum, Zeit, Materie* (Springer, Berlin).

MANFRED STOECKLER

SYMMETRIES AND EXPLANATIONS: THE LESSONS OF ELEMENTARY PARTICLE PHYSICS[1]

Symmetry, as wide or as narrow as you may define its meaning, is one idea by which man through the ages has tried to comprehend and create order, beauty, and perfection.

Hermann Weyl[2]

1. INTRODUCTION

Symmetries play an important role in the history of science and art.[3] In the development of modern physics the use of symmetry arguments changed radically in the beginning of the 20th century. To a considerable extent the new rise of symmetry in theoretical physics is due to A. Einstein, H. Weyl, and E. Wigner.[4]

The use of symmetries in quantum field theory follows the tradition of the ideas of Weyl and Wigner. Of course there are some special features coming from the mathematical structure of the quantum field theories of the fundamental interactions in elementary particle physics, for example the method of spontaneous symmetry breaking and the idea of local gauge theory.

Bas van Fraassen discusses two forms of symmetry arguments.[5] The first one relies on a meta-principle: Essentially similar problems have essentially similar solutions. The second form is a sort of metaphysical assumption: An asymmetry can only come from a preceding asymmetry. There are many examples of both forms of symmetry arguments in elementary particle physics.

I am going to emphasize two aspects of symmetry arguments, which can be studied in elementary particle physics:

- Symmetries are instruments of provisional physics. They are used in not yet fully developed theories.
- Symmetries supply some sort of "justification". They are often used in the beginning of a chain of explanations.

From a philosophical point of view the upshot will be that the role of symmetries in particle physics can be well understood if we apply the theory of explanations to the special conditions of modern theoretical physics. A special ontological role of symmetries, as Heisenberg seems to assume in some places, must not be invoked. Nevertheless, there are some heuristic and justificational uses of symmetry arguments, which are similar to the employment of metaphysical principles.

But first I should give a rough sketch of the mathematical role of symmetries in quantum field theory, which is the language of theoretical particle physics.

D. Ginev and R. S. Cohen (eds.), Issues and Images in the Philosophy of Science, 343–355.

2. SYMMETRIES IN ELEMENTARY PARTICLE PHYSICS: DEFINITIONS AND DISTINCTIONS

"A symmetry is a transformation that leaves all relevant structure the same."[6] In order to characterize a special symmetry one has to give *transformations* T and features, which remain unchanged during these transformation (*invariants I*).[7] So a symmetry can be explicated as an ordered pair ‹I, T› of two sets T and I. In contemporary elementary particle physics there are various types of symmetries.[8] The most important distinction is between space-time transformations and so called internal transformations (Isospin, SU(3) etc.). Whereas the space time symmetries are universal, i.e., they are valid for all interactions, the internal symmetries are characterized by the special sorts of interaction (strong, electromagnetic or weak interaction). A geometrical language is used for the internal symmetries although a space-time interpretation is not possible. This is a general tendency in modern physics: the formal development of mathematics is used to construct theories, even though there is not an intuitive physical interpretation at the beginning.

Let me give an example of an internal symmetry: The transformations T may be given by a matrix M_g, which mixes and rearranges the three components of a wave function (or field operators in quantum field theory). The matrices may be representations of the group SU(3). The invariants are Lagrangians,[9] S-matrix-elements, and expectation values built up by field operators and state-vectors. In a sense, the transformations change elements of the mathematical description whereas the empirical content of the theory is unchanged.

$$M_g \begin{pmatrix} \psi_1 \\ \psi_2 \\ \psi_3 \end{pmatrix} = \begin{pmatrix} \psi_1' \\ \psi_2' \\ \psi_3' \end{pmatrix} \qquad L(\psi_1, \psi_2, \psi_3) = L(\psi_1', \psi_2', \psi_3')$$

Further examples of internal symmetries are isospin, which is linked to the group SU(2), and colour (linked to SU(3)). The symmetry transformations define properties of particles (of the quantum field), which are conserved if the symmetry is not broken. Usually the names of the symmetry transformations have been used to label conserved properties before the mathematical expression could be constructed.

Although symmetries play a very important part in modern particle physics and the search for symmetry is a central motive in research, it does not mean that nature is always symmetrical. Symmetries like parity have been broken down on the level of *fundamental laws* (the Lagrangian is not an invariant under the transformation). The mass of electrons can be explained by *spontaneous symmetry breaking*, which is connected with the asymmetry of a special state. And thirdly, in actual processes, corresponding constraints lead even in symmetrical laws to *asymmetrical situations*. The fundamental laws of nature are symmetrical, but the manifold phenomena can only be explained by contingent and asymmetrical boundary conditions.

The first and the third way of breaking symmetry is well known from classical physics. The concept of spontaneous symmetry breaking is especially interesting for elementary particle physics. The rough idea is the following: The Langrangian is invariant under a transformation, but the state of minimum energy is not invariant. The physical states do not provide a representation of the symmetry, which is the usual case in atomic physics.[10]

The idea of spontaneous symmetry breaking was a very useful mathematical guide to the theories of weak and strong interactions, because it provided a way to work with massive vector-bosons and to retain gauge invariance and renormalizability. This idea depends on a special relation between field operators and state vectors, which is typical for quantum field theory. But spontaneous symmetry breaking is a highly technical matter, and the mathematical presentations in the usual textbooks are not very rigorous. We will omit these problems as not being too important from a philosophical point of view.

3. SYMMETRIES AND PROBLEM SOLVING

Problem solving is a classical field of application for symmetry principles. Bas van Fraassen gives a wonderful example of a symmetry argument. He shows how Farmer Able can find the shortest way from the farmhouse to the creek (where he fills a pail with water) and to the hen-house. The central idea is mirroring the hen-house on the creek as a line of reflection. Then there is a straight line from Able's house via the creek to the reflected hen-house. This determines the point where Able should fetch the water for the hens if he wants to walk the shortest distance.[11] If we work with symmetrical theories, sometimes it is possible to transform a problem that is very difficult to solve into a simpler, solvable problem and to retransform the solution to the original problem. The symmetry of the theory guarantees that the essential structure of the problem is conserved.

This is one of the reasons for the advantage of the so-called gauge theories in elementary particle physics. By using a gauge transformation you can bring the Lagrangian in a form which makes it more easy to show the renormalizability. For working with the theory, it is very helpful to skip mathematical complexities which are not essential for the given problem.

In quantum field theory only a few problems really can be solved. But even if such a theory is too complicated to cope with, at least some special problems can be solved. Very often the most symmetrical systems can be figured out without using the whole mathematical formalism. For example, in relativistic cosmology symmetry assumptions can be used to reduce the number of parameters which characterize a system. In this case the symmetry of the objects of the theory helps to *apply* the theory. Usually the symmetrical objects are idealized models which represent reality only by strong approximations. This is a form of preliminary physics which is wide-spread in the contemporary research in particle physics and cosmology, but widely neglected by philosophers of science.

Such methods of problem solving are important in theories like quantum field theory or general relativity which have a rather complex mathematical structure. E. P. Wigner showed in the twenties, that in quantum mechanics of the atomic shell lots of spectroscopical phenomena follow from rotational symmetry. That means that to understand these features you need not know all the peculiarities of dynamics. So people can use a theory without being able to solve all the detailed problems of the dynamical laws. An important consequence of symmetries is conservation laws. In electrodynamics and mechanics many phenomena can be easily explained by using symmetry arguments:[12] "This is in the nature of a conservation law. We do not need to watch the insides to know at least something about the game."

Working with theories which have a complicated mathematical structure requires additional abilities of the theoretical physicists. Deducing in a logical sense is not enough. One has to find models which are simple enough to be manageable, but sophisticated enough to be similar to real objects. Here symmetries both of the theory and of the model object are very helpful.

From a pragmatic point of view these symmetries are preconditions of working with the theory, but it is a contingent fact that symmetrical theories are empirically adequate and that symmetrical models are "near enough" to reality.

4. SYMMETRIES AND HEURISTICS

Symmetries not only facilitate solving problems of a given theory, they also help to find new theories. In the first case the symmetry is fixed, in the second case some symmetry assumptions have to be employed. What reasons can be given for such assumptions? Must we assume not only symmetries in the description but also symmetries in nature?

Symmetry considerations help to construct new theories, they have a large heuristical potential.[13] E. P. Wigner emphasized the change in the usage of symmetry:[14] today physicists try to deduce the natural laws from symmetry principles and not, as in the 19th century, the symmetry principles from the natural laws. This is a renaissance in the use of symmetry which can be traced back to Anaximander and Archimedes.

In order to understand the special role of symmetries in particle physics one has to consider the interplay of two types of research.[15] On one side, there is what physicists call phenomenology: In this area people work on concrete quantitative problems. They use theoretical models with a limited sphere of application closely related to experimental results. Typically the theoretical assumptions are not yet derivable from a general theory. An example from a more familiar area of physics is Kepler's laws.

On the other side, there are people working on the general theory. This construction of fundamental theories in mathematical physics is fairly remote from experiment and mainly guided by mathematics. The division of labour sometimes leads people to construct general theories even if no empirical claim follows from these theories (at least for many stages of the development). Later on, it is the business of phenomenology to find consequences which could be tested by experience.

Phenomenology and general theory sometimes develop in phase, sometimes there are only few links between the two areas. Having settled this distinction we can come back to symmetries.

Symmetries can bridge the gap between phenomenology and general theory. This is the most important reason for their heuristic force.[16] The analysis of many high energy collision experiments led to the assumption of special conservation laws for abstract properties like baryon number or strangeness. Evaluating the experiments in this way made possible a classification of the particles. The phenomenological classification was good enough to predict new particles which could be found in the experiments.[17] So the free places in the particle multiplets could be filled even if the dynamics of the theory (for example the Lagrangian of strong interaction) was unknown.

Considerable constraints on the construction of dynamics follow from the structure of the multiplets and in general from the symmetry of phenomenological description: The Lagrangian can not exhibit less symmetries than the phenomenology.

Of course the dynamics can not be deduced only from symmetries. There is a mathematical background, which gives a general scheme teaching how to construct a quantum field theory. This general scheme is partly taken from classical mechanics, partly it is an extension of quantum electrodynamics, the first successful quantum field theory.

We see that the relation between theory and experiments has become rather involved. Phenomenology grew as a third level between theory and experiments. Symmetries allow the transfer of the experimental experience into general theory, because symmetries can be discovered without knowing the details of dynamics.

There are additional mathematical structures enhancing the heuristic power of symmetries. As we will see in the following section, gauge theories include a general recipe to find the fields mediating the interaction. Combining the symmetries of different theories proposes a concrete way to their unification.

The fundamental work in general theory sometimes seems to be far from experience. But the phenomenological models bring an essential empirical impact. The symmetries found in experiments, together with other principles of a theory construction (for example simplicity), determine the Lagrangian, that means the fundamental law of the theory. This heuristic way is complicated. But should we expect a direct connection between experiments and the general theory in a field like high energy physics, where only a small number of different types of experiment is possible and where it is extremely difficult to use the general theory for figuring out concrete examples? There are no *a priori* assumptions of symmetries, but from a pragmatic point of view only symmetries provide hope for progress. And as a matter of fact, it seems that nature can be described by highly symmetrical theories.

5. SYMMETRIES AND JUSTIFICATION

In modern particle physics, symmetries have more functions than problem solving and helping to construct new theories. They have to provide something that I call "justification". Claiming a justificational role of symmetries could be misleading. I do not think that symmetry arguments could justify a theory in an *a priori* manner. But even if one agrees with the empiristic intuition that experience is a constraint on theories, there is a lot of reasoning in science which is not covered by a bold falsificational method.

Explaining is more than deducing from general laws. In some contexts explaining means unifying knowledge. The logical form of explanations does not determine the quality of an explanation. In fundamental theories, there are special criteria characterizing good explanations. In applied sciences, the general laws and many contingent assumptions can be adopted without further explanations, because in this area people are more interested in special mechanisms. In fundamental theories as much as possible should be explained: Motivate the special form of the Lagrangian! Avoid free parameters which can be determined only by experiment! Give reasons why proton and positron have the same charge!

One way of minimizing the unexplained premises is unifying the fundamental interactions. We shall discuss this topic in the following chapter. Here we shall prepare the discussion and show how symmetry arguments play a justificational role and how this role evolves from the general aims of explanations in science under the conditions of a fundamental theory expressed by a mathematical formalism.

The aim of elementary particle physics is not only to describe how matter is, but also to explain why it is like that and not otherwise. People like to have arguments, and not just start with a complicated Lagrangian. To a remarkable degree the present theories of elementary particle interactions can be understood by deduction from general principles. On the one side, there is the principle of renormalizability, which is often used as a constraint on the possible forms of Lagrangians. On the other side, there are special symmetry principles, which help to determine the Lagrangian. For example, the only Lorentz invariant and gauge invariant renormalizable Lagrangian for photons and electrons is precisely the original Dirac Lagrangian.[18]

Symmetry arguments help to minimize the unexplained basic assumptions of a theory. The same symmetries work in different theories. Gauge theories are good examples for deriving the dynamics of a theory with the help of symmetry principles. I shall give a rough sketch of this mechanism.

The Lagrangian for a free electron is invariant under the following transformation (in this example it is a phase-transformation):

$$\psi(x) \rightarrow \psi'(x) = e^{i\beta}\psi(x)$$

To gauge a theory means to require that the parameters of the transformation may depend on space and time (local gauge symmetry):

$$\psi(x) \rightarrow \psi'(x) = e^{i\alpha(x)}\psi(x)$$

The Lagrangian of the free election is no longer invariant under this transformation. To restate the symmetry you must add an additional term, which happens to be the term which describes the interaction between the quantum fields of electrons and photons. In this way the requirement of invariance under a group of transformation determines in some sense the explicit form of interaction. This concept can be extended to other fundamental interactions.

This does not mean that elementary particle physics is fixed by *a priori* arguments. There are a lot of premises: Space time symmetries are assumed. Before the start of the procedure of gauging a theory you must have some ideas of the structure of the Lagrangian (very often the analogy to classical Lagrangians give the essential idea). And, of course the symmetry requirements may be wrong, which is what happened to parity.

In the next stage, one can ask the question whether it is possible to justify why one particular symmetry is realized in nature and not another one. Usually people believe that this cannot be explained. There are however attempts (for example by C. F. von Weizsäcker) to deduce symmetries from even more general claims.

Often chains of explanations start with a symmetry assumption. Perhaps this is the effect of a metaphysical tradition, in which the symmetrical things are more perfect and less arbitrary than the asymmetrical states of affair. "An asymmetry can only come from a preceding asymmetry." So asymmetry seems to call for explanations, while symmetry is more intuitively acceptable.

But we should not forget the role of symmetry breaking. The case of parity violation has been mentioned before. The mechanism of spontaneous symmetry breaking was an important step in the development of a unified theory of weak and electromagnetic interaction. This shows again the physicist's pragmatic way of using symmetry arguments. They try to find theories which have many symmetries, but the theories must work, they must be fruitful and they should not be in conflict with experiments.

6. EXPLANATION, SYMMETRIES, AND THE UNITY OF PHYSICS

Even if all the fundamental laws were known, the symmetry principles would not be superfluous, because they increase the coherence of the theory. There are two ways of unifying particle physics by using symmetry arguments.[19] First there are similar ways of solving problems in different theories. General principles (like Lorentz-invariance) can be applied to quite different theories. The same group of transformations can be realized in different representations and can in this way connect different multiplets of particles.

Secondly, mathematical relations between the symmetry groups used in the different theories provide a recipe to the unification of these theories. The unified theory encloses the old theories as special cases. In addition, they have a surplus meaning. In grand unified theories it can be explained why the proton and the positron have the same charge. It is also possible to determine a value for the so-called Weinberg angle, which is an unexplained free parameter in the Salam-Weinberg-model.

In this way, symmetries help to unify the fundamental laws of particle interactions and to make progress in the unification of physics. This can be considered as a consequence of the most important aim of explanation in science which is unification of knowledge.[20] Symmetry arguments have a unifying power, and this characterizes good explanations.[21] So S. Weinberg is quite correct in claiming: "Our job in physics is to see things simply, to understand a great many complicated phenomena in a unified way, in terms of a few simple principles."[22]

Again it is a matter of fact that the world is such that symmetries are useful tools in unifying knowledge. But from this does not follow that in particle physics symmetries have got a special ontological dignity. This is the topic of the following section.

7. PARTICLES AND SYMMETRIES

There is an aspect of the justificational role of symmetries which is special to elementary particle physics. The development of particle physics since the thirties led some people to an idealistic interpretation, so that particle physics seemed to be at variance with the atomistic tradition.

In the atomistic approach, properties of composed objects are explained by the properties of their constituents. The variety of phenomena are reduced to the changing combinations of unchangeable atoms.

The atomistic strategy faces some problems in particle physics. A detailed discussion of scattering processes and radioactive decay showed a new feature which is not known from classical atomistic theories. It is no longer possible generally to assume that the particles to be seen in the end of the experiment existed already from the beginning as parts of the decaying compounds. For example, the electron is not part of the nucleus before Beta-Decay. As distinguished from the old quantum mechanics quantum field theory can describe the creation and annihilation of particles. So according to the new theory the basic building blocks of matter are not unchangeable.[23]

Especially W. Heisenberg drew philosophical consequences from this feature of particle physics.[24] He emphasized that, already on the experimental level, the concept of constituents lost its usual meaning. "Y consists of X" cannot any longer be explicated by "Y can be divided into X", if "dividing" is understood in an experimental sense as dividing by an energy which is small compared to the rest mass of the particles.[25] Using an operational criterion W. Heisenberg got the result that the concept of "constituent" (the concept of 'consisting of') does not make sense in particle physics.[26] Similar to Bohr Heisenberg followed the general idea that our language is limited. "We must use words like 'divide', or 'consist of', or 'number of particles', and at the same time we learn from observations that these words have only a very limited applicability."[27]

Werner Heisenberg saw the consequences for the concept of atomistic explanations. In elementary particle physics it is no longer possible to find a hierarchy of particles so that some basic building blocks can be used to explain all the other composite particles. In atomism there is a natural end of dividing particles. The fundamental particles, which cannot be divided, are a natural beginning of a chain of explanations and can guarantee the unity of physics.[28] But in modern particle physics, atomistic explanations have lost a basic precondition, namely a clear conception of "constituent".

If no other explanatory structures can be found, the failure of atomistic explanation could lead to a pure descriptive form of physics. Werner Heisenberg objected to an approach in which particle physics became a set of tables and mathematical tools.[29] So something has to replace the atomistic strategy.

Here the symmetries come into the game. Whereas the elementary particles transform into each other, emerge and vanish, symmetries provide the stabilizing factor in the ever-changing phenomena and, as we have seen, a unifying principle in explanations. The symmetries play their main role in understanding and explaining the empirical data. "What had to be added to the experimental material and to the corresponding mathematical forms in order really to understand the phenomena? I think there is only one answer: One has to add the conviction that behind the infinite variety of phenomena there is a single unifying principle of form."[30]

Heisenberg discussed symmetries mainly in their role as replacements for atomistic explanations, which relied on the composition of particles. "In order to find the way to a real understanding of the spectrum of particles it will therefore be necessary to look for the fundamental symmetries and not for the fundamental particles."[31]

This is quite correct if understood in the following way: In order to understand particle physics and in order to find good explanations it is more useful to look at the mathematical structure of the theory, especially at the symmetries, than to look for the constituents.[32]

Symmetries determine the structure of the fundamental particle interactions and the range of possible processes. Heisenberg hoped to find even the possible sorts of particles by symmetry considerations. In any case, symmetry groups define to a large extent the structure of mathematics which is used for describing elementary particles. "In that sense one could say, the symmetry groups are the most 'fundamental' level as far as we can see now."[33]

Heisenberg interpreted this opposition of particles and symmetries as a turn from the philosophy of Democritus to the philosophy of Plato. "What is really needed is a change in fundamental concepts. We will have to abandon the philosophy of Democritus and the concept of fundamental elementary particles. And, we should accept instead the concept of fundamental symmetries which is a concept out of the philosophy of Plato."[34]

This was interpreted as a turn to an idealistic philosophy in the sense that matter can be reduced to symmetries. Indeed, especially in his last years, Heisenberg relied on Plato's philosophy, especially on the ideas in the dialogue *Timaeus*. The role of symmetries in explanations, their unifying power, and affiliated conservation laws fit quite well in a Platonistic program, where, according to Heisenberg, order is guaranteed by mathematical form, not by the material bodies of atoms.[35]

But does particle physics really exclude a materialistic ontology? It is hard to characterize Heisenberg's own idea of Plato's philosophy. Does he think of the "fundaments of reality" in an ontological or in a methodological (explanatory) way? Consider the following quote from his famous paper "The nature of elementary particles": "If we wish to compare the results of present-day particle physics with any of the old philosophies, the philosophy of Plato appears to be the most adequate: The particles of modern physics are representations of symmetry groups and to that extent they resemble the symmetrical bodies of Plato's philosophy."[36] Obviously this is fully consistent with a materialistic ontology. Symmetries are invariants in transformations, but they are not purely mathematical entities. The operators and wave functions, the transformed objects, have a physical meaning; they refer to physical, material objects. I am not sure whether Heisenberg was interested in ontological questions at all. There are some hints that Heisenberg really believed in some form of foundations of matter in mathematical concepts,[37] but in his writings there is no convincing proof. We do not know whether he had changed his mind when he could have learned the success of quark physics which finally met his requirement for a dynamical formulation of the theory.[38]

In any case, from a systematic point of view particle physics does not give any special reason to skip a materialistic ontology. Elementary particles are not made out of mathematical forms, even if mathematical forms are more important in explaining the fundamental processes than the material objects.

Concluding this section, it may be emphasized again that the main tool for understanding high energy physics is not unchangeable building blocks but laws and symmetries. The symmetries are the unifying principles. They are used in explanations, but not as a basic set of fundamental particles. Heisenberg correctly characterizes the growing importance of mathematical structures in physics. But this does not justify the departure from a physicalistic ontology, in the sense that matter can be reduced to mathematical concepts. The new role of symmetries is not a question of ontology, but a problem of types of explanations. So maybe neither Democritus nor Plato, but Aristotle should be the godfather of particle physics.

CONCLUSIONS: SYMMETRIES BETWEEN METAPHYSICS AND METHODOLOGY

The role of symmetries in physics has been discussed mainly by scientists. The methodological problems which are involved in the use of symmetry arguments can be discovered only by people who know the structure of the theory. Symmetries are examples of a topic of philosophy of science, which is special to some branches of science, because it depends on the content and structure of the theories used in this branch.

Nevertheless the role of symmetries in particle physics can be understood if the general aims of science like problem solving and explanations are considered in the given context of a fundamental theory, which uses a highly developed mathematical formalism and is characterized by a small basis of experience.

So in other parts of science symmetries may play a different role (or may be quite unimportant). But in particle physics symmetry arguments are extremely fruitful. Nature seems to be such that on the fundamental level symmetry theories are successful. Metaphysical motivations were useful as heuristic tools. But from an ontological point of view symmetries do not play an autonomous role in particle physics. Nevertheless, nature must have special properties so that the methodological advantages of symmetries can be used in its explanation.

University of Bremen, Germany

NOTES

[1] I had the privilege to discuss these topics with Professor Azaria Polikarov, who stayed at the University of Heidelberg as a visiting professor when I worked there on my "Habilitationsschrift" (Stoeckler 1987). In many tea hours he provided cake, valuable hints and encouraging criticisms. It is my pleasure to dedicate this paper to the 75th anniversary of my fatherly friend.

[2] H. Weyl (1952), p. 5.

[3] See Weyl (1952).

[4] Cf. Radicati (1979), p. 524.

[5] See van Fraassen (1989), pp. 233–43.

[6] van Fraassen (1989), p. 262.

[7] See (also for many other parts of this paper) Redhead (1975).

[8] For a first informal introduction see Dodd (1984). Pickering (1984) gives an extended historical account.

[9] In classical physics the Lagrangian determines the dynamics (similar to the formula for the energy of the system). In quantum field theory, the Lagrangian characterizes a fundamental interaction. So the Salam-Weinberg theory of electroweak interaction is determined by a special Lagrangian.

[10] Cf. Aitchison (1982), chapter 6.
[11] van Fraassen (1989), p. 234.
[12] Feynman (1967), p. 60.
[13] Polikarov (1983), part 2, emphasizes the philosophical aspects of heuristics in scientific research.
[14] Wigner (1979), p. 5; Redhead (1975), p. 80.
[15] Falkenburg (1988) emphasized the decisive role of this interplay in the development of particle physics (p. 114, p. 125). In characterizing phenomenological and fundamental research I follow her conception.
[16] See Falkenburg (1988), pp. 121 ff.
[17] Pickering (1984), p. 60.
[18] Weinberg (1980), p. 517.
[19] Polikarov (1983), p. 124, distinguishes different forms of unification in scientific research.
[20] Cf. Friedman (1974).
[21] In a similar way Falkenburg (1988) interpreted the unifying role of symmetries with the help of Kant's concept of the systematic unity of nature.
[22] Weinberg (1980), p. 515.
[23] Analyzing the ontology of quantum field theory would require a more sophisticated technical discussion. In some sense the quantum fields could be taken as the fundamental objects of the theory. These fields can change their state, but they cannot really vanish. See Redhead (1988).
[24] See especially Heisenberg (1976a).
[25] W. Heisenberg used the following experimental criterion for 'consisting' of: "When a particle, by a small amount of energy, can be disintegrated into two or several parts, the restmass of which is very large compared with this small energy, then and only then may one say that the particle consists of these parts, can be divided into these parts" (quoted from Heisenberg (1976b) p. 64). Cf. Heisenberg (1976a) p. 38.
[26] Heisenberg did not consider that the theory can provide a clear concept of "constituent". See Stoeckler (1987), p. 298.
[27] Heisenberg (1985), p. 446.
[28] Cf. Heisenberg (1985), p. 395.
[29] Heisenberg (1976a).
[30] Heisenberg (1985), p. 504.
[31] Heisenberg (1985), p. 505. Cf. Heisenberg (1971), chapter 20. Cf. also Heisenberg (1985), p. 414: "What then has to replace the concept of a fundamental particle? I think we have to replace this concept by the concept of a fundamental symmetry." Heisenberg emphasizes the methodological side of the problem. In his opinion, asking "Of what does this object consist and what is the geometrical or dynamical configuration of the smaller particles in the bigger object?" is asking the wrong question (1985, 414).
[32] Heisenberg (1985), p. 494.
[33] Heisenberg (1985), p. 462.
[34] Heisenberg (1985), p. 446.
[35] Cf. for example Heisenberg (1985), pp. 395 ff. and pp. 496 ff., and Stoeckler (1993).
[36] Heisenberg (1976a), p. 38.
[37] See Liesenfeld (1992), chapter 6.
[38] The development of a unified field theory is not bound to a Platonistic ontology.

REFERENCES

Aitchison, I. (1982), *An Informal Introduction to Gauge Field Theories* (Cambridge).
Dodd, J. E. (1984), *The Ideas of Particle Physics* (Cambridge).
Falkenburg, B. (1988), 'The Unifying Role of Symmetry Principles in Particle Physics', *Ratio (New Series)* **I(2)**, pp. 113–34.

Feynman, R. P. (1967), *The Character of Physical Law* (Cambridge, Mass.)
Friedman, M. (1974), 'Explanation and Scientific Understanding', *Journal of Philosophy* **71**, pp. 5–19.
Heisenberg, W. (1985), *Heisenberg, Collected Works*, Series C, Volume III, Munich 1985 (In the quotation from this Volume, I omit all information about title and original place of publication).
Heisenberg, W. (1971), *Physics and Beyond* (New York).
———. (1976a), 'The Nature of Elementary Particles', *Physics Today* **29**(3), pp. 32–9.
———. (1976b), 'Cosmic Radiation and Fundamental Problems in Physics', *Die Naturwissenschaften* **63**, pp. 63–7.
Liesenfeld, C. (1992), *Philosophische Weltbilder des 20. Jahrhunderts* (Würzburg).
Pickering, A. (1984), *Constructing Quarks*. A Sociological History of Particle Physics (Edinburgh).
Polikarov, A. (1983), *Methodological Problems of Science* (Sofia).
Radicati di Brozolo, L. (1979), 'Einstein and the Role of Symmetry in Modern Physics', in *Relativity, Quanta and Cosmology*, F. de Finis (ed.) (New York), pp. 523–35.
Redhead, M. (1975), 'Symmetry in Intertheory Relations', *Synthese* **32**, pp. 77–112.
———. (1988), 'A Philosopher Looks at Quantum Field Theory', in *Philosophical Foundations of Quantum Field Theory*, H. R. Brown and R. Harré (eds.) (Oxford), pp. 9–23
Stoeckler, M. (1987), 'Philosophische Probleme der Elementarteilchenphysik', Habilitationsschrift, Gießen 1987 (unpublished).
———. (1993), 'Hat sich Werner Heisenberg zu Recht auf Platon berufen?' in *Werner Heisenberg-Physiker und Philosoph*, B. Geyer *et al.* (eds.) (Berlin), pp. 335–43.
van Fraassen, Bas C. (1989), *Laws and Symmetry* (Oxford).
Weinberg, S. (1980), 'Conceptual Foundations of the Unified Theory of Weak and Electromagnetic Interactions', *Reviews of Modern Physics* **52**, pp. 515–23.
Weyl, H. (1952), *Symmetry* (Princeton).

LADISLAV TONDL

COGNITION AS A SYSTEM

1. INTRODUCTION

This study does not aspire to classifying the issues traditionally studied by philosophy, problems which are most frequently termed as epistemology, noetics or gnoseology. The term theory of knowledge or theory of cognition does not seem to be very suitable, particularly so as the word "theory" should meet some much stricter requirements. While attempting to solve these problems, we probably cannot bypass some features of the connection between a thorough analysis of cognitive process, modeling of some cognitive procedures (a case in point in these contexts are models of explanatory procedures, predicating proceedings, formulation of generalisations etc.), on the one hand, and global reasoning pertaining to cognition as a whole on the other. The other side of the coin – or perhaps a parasitic by-product of such global reasoning – is an attempt at cramming what is a considerably heterogeneous complex of such activities, whose entirety as well as comprehensive results are called "learning", into the Procrustean bed of an all-encompassing scheme, such as, e.g. the pattern of "reflection", the pattern of simple reductions to elements of empirical activities etc. Attempts at delineating a single, universally conceived source of cognition, in both empirical and rational terms, can be dismissed as equally one-sided.

The author of this paper is far from disparaging the merits of the analytical approach to cognition, among other reasons also because for many years he himself had been involved in developing that approach or utilising its stimulating activities. Studies of the formation of terms and conceptual frameworks, models of scientific procedures, generation of hypotheses and other forms of expressing the results of cognition have proved their viability in areas, such as the so-called cognitive sciences, artificial intelligence and some other branches. The same applies to the results of analyses and functions of what is termed "social memory" in cognition, to the irreplaceable role of the relatively *a priori* cognitive and value-related equipment or cognitive competencies. The perception which also views cognition as a system of interactions of different steps, series of steps, activities, decisions as well as creation by no means diminishes the value of analyses of individual steps or attempts at their modelling.

Accent placed on interactions of individual partial components, their interlinking and interplay is of considerable importance also because each partial component or initiative towards other steps is affected by what could be called a relatively *a priori* equipment of the subject of cognition, the subject's cognitive and value-related equipment in particular, or also by what can be

D. Ginev and R. S. Cohen (eds.), Issues and Images in the Philosophy of Science, 357–375.

termed an epistemic competence. Therefore, a substantial feature of this competence is the capacity or skill to tie on to the preceding steps, and not only to one's own steps but also to those taken by others, elsewhere or in the past. What is known as social memory has a significant role to play in this respect. That is why cognition has its social and cultural dimensions too; that is why the best results can be achieved by erudite teams in advanced research communities.

The author of this study, which is of a predominantly philosophical and methodological nature, had no intention of writing about the views, notions or standpoints of other philosophers, even though this seems to be a very frequent, popular and probably also legitimate mode of philosophising. What is more, this paper makes still smaller claims to presenting the only acceptable or even the only correct viewpoint of the problems under scrutiny. Instead, the underlying purpose was to outline key problems and their interrelations, to stress the importance of what can be called the creation of well-functioning networks, linkages, and simulating interactions in cognitive processes or when solving comprehensive research projects. In actual fact, the processes of creating purposeful modifications of entire large networks when tackling complex problem situations are not characteristic solely of cognitive roles and research projects. This applies to other walks of human activities as well, for instance to different areas of technical assignments and technical projects, an area in which the author had worked for almost two decades.

2. THE DILEMMAS OF EPISTEMOLOGICAL TOPICS

Everyone who seriously contemplates the issues of cognition, sources of human learning, the nature of its results, their functions in other spheres of human activities, the criteria of decision-making, evaluating the products of cognition, and many other problems studied by epistemology (or by the philosophical disciplines with similar or analogous objectives under other names) will hardly be able to avoid many dilemmas. These dilemmas, which have accumulated as attendant phenomena of the centuries-long development of philosophical and scientific thinking, have usually originated as a result of attention being focused on some partial aspects of a very intricate cognitive process, as a hypostasis of those aspects, i.e. assignment of a completely independent role to some, usually abstractly conceived, components of highly complex cognitive process. Needless to say perhaps, this is not meant in any way as a detraction from the stimulating importance of the trends that have been focused on analysing the significance and function of such partial components as empirical evidence, role of conceptual equipment, *a priori* knowledge equipment, role of linguistic means, and of sign systems in general in cognition, the importance of interpretation, as well as other tasks ensuing from the interpretation of obtained findings.

The traditional topics of epistemology, even though specifically philosophical subjects are involved, have been given major stimuli and initiatives as a result of significant shifts in science and scientific learning, especially as a result of progress in modern science, changes in the basic conceptual models of science or changes in fundamental scientific paradigms. In this respect, it is certainly true that any philosophy of science (and also the philosophy of technology or a philosophy examining human creative activities, the sphere of artefacts) can hardly afford to ignore epistemological issues. In a sense, epistemology is a "self-reflection" of cognition (and hence the learning of learning and also the learning of findings and their functions). In these contexts we come up against possible variant interpretations of what we view as a "self-reflection" of cognition. (Such possibilities of the "self-reflection" of cognition were explicitly pointed out by W. V. Quine in replies to the ideas expounded in his book "Word and Object") [4], [5]: Are these merely analyses of cognitive activities, results of such activities, which represent artefacts *sui generis*, analyses of their functions in other walks of human activities and also (or even predominantly) the self-critical reflection, which, besides cognitive aspects, also encompasses value-related, ethical, aesthetic, emotive etc. aspects? In all probability, a universal answer to those questions, which would be satisfactory in all the situations or contexts, can hardly be found.

We will find ourselves in an analogous situation, if we overlook possible answers to some classical questions, the solutions of which epistemological subjects are associated. Figuring prominently among these issues is the question of the *sources* of cognition. The traditional dilemma of empiricism and rationalism (while each of these trends may be further specified, as manifested by the additional series of possible modifications and attributes, e.g. "logical", "critical" etc.) presupposed the existence of alternatives highlighting the role of empirical evidence and the role of rational creativity. Having admitted that the results of cognition are always bound up with some operators whose analyses are submitted by contemporary philosophy or intensional logic (while these are not only epistemic operators of the type "somebody knows that" etc. but also other types of operators, such as, e.g. operators of propositionary procedures, levels of conviction, belief, preferential attitudes etc.), we have to recognise other sources of cognition as well. But this does not imply that the role and value of these different sources are identical under any situation, that their meaning is equivalent for all the cognitive roles. If we thus accept the plurality of sources, we have to accept their different roles in various thematic spheres, under different problem situations. Such is the case of situations in which our cognition is based on what Bertrand Russell called cognition through immediate acquaintance and on what he called knowledge mediated by descriptions.

Basically, there is no dispute as to the fact that the results of cognition are expressed in sets of communicable means, most frequently in *sets of linguistic statements* which may be shared, recorded, interpreted, and stored, in memory facilities. What has been or could be a bone of contention are the procedures, methods or criteria stating how these statements should be differentiated, typified, and especially interpreted, i.e. assigned what is very generally and very vaguely called their meaning. The philosophical and also logical tradition used to be focused on the fact that to such statements we assign various states, which always form elements of a specific world, specific parts of the world, domains, phenomena or events or – what cannot be excluded either – a specific consciousness. (This does not concern only "states of objects" but also "states of consciousness" or rather "states of minds".) Considering the elements or states of a specific world, we do not and definitely cannot be referring to a momentarily or immediately available world, to what we call the actual world, but also other worlds to which we assign other attributes, setting them apart from the actual world. Therefore, this involves possible worlds, future and practicable worlds, constructive worlds, worlds originating as outcomes of different types of human activities and hence also imaginary worlds or worlds of human thought constructs.

If there is no doubt that the results of cognition are expressible in sets of statements (while the use of dispositional predicate, i.e. the fact that it can be expressed that way, is certainly not immaterial), then discussions and disputes revolve around specification and typisation of such statements, modes or means of their decision-making. In modern philosophy it was mainly Leibniz and Kant who – although each spurred by different motives – differentiated analytical statements and synthetic statements or – if we are to use Leibniz's terminology – "vérites de fait" and "vérites de raisonnement". At first sight, there is an evident and substantial difference between the statements of a historian on the date and course of the battle of Austerlitz and the statements explaining the Pythagorean theorem or other theorems of pure mathematics. Proceeding from such a relatively simple comparison, we can formulate a seemingly simple conclusion that synthetic sentences may be decided, i.e. assigned truthfulness, namely in view of a specific world, usually in view of the actual world, generally on the basis of empirical procedures. On the contrary, analytical sentences, i.e. sentences in mathematics and logic, may be decided, and thus their (logical) truthfulness stated, in view of all the possible worlds. But a detailed scrutiny of these differences and also of the applicable decision-making criteria will show that these problems are far from simple and explicit. The complexity and also the great demands posed by a clear-cut delineation of formal and contextual components in sets of statements, which may regarded as the results of cognition, are illustrated by changes as well as disputes relating to the foundations or conceptual starting points of mathematical thinking. For centuries it was regarded as a matter of course that the sum total of the inner angles of a triangle made up two right angles. Only the continued development of non-Euclidean geometries revealed the need of considering other prerequisites, which could change the scope of the originally assumed competence. Equally remarkable is the development of concepts and standpoints pertaining to the consistency of some formal systems to which the attribute of validity was originally assigned – if we can again use the terminology derived from Leibniz's traditions – in all the "possible worlds".

A somewhat different sphere of dilemmas and potential disputes may emerge if we consider the scope of the competence of statements resulting from cognition. Also in these contexts one should not ignore the importance of philosophical traditions linking up to Kant and distinguishing two basically different types of statements or rather two types of scientific knowledge: The first type, which originated in modern natural sciences, was associated with the term "scientific law", i.e. a statement which is of generalising importance, which can be expressed in formal terms as a statement operating with variables from a well-defined domain of entities and recorded with a general quantifier. (Expressed in words, these are statements of the type: It is valid for all the entities, which have limited attributes that are characterised by certain relations and functions, that these entities have other attributes or functions as well, that they cause certain changes, that they start certain processes etc.) The fields of cognition leading to thusly formulated statements as the results of cognition, which, in fact, represent generalisations pertaining to a well-defined domain, are traditionally called nomothetic branches or disciplines. On the other hand, areas of cognition which are instrumental in generating individualising statements that express a specific and unique state, a specific event or a sequence of events, are traditionally called idiographic branches or disciplines.

It should be stressed that this particular distinction can be made neither quite consistently nor unequivocally, that the interlinks between what we call the "theory" and "conceptual framework" of a given domain or a problem area and what is expressed by descriptions or specific characteristics of individual states or situations are more profound than generally meets the eye. True enough, a decisive position is occupied in some disciplines by creation, verification and confirmation of hypotheses, generalisations, scientific laws or "rules" (in the sense of expert systems), while statements related to specific situations, whether they are topical, past, anticipated, requested or projected situations, prevail in other disciplines. But no historian who is called upon to provide the most complete characteristic of a sequence of events, to show full respect for sources necessary for such a characterisation, can ever work without a conceptual framework, without a system of terms, models or generalisations, expressed not only during a global view of the overall situation in which the given sequence of events takes place, and – of course – without those, terms, knowledge and indisputably also value-related structures resulting from a long-term intellectual and cultural development, without the impact of that "social, intellectual and cultural memory", which is actually projected in each step taken by those who have mastered part of that memory. This also applies to statements with an analogous function in all other branches in which we have something to do with unique irreplaceable phenomena, in which we strive for analysis, explanation of the smallest droplets in which the knowledge of the entire ocean are always projected.

It is also necessary to reject the ambitions of "nomethetic disciplines" claiming that only they are competent enough to move the boundaries of human knowledge forward or to enrich what we have called "social, intellectual and cultural memory". These ambitions have been called forth by the great achievements of the technical applications of classical mechanics and Newtonian physics, which – although within a limited area – could become a prerequisite for control, predication and practical application of some processes in these branches. The illusion that such a pattern of relations between knowledge based on generalisations (or "rules" as statements with an analogous function are called in knowledge or expert systems) and the whole spectrum of anthropogenic activities leads to an erroneous and highly one-sided perception of the function of knowledge. The core of this one-sided interpretation lay in the notion that for all these spheres of cognition it is possible to create an analogous relation between knowledge and the sphere of anthropogenic activities, primarily practical mastery, initiation of quite unbounded changes, unlimited possibilities of control, prognoses or creation of technical artefacts. Since this illusion or one-sidedness was also related to the social fabric, ideas emerged concerning the unlimited possibilities of changes in the nature of man and society, illusions concerning the possibilities of social engineering, which is a mere analogy to those engineering activities that are based on the application of classical mechanics.

If the results of cognition (at least in its substantive components) are statements, a dispute may easily arise as to the content of statements, to what extent does this content – or to put it more precisely – the meaning of these statements correspond to actual or possible states, or – as is most evident – to what extent does the meaning of statements expressing empirical evidence correspond to empirically determined states. It is known that an attempt was made in the development of philosophical thought, when considering the nature of empirical statements, to distinguish the so-called primary and secondary qualities. Primary qualities were assigned to observed entities which, in the spirit of Cartesian traditions, are marked by expansiveness, thus bearing spatial and temporal attributes. But having noted that roses given to a friend for a birthday are red, it then depends on our capacity to perceive and distinguish those parts of the colour spectrum which are, within the framework of an interval, optically identifiable by the human eye or rather in the perceivable part of the spectrum, thus named as red. This kind of argumentation is fully justifiable. But no less justifiable is the standpoint that our determination, scaling or quantification of spatial and temporal parameters also depends one man-made instruments and yardsticks for measurement and quantification.

This problem of empirical knowledge and the reliability of such knowledge has passed through philosophical thought in modern philosophy like a red thread. Traditionally, this problem is characterised as a continuation of the scepticism which was explicitly formulated by D. Hume. Continuation and further substantiation of this scepticism formed the stance of 20th century positivist-oriented philosophers, starting with the conceptions of E. Mach, A. J. Ayer, R. Carnap, C. I. Lewis and many other thinkers.

Within this context, we cannot question those traits of the above-mentioned scepticism that lead to the conclusions that the results of human cognitive activity, including results of observation, measurement and experiments in science, bear the nature of artefacts. The content and substance of these artefacts doubtlessly have the character of human creations, which are the results of an organised and target-oriented interaction of certain sources, starting situations or stimuli and human cognitive activity equipped with certain distinguishing means, a certain conceptual equipment and also a certain level of knowledge. Used in such contexts, the term "artefact" does not designate a space of arbitrariness and an unlimited choice of goals and means, which is, to a smaller or lesser extent, possible in other cultural spheres. Having claimed that science is an important and utterly consequential part of culture and cultural activities, we should add that this sphere is controlled by relatively strict rules enabling testing, verification, confirmation and application of other decision-making and evaluating procedures. Thanks to respect for such rules, generated findings are universally communicable, and on the basis of some of them, other target-oriented activities can be implemented, securing approximation to desirable objectives. Thanks to respect for such rules, findings which do have the nature of artefacts can be assigned an intersubjective nature and therefore an objective content as well.

3. CRITICAL REMARKS CONCERNING ONE-SIDED AND SIMPLIFYING NOTIONS OF COGNITION

An overview of some of the most pronounced dilemmas of the epistemological problems was aimed at showing the multi-dimensional and not always homogeneous substance of these problems. Also due to these reasons, attempts at reducing these problems to simple patterns or models seem to be hardly justifiable and acceptable. Such attempts are usually based on an isolation and overestimation of some partial aspects of cognition which can, doubtlessly, be important in some contexts, but which cannot be taken as a universal tool for the clarification of cognitive activities and their results.

One of the relatively simple and often very primitively interpreted versions of such attempts are concepts of learning and its results operating with the term "representation". One cannot ignore the fact that the original term "representation" ("*Abbildung*" in German) and its interpretation were further simplified by the incorrect translation of the German original by the term "reflection" (taken over by Marxist philosophers from the Russian word "*otrazheniye*"). This term denotes quite primitive associations, comparing the function of the cognizant subject to that of a mirror, regardless of the state or equipment of such a mirror. With a certain degree of irony we might claim that in this way the represented entities or states are placed in front of numerous highly different mirrors of different properties, distinguishing abilities and different linguistic equipments necessary for the expression of the results of representation. The notion that a mirror image is a copy, a model or even another version of the original, just like a photographic reproduction of a painting, is probably justified in some special situations, but it is definitely insufficient for the explanation of the whole spectrum of activities and their results which we call cognition, findings, or knowledge.

But the terms "representation" or "models" hold an important place in the analysis of some major aspects of human creative activity, i.e. not only in spheres of cognitive processes but also in the field of technical activities, projecting, or designing. Representations or models are invariably renderings of something, i.e. of a specific real or possible situation, specific states, processes, sequences of changes, and they are renderings in a system of applicable representational means that make it possible for the generated artefacts to be communicable and interpretable. As a result, this involves a relationship between a state or a situation in one system of entities and states in another system which is usually accessible to other subjects, i.e. being communicable and interpretable. This semiotic role makes a substantial function of each representation or each model. From a formal point of view, this is a relationship of isomorphism (which is virtually only a theoretical possibility) or a relation of homomorphism whereby only some attributes or relations of the represented set of entities are depicted, or figuratively speaking that which is to be rendered, regardless of the fact whether these are real, empirically identifiable entities, theoretically considered entities, possible states, prepared technical artefacts or considered technical solutions. A set of representational means should, however, be prepared beforehand, sometimes it should even be purpose-built, for instance by adopting certain rules, conventions etc. This is certainly not a universally created mirror which is at the disposal of any subject of cognition (or, generally speaking, any creative activity).

The relationship of homomorphism presupposes that the core of homomorphism should be specified, i.e. those attributes or relations of sets of depicted entities, which are dependent on contexts of a pragmatic nature, on the possibilities of differentiation, identification, and qualitative as well as quantitative specification of selected attributes or relations. In other words, the subject of cognition invariably makes some kind of selection in terms of what he wants or needs to select, what he is capable of representing sufficiently reliably, or what he can – in view of his knowledge, epistemic competencies and capacity possibilities – depict. (While the proponents of the so-called "theory of reflection" claim they do not rule out what they call "active participation of the subject of cognition" in the results of cognition, they are usually unable to specify such an involvement any further.) A set of these available possibilities of the subject of cognition can be characterised as an "epistemic competence" which is however always rendered relative also due to the given problem situation, entities or situations under scrutiny. If homomorphic representation is, indeed, conditioned by selection, affected both by the claims laid to the function of representation and also by epistemic competence (while both given sets of prerequisites can be mutually conditioned), it is only natural that we cannot rule out the possibility of various representations of the same set of depicted entities or processes, notably for different situations or differently equipped subjects of representation. That is also why the spectrum of forms and types of models displays such a great variety, covering different types of pictorial (iconographic) and graphic models, formal models etc. Furthermore, the pluralism of forms and types of representations or models is strengthened by the possibilities of mutual transformations. A marked expansion of these transformational possibilities has been brought about by progress in information technologies, widespread possibilities of digitalisation and their subsequent transformations into their original forms. Also these features of the plurality of the possibilities of representation show that the term "reflection" can offer only a minimum explicatory value in expounding cognitive processes and their results.

As a matter of fact, the homomorphic model of a certain real or possible situation, a certain process or a sequence of events always represents an artefact conditioned by selection in the given sense, i.e. with a view to the epistemic competencies of its authors and with a view to the possible, considered or anticipated functions of representation. In a number of spheres of creative activities, artefacts with these functions should take into consideration not only the abilities, possibilities and competencies of the potential interpreters and users of such models but also their wishes, right, and ability actively to participate in what is an additional, expanding, and sometimes enriching interpretation. That is why selection of the core of homomorphism does create, in many cases, a necessary framework of given representation, while leaving sufficient scope for creative participation of interpreters or users. Such is the case of technical models, especially processes in the shape of project documentation, i.e. a model of what, how, where, when and under what conditions should the project be technically carried out. It is only natural that this space for creative co-participation of interpreters and users has different scope and different limitations and hence different cognitive activities in different spheres of human creative activities. One can well imagine different interpretations of artistic artefacts, much greater limitations are imposed in the spheres of technical, projecting and designing activities, while relatively strict limitations are justifiably required by science. But even in science, some findings may be associated with different interpetations, while a change of interpretation can sometimes lead to a major knowledge-related shift. (A case in point is the change of interpretation of the well-known Michelson-Morley experiment, originally designed to prove the non-existence of "ether wind".)

Another example illustrating attempts at shaping the concept of a universal pattern of cognition and the results of cognition is the attention focused on analysing statements and their interpretation, i.e. what can be described as a semiotic approach, semiotic evaluation and decision-making. This approach was launched by the classical positivistic attitude to cognition, as represented by the works of E. Mach whose pioneering study associated cognition with an analysis of sensations [3]. A characteristic feature of this approach is accent laid on the analytical method (which, according to the pioneer of this approach, is typical of scientific cognition) and subsequent interpretation.

It is not an objective of this study to describe Mach's analysis in greater detail or to characterise various other standpoints developing that approach. We will try to characterise that approach through the conception of the leading American philosopher C. I. Lewis who developed the heritage of C. Pearce, the distinguished proponent of pragmatic concepts at Harvard University. His underlying work bears the characteristic title "An Analysis of Knowledge and Valuation" [2]. Lewis describes knowledge as an assertive state of mind which is, however, called forth by a specific activity. This accentuates the activity-related context of knowledge. It also stresses the practical significance attributable to knowledge. Another substantial feature of a state of mind characterised in this way is what is described as the "believing attitude", i.e. an attitude which always claims to truthfulness, to what brings justification. Pragmatic dimensions of such an approach are then spelt out in the statement claiming that epistemology is not a psychological description of states of mind or attitudes in the given sense but rather an analysis and criticism of their cognitive claims, an evaluation and assessment of their validity and truthfulness, a specification of criteria for such forms of decision-making. Based on such attitudes, as Lewis continues to assert, one can come to the intentions of knowledge, to the types of meaning, to the semiotic analysis of propositions or statements expressing states of mind or attitudes in the given sense, to the differentiation of analytical and empirical (synthetic) truthfulness. This particular approach then does not claim any ontological commitments (for instance in the sense, as specified by W. V. Quine), it can be satisfied merely with the claims to success or positive experiences grounded in knowledge in the given sense, i.e. states of mind, attitudes and propositions, or statements proceeding therefrom.

We have demonstrated that attitude which actually has many common denominators or analogous features with some other epistemological conceptions, developed within the framework of pragmatism, operationalism or some empirical and inductionist theories primarily in a bid to prove the possibilities of hypertrophy of some, often very important, features of cognitive processes and the creation of universal patterns on the basis of these features. This certainly does not question the greater or smaller importance of the practical aspects of cognition, the application of its results in various spheres of active components of such an activity, the creation of technical artefacts, the safeguarding of desirable change, and the prevention or blocking of undesirable changes. The limited nature of such an attitude can be demonstrated by the following example: The success of a certain therapeutic procedure in medicine by itself is not and cannot be a guarantee that the previous diagnostic decision represents the sum-total of the knowledge concerning the nature, causes, sources and other contexts of the illness that is being treated. Thus some partial successes scored by charlatans, back-street healers and shamans, which these people usually associate with a certain level of justification or verbal substantiation of their own procedures, could be passed off for a legitimate alternative to one's own knowledge. In other words, the success of a certain practical approach (which is often only a seeming or loudly proclaimed accomplishment) is not and cannot be a criterion for recognising and appreciating the knowledge-related starting points of such a procedure. Such reservations are also fully applicable to what is known as the criterion of praxis, which was so vociferously proclaimed by Marxist ideologues.

Some other examples could be added to the afore-mentioned cases illustrating one-sided concepts of cognition, which accentuate only come components of the cognitive process (even though in many situations the role of such components can be quite substantial). These are primarily procedures which reduce epistemological problems to semiotic analysis, to language analysis, especially of the language of science, to inductive procedures stemming from empirical evidence etc. Basically, it should be stressed that cognition is always a complex of several types of human activities, not only outright cognitive but also decision-making, evaluating, designing and creative activities which do, or at least could, lead up to communicable results of cognition. At the same time, it should be added that this complex is practicable in a specific problem situation, given a number of necessary prerequisites which we have summed up as an epistemic competence.

4. SYSTEM FEATURES OF COGNITION

The criticism of one-sided or simplified attitudes to cognition and especially to highly qualified, target-oriented knowledge – and technology-laden cognitive activity, i.e. an activity realised within the framework of science, research and development, should respect, or at least take into consideration, interlinks between various aspects of the cognitive process, sources, stimuli and motivations of that process, the nature of attained or attainable results, as well as a large but not always easily integratable space for manipulation of those results. Following a thorough appraisal of all the reasons and impacts of the above-mentioned criticism, one can accept the view that cognition may be seen as a complex system of interactions between initial problem situations, immediate reactions to these situations, the impact of accumulated intellectual and cultural memory, complex of rational activities and resultant artefacts as products of these activities. While considering the main sets of phenomena or subsystems which, in their entirety, make up cognition as a system, we should single out some dangers posed by generalisations or something which is variegated, which is sometimes hard to compare or which leads to very different results. At the same time, it is necessary to bear in mind that cognition is always the cognition of somebody, the cognition of something, and cognition under specific situations. Therefore, if we understand these contexts as arguments of a complex function, such circumstances or connections can substantially affect the nature of cognition, viewed as a complex function. Also due to these reasons it is expedient to regard any generalisation, including that which outlines the system features of cognition, only as an approximation to understanding and explaining those facts whose sum-total makes up cognition and findings, while the weight of such an approximation can be quite different in different contexts.

Reflections examining system features traditionally begin with what we call a system analysis, i.e. specification of constitutive subsystems or system elements. As for the system analysis of cognitive processes, it is useful to distinguish and specify at least the following subsystems:

- a problem situation which represents a starting point, a source or stimulus for learning;
- a complex of activities associated with target orientation *sui generis*, i.e. efforts to seek a solution of an identified and specified problem situation, to lower the level of entropy in relation to that problem situation, with efforts to expand or specify spheres of available knowledge;
- generation of findings, i.e. artefacts with a cognitive function, which meet the acknowledged requirements for acceptable or recognised claims to the results of cognitive activity, notably communicability and interpretability;
- analyses of findings, their normalised types and forms, interpretation of findings, sets of decision-making procedures, namely procedures of verification, confirmation, testing, interpretation, semiotic analysis of artefacts with a cognitive function;
- forms and means of manipulation of artefacts with a cognitive function, especially procedures of application, transformation, of arranging and integrating findings.

The situation we have readily characterised as a *problem situation* is probably marked by a whole range of substantial features. A man who feels pain, who does not feel good or who believes that he is running a temperature, thinks he is ill. If we realise that we are unwell, that we need something or lack something, that the situation we find ourselves in should be changed, we have come to identifying such a situation as a problem one. A physician who is called upon to help a patient, who is asked for therapeutic action, finds himself in an analogous situation. A problem situation must therefore be stated, identified and recognised as a problem situation, as one that requires an intervention, usually by incorporating a certain human activity which is in a position to change the given situation in a desirable direction. This, however, requires that the given situation should be specified, identified, so as to be recognised, as in the case of an illness and possible action by a physician, as a situation requiring necessary action by a human activity. If a man is not to express his helplessness or indifference in the face of an identified situation, such a situation requires that efforts be launched, which can be generally characterised as a search, a quest or an investigation. As a rule, it is an effort to specify the characteristic or nature of the given situation, its identification, to stipulate goals and means of possible interventions, to seek, find or specify such means. In this context one cannot ignore that many leading thinkers view the terms such as "search" or "quest" as starting points for an understanding and explanation of an active anthropogenic activity, i.e. an activity focused on the attainment of desirable changes, the creation of artefacts, a more perfect and efficient use of available means or resources. It was certainly not by chance that the term "search" was made a key notion of the well-known conception of the creation or genesis of artefacts and also of the projecting or designing activity in technical spheres by H. Simon in his famous monograph on sciences of the artifical [7].

If it is true that recognition, identification, or specification of a problem situation stimulates a set of efforts, ambitions or activities which can generally be characterised as a search, then the very start of these activities, the will, or an effort to carry out such a search, by themselves do not provide any guarantees that an acceptable, or at least satisfactory, solution of the pertinent problem situation will be found. This applies to all walks of human activities which have the nature of a "search", i.e. research and development, activities in the technical sector and probably also some other activities instrumental in creating artefacts. That is also why a team of prominent authors engaged in studying the current problems of science, technology and innovation policy at the end of the 20th century headlined its studies with the phrase "uncertain quest" [6]. Such procedures, associated with the solution of problem situations in the fields of research, development and technical activities, may be characterised also by using the terms "uncertainty" or "entropy". If cognitive activity can be described as a search whose start will not provide any *a priori* guarantee that a solution will eventually be found, or at least an acceptable or satisfactory solution will be found, then this particular endeavour can be characterised as an effort to lower the original uncertainty or to reduce the original level of entropy. As a result, specification, recognition or identification of the given problem situation has the nature of becoming aware of such an uncertainty or the unacceptability of such an entropy level.

If it is true that recognition, specification or identification of a certain problem situation motivates or stimulates specific activities, then it is understandable that not every situation gives the ground for the start of cognitive activities. In many cases, identification of a problem situation, if we already know a suitable , possible or available solution, may directly stimulate such a solution. Only in cases in which not all the aspects of a solution are known, in which the knowledge usable in the given problem situation (regardless of whether a solution has a practical nature or that of intellectual activities) should be expanded, specified or improved, is it vital to start *a complex of activities* connected with a target orientation *sui generis*. In this respect, the ultimate goal need not be an immediate solution but, for instance, identification of possible risks, not only an implemented change but also an identification of possibilities, boundaries, or risks associated with such a change. That is why the purpose of the complex of activities associated with a cognitive orientation is, first and foremost, the establishment, specification, or expansion of the available knowledge. In this sense, the purpose of the complex of these activities is the production of data with a cognitive function.

The claim that cognition also forms a complex of target-oriented activities could lead us to the simplifying idea that these are primarily cognitive procedures in the narrow sense of the term, i.e. especially empirical procedures, starting with simple observation, measuring procedures using sophisticated measuring devices, and ending with experimental activities of the most diverse types. In addition to empirical procedures, we usually also distinguish theoretical procedures, which operate with the already generated data, making it possible to create generalisations of a greater number of data, for instance scientific laws, hypotheses or empirical generalisations, the creation of data which have the nature of rules, instructions, guidelines, algorithms etc. The sphere of theoretical procedures also contains those procedures that make it possible to use derivative (inferential) procedures, safeguarding the possibility of creating new data from the already available ones [8].

The sphere of activities, which make up complexes of different activities, also covers evaluating activities, largely connected with the evaluation of the acquired, partial or total results, with the procedures of verification, testing, confirmation, evaluation of possible applications, and with the efficiency and effectiveness of such applications. Evaluation is also interrelated with the activities which have the nature of decision-making procedures, especially in the process of selecting feasible procedures, applicable methods and means, in opting for the continuation of the complex of launched activities. As a result, evaluation and decision-making should be regarded as an integral component of cognitive processes.

Cognition is therefore also a system of interactions of different activities that lead to the creation of data with a cognitive function, i.e. generation of intellectual artefacts. This also highlights the creative, active, designing or inventive component of the complex of activities which, in their entirety, lead to cognitive changes. It is natural that no cognitive change can be isolated or seen as a result of a sudden flash of genius. The memory of past steps and their results is projected in every shift taking place within the sphere of our knowledge.

If we understand the cognitive process as a system of interactions of different stimuli, prerequisites, and sets of target-oriented activities, then a substantial component of the target structures is *the creation of findings*, i.e. artefacts with a cognitive function which are communicable, reproducible, and interpretable. This means that from an information point of view, such a process of creation calls for the application of specific information carriers, i.e. means enabling information transfer in time and space. Needless to add, such creation makes it imperative that certain acceptable, recognised claims or rules be met. To put it in other words, the process of generating findings, and not only in the sector of science and research, where such process of creation constitutes a key goal, carries social dimensions as well. The author of findings or, generally speaking, the subject of cognition is, therefore, never totally independent in his choice of means of expression, i.e. as a rule, linguistic or other sign systems.

If we realise that a substantial part of the results of cognitive activities is expressed by linguistic means, the pride of place among these prerequisites is held by what we usually call a conceptual system. As a matter of fact, it is symptomatic that many authors who were engaged in analysing cognitive processes devoted considerable attention to creating terms, distinguishing observatory and theoretical terms in empirical spheres, interrelations in these and other conceptual types (for instance in the shape of the so-called reduction sentences), criterion of distinction of logical and extra-logical terms etc. A major problem in these contexts is the interpretation of terms, and this usually applies to accepted contexts. (Such a context can be demonstrated against the background of a given conceptual system and a thematic sphere as follows: Talking about a knight or a rook in chess, we are actually using a different interpretational framework than in cases in which the same terms are used in different contexts.)

The results of cognitive activities are not expressed solely by terms alone but by meaningful statements using a specific conceptual system. An analysis and typology of statements (or sentences) and systems of statements forms a substantial factor in analysing findings. Of decisive importance for the analysis and interpretation of statements is the delineated context, i.e. linkage to the delineated conceptual system and specified domain. In other words, substantial features of these problems are not only formal but primarily semantic criteria. This concerns the typology of statements which lays a claim to having cognitive function, for instance the differentiation of singular and generalising statements. One can hardly ignore that the tradition of epistemological interests underlined the distinction of different types of generalising statements, the distinction of confirmed scientific laws and hypotheses, empirical generalisations and random generalisations. (The difficulties associated with a clear-cut delineation of these types have been singled out by the author of this study in chapter on cognition and scientific cognition in [9].)

An overwhelming majority of statements which lay claim to having cognitive functions concerned the above-mentioned areas and, naturally, also the decision-making issues closely connected with them. But statements expressing results of a cognitive activity are not only statements relating to something topical, possible, something that is valid only in a delineated world or in all possible worlds. Such statements also represent assertions made by somebody, uttered or otherwise expressed with a certain degree of conviction, certainty, confirmation etc. Claiming this, we are actually touching upon the role of intensional operators in the process of formulating findings. By the way, the very statement that we know that we know, why do we know and how, is, in itself, an expression of the irreplaceable role of intensional operators. Also the fact that epistemic logic has been developed (whose prototype is, for instance, the work of J. Hintikka [1]) confirms that some recent trends in contemporary epistemology take into account the relationship between the author or recipient of a finding and its content, that such an attitude which can be characterised as an epistemic attitude is given due attention and explanation.

While considering the generation of findings, we cannot ignore those forms of expressing findings that use the means of pictorial, graphic or cartographic communication. Such means too are known to have quite a legitimate nature of information carriers: they can be communicated, remembered, interpreted etc. It is therefore logical that scientists talk of graphic languages, the language of cartographic representation, iconic models and other analogous means of communication. In this context, it is not only possible interpretation that is of importance, this naturally always happens under the assumption that the addressee of a statement of this kind has at his disposal an adequate competence, equally important is the possibility of having various forms of transformation, including transformation into traditional verbal forms. The fact that while expressing some findings, we tend to select pictorial or graphic forms is based primarily on semiotic grounds, particularly on the ease and great accessibility of interpretation. (The author of this paper has examined the semiotic aspects of graphic communication in greater detail in [10].)

In cognitive processes as systems of interlinks of interactions, as processes of mutual connectivity among various activities with a target orientation, individual partial results must be decided, evaluated or assessed, which, in turn, affects the selection of other procedures, of the possible trajectories of follow-up activities. Also these results of the complex of cognitive and other interrelated activities, which we regard as coherent and acceptable findings, are usually further analysed, evaluated and verified. It is therefore possible to speak of *an analysis of findings* as a legitimate subsystem of the entire cognitive system. Since the nature and forms of findings cover a relatively wide spectrum, and not only in terms of formal nature but also in terms of content as well as cognitive function, an analysis of findings includes a large and comparatively variegated set of procedures. Basically, these procedures can be divided into the following two groups:

- procedures of epistemic decision-making;
- procedures of semiotic decision-making or – as it is more commonly called – procedures of interpretation.

Epistemic decision-making means procedures making it possible to attribute adequate epistemic values, i.e. to assign truthfulness, validity, justified acceptability etc. These involve procedures of verification (most frequently verification of empirical findings or singular statements derivable from acknowledged generalisations or rules), the testing, confirmation and falsification of hypotheses, determination of validity, derivability of sentences or theorems from acknowledged or accepted axioms etc.

The term semiotic decision-making or interpretation of findings means various procedures of assignment, the most frequent being the assignment of sense and denotation (or at least one of the afore-mentioned modes of meaning). Semiotic decision-making also means transformation of one form into another which is better communicable, interpretable, or faster interpretable. Semiotic decision-making presupposes the availability of such a set of prerequisites, knowledge, and competencies which make it possible to assign meaning and thus to understand. To be true, semiotic decision-making is a major prerequisite for what can be called ontological decision-making, but it does not provide any guarantees for such a kind of decision-making or – if we can use Quine's terminology – does not give any "ontological commitments". (The relationship between epistemic, semiotic, and ontological decision-making is examined in greater detail by the author in [11].)

If we understand the cognitive process as a system of interlinks among the given systems, encompassing a problem situation, complex of activities, the generation of findings, it is also expedient to consider what happens to findings, what is the functioning of a subsystem that can be called *manipulation with acquired and generated findings*, this applying not only to manipulation in the form of arrangement but also to the application of findings. This subsystem, in turn, affects the conception of problem situations, the creation of goal structures, and generally the entire knowledge – and value-related equipment which co-participates in the entire cognitive process. The traditional and – as seen from the present-day angle – highly limited interpretation of such a subsystem has been confined to a concept which saw knowledge as an instrument of control, as an intellectual prerequisite for using resources and an opportunity provided by nature and an instrument of manipulation with these resources. Proceeding from that context, which has its roots in the foundations of modern science dominated by classical mechanics, so aptly depicted by Bacon's "scientia est potentia", not only science but also philosophy and value-related perception of knowledge have proceeded to a more limited and, at the same time, more accurate understanding of knowledge. The forms of these limitations and accuracy are highly diversified. They can have the nature of well-defined horizons of opportunities and validities of identified findings, and also the character of limitations proceeding from the acknowledged value structures, ethical principles, and an awareness of responsibility. The thematic sphere which could be called "the boundary of the possible" or "the limits to freedom in research and mainly in the application of its results" will probably grow to be a still greater component of the entire cognitive activity. The spectrum of shapes and also both knowledge- and value-related grounds for specification of these boundaries is very broad and certainly deserves a more thorough analysis.

The forms of manipulating with acquired and generated findings can be divided into two different groups:

1. registration, publication, incorporation into normalised memory formats, and
2. sphere of application of findings, both in terms of further orientation of cognitive activity and analogous processes, and in various forms of practical applications.

(1) Registration, publication and incorporation into adequate memory formats are marked by a significant trait: The chief purpose of these forms of manipulation is to safeguard accessibility and efficient communicability. These objectives are served by abstracts, key words, suitable thematic structures for the purpose of data storage, for instance in libraries, catalogues or other memory media. Parallel with these, there are different forms of semiotic integration, comparison with findings acquired in the past or elsewhere, decision-making concerning the integratability into certain systems, decision-making relating to the arrangement of findings acquired in the past or elsewhere, or appraisal of forms of such an arrangement, for instance decision-making relating to the consistency of the generated systems of finding etc.

(2) The sphere of application can also assume different forms. It would be a great simplification to suppose that all the findings provide the possibility of practical implementation of desirable or necessary changes that have the nature of an algorithm or general instructions for the purpose of solving tasks of a given type. Although this relationship between knowledge and the possibility of practical implementation of desirable changes is typical of a part of acquired findings, it is not the only type of application. Some kinds of knowledge make it possible to launch such decision-making and follow-up activities whose goal is to maintain the existing stereotypes and prevent undesirable changes. Other findings are marked by the fact which is sometimes characterised as predicative or also as a systematisation capacity, i.e. the possibility of deriving potential states, possible sequences of other states, the possibility of estimating anticipated impacts, risks, and otherwise characterised effects of specific natural processes as well as changes initiated by man. That is also why the broad, multifaceted, and considerably heterogeneous sphere of applications of knowledge cannot be crammed into the straitjacket of simple patterns of knowledge as an instrument of control, especially an instrument for mastering nature or for mastering man and society.

Academy of Sciences, Prague, Czech Republic

REFERENCES

1. Hintikka, J. (1962), *Knowledge and Belief.* An Introduction into the Logic of Two Notions (Cornell Univ. Press, Ithaca).
2. Lewis, C. I. (1950), *An Analysis of Knowledge and Valuation*, second edition (The Open Court Publ. Comp., La Salle, Illinois).
3. Mach. E. (1897), *Beiträge zur Analyse der Empfindungen*. Jena, 1886 (English translation *Contributions to the Analysis of Sensations*, The Open Court, Chicago, 1897).
4. Quine, W. V. (1960), *Word and Object* (MIT Press, Cambridge, Mass.).
5. Quine, W. V. (1969), 'Reply to Smart', in *Words and Objections*, D. Davidson, J. Hintikka (eds.) (D. Reidel Publ. Comp.).
6. Salomon, J. J., F. R. Sagasti and C. Sachs-Jeantet (eds.) (1994), *The Uncertain Quest: Science, Technology and Development* (United Nations Univ. Press, Tokyo, New York, Paris).

7. Simon, H. A. (1969), *The Sciences of the Artificial* (The MIT Press, Cambridge, London).
8. Tondl, L. (1972), *Scientific Procedures* [Boston Studies in the Philosophy of Science, Vol. 10] (D. Reidel Publ. Comp., Dordrecht, Boston).
9. Tondl, L. (1994), *Věda, technika a společnost* (Filosofia, nakl. Filosofického ústavu AV ČR, Praha). (English, Science, Technology and Society.)
10. Tondl. L. (1997), *Mezi epistemologií a sémiotikou* (Filosofia, nakl. Filozofického ústavu AV ČR). (English, Between Epistemology and Semiotics.)
11. Tondl. L., *Sémiotické a ontologické rozhodování* (V tisku, Filozofický časopis). (English, Semiotic and Ontological Decision-Making.)

A. I. UYEMOV

SYSTEM APPROACH TO THE PROBLEM OF THE CLASSIFICATION OF SCIENCES AND SCIENTIFIC RESEARCHES

The problem of the classification of scientific knowledge is one of the most important problems of the methodology of science. A. Polikarov has given considerable attention to this problem. The author of the present paper is greatly satisfied by the fact that he used to cooperate with A. Polikarov in the development of that problem. He hopes that the joint efforts will result in drawing attention to the new fundamental divisions which correspond to the scientific spirit of the second half of the twentieth century.

Up to the sixties, for philosophers in the Soviet Union, Bulgaria and other countries of Eastern Europe, the classification of sciences according to the forms of matter and motion seemed the only correct one. The fundamentals of such a classification were drawn by F. Engels [1, 2]. Engels' views were systematised and presented in detail by B. M. Kedrov [3]. Later on they were developed and thoroughly illustrated with the help of materials from modern science in three volumes written by the same author [4–6]

The main idea of such a classification was as follows. There are different forms of motion which differ from one another as to their degrees of complexity. A more complex form – a higher one – includes the lower form but is not reduced to it. Each science studies a definite form of motion. The sequence of sciences in their classification comes from the development of forms of motion in the historical course of the solar system. The simplest form of motion is the mechanical one. Further, in the order of growing complexity, there come physical, chemical, biological (life) forms of motion. This sequence defines the respective order of the sciences: 'Mechanics', 'Physics', 'Chemistry', 'Biology'. Each form of motion has its corresponding material substratum. That is why sciences may be defined by means of this substratum. Hence Mechanics is the science of masses, Physics is the science of molecules, Chemistry is the science of atoms, and Biology is the science of albumen bodies. This is very important from the logical point of view, as it permits us to avoid a vicious circle in the definitions of sciences. The historic foundation of such a classification is connected with the development of both Nature itself and knowledge of it. The simplicity of the given scheme is quite obvious. However, it is also clear that it does not cover all the sciences, not even the natural ones.

This is why when F. Engels had completed it, he came to a classification which reminds us of that of St.-Simon and Kant:

I. Sciences of nonliving nature: Mathematics, Astronomy, Mechanics, Physics, Chemistry, Geology

D. Ginev and R. S. Cohen (eds.), Issues and Images in the Philosophy of Science, 377–390.

II. Sciences of living nature – Biology
III. Sciences of Society – History

The consistency of the classification according to the forms of motion is broken here into quite a number of items. Firstly, Mathematics does not correspond to any particular form of motion and has no special material substratum. F. Engels and B. Kedrov try to overcome this difficulty, considering Mathematics to be an auxiliary discipline giving only a tool for the study of the quantitative aspect of every motion, primary a mechanical one. One can, of course, relate Astronomy to some form of motion and define its substratum, but 'astronomical motion' would not be a simpler form of motion than a mechanical one. Moreover, one cannot include astronomical motion into physical, chemical motions as an aspect of them, which can be done with mechanical motion. And, finally, biological motion evidently does not contain a geological motion as one of its aspects.

It is obvious that Engels' classification reflected the state of science of his time. As long as science develops, the concrete character of such classification must become obsolete. However, B. Kedrov insisted that ". . . a concrete solution of a problem might be obsolescent, but then principles lying at the basis of its solution must remain intact" ([3], p. 443). One could have objected to this statement even at the time it was formulated. The explosion of the atomic bomb eloquently proves the validity of the use of the verbal combination 'atomic physics'. But from Engels' point of view, Chemistry studies atoms, and Physics is the science of molecules. Therefore the combination 'atomic physics' would be meaningless.

The existence of atomic physics as well as that of the molecular chemistry makes the correlation between the material substratum and the form of motion rather ambiguous. One and the same substratum can correspond to different forms of motion. The practical question arises: how to define, namely what in the motion of atoms and molecules refers to the physical and what to the chemical form of motion? To answer this question, one should know what Physics studies and what Chemistry studies. Thus in defining these sciences, a vicious circle arises. Physics is the science of the physical form of motion and the physical form of motion is those aspects in the motion of atoms and molecules which are being studied by Physics. The same *mutatis mutandis* may be said about Chemistry.

The described situation witnessed to the fact that it is necessary to change not only the concrete details but also the principles of the classification. Another ground of such a transformation is still more important. It is connected with the origin and the acknowledgement of Cybernetics as a respectable science deserving to take its place in the classification of sciences.

In the Soviet Union it happened in the middle of the fifties after the publication of quite a number of articles in defence of Cybernetics [7–9]. Alas! There is no place for Cybernetics in the classification of the sciences according to forms of motion. Moreover, Cybernetics has broken this classification! It should have been referred simultaneously both to the sciences of living and nonliving nature and to the sciences of society! Nevertheless, B. M. Kedrov made an attempt to find a place for Cybernetics. It is most thoroughly presented in the last volume of his trilogy ([6], pp. 309–20).

B. M. Kedrov spoke about the discovery of a new, cybernetic, form of motion. This consists of the control processes assuming a number of joint movements, directed in a definite way. As with any other form of motion, cybernetic motion has its 'quite definite' material substratum. This is "a certain mechanism, which is a living organism, society, human brain, machines created by the man, which are quite different as to their qualitative content, as they are built of different material" ([6], p. 344).

The cybernetic form of motion has its peculiarity – "it does not act directly on our sense organs" ([6], p. 314). In this respect it is similar to the quantum mechanical form of motion and to our mind. This is a very complicated form of motion and therefore here the independence of the form of motion from the character of its material substratum becomes more vivid ([6], p. 315).

This way of the placement of cybernetics into the classification of the sciences caused a protest on the part of many philosophers and scientists. It is quite natural. The founders of cybernetics emphasized the independence of cybernetic relations from the material substratum. The application of the notion 'a form of motion' to the subject of cybernetics is hardly more reasonable than using such a notion for the definition of the subject-matter of mathematics. But it is possible to some extent to understand mathematics as a propaedeutic to mechanics. This possibility absolutely fails when applied to cybernetics.

One can radically overcome the above-mentioned difficulties only by revising the very idea of sciences' classification according to the forms of motion. As Bogdan Dyankov pointed out ([10], p. 78), A. Polikarov, in his talk at the international philosophical seminar in 1958, had put the question that the form of motion is not a sufficient principle for the classification of the sciences. From Polikarov's point of view, Kedrov is not right to make this principle absolute, when considering it as the only and sufficient one for constructing a classification of the sciences. It is necessary to look for some additional basis for the classification.

Our paper of 1961 [11] offered such a basis. All sciences may be divided into three types. The sciences *of definite things irrespective of the properties they possess and of the relations in which they stand* comprise the first type. Thus, for example, in the last century one may consider that physics studies molecules, discovering their properties, no matter which; in the same way chemistry studies atoms, astronomy studies celestial bodies, biology studies living organisms, etc. Astronomy and biology despite their successes have preserved their subject-matter in the twentieth century, but this cannot be said about physics and chemistry. The difference between these sciences is defined not by boundaries between certain classes of objects, but by differences of the properties of complexes which are of interest in both sciences.

One set of complex properties is studied by chemistry, another is studied by physics, no matter where these properties are discovered, in molecules or atoms. Thus both of them are *sciences of definite properties, no matter what things possess them*. It is easy to notice an interesting fact. The definition of the second type of the sciences is derived from the first part of the first type's definition by a dual transformation. It consists of replacing the notion of things by the notion of properties, and the replacing of the notion of properties by the notion of things.

The third type of the sciences is sciences of definite relations irrespective of the fact, what the things are which are involved in these relations. Here we have the result of the dual transformation of the second part of the first definition, when 'things' are replaced by 'relations' and 'relations' by 'things'. The classical example of a science of this type is mathematics. "The subject-matter of pure mathematics is spatial forms and quantitative relations of the real world" ([1], p. 43). Here two types of relations are the subject of mathematical studies. The first one is spatial relations, the second one is quantitative relations. Modern mathematics begins to study more abstract types of relations, but it does not cease being the science of relations.

Cybernetics belongs to the sciences of this type also, as it studies relations of control and other relations connected with control irrespective of their material substratum. The discovery of a new material substratum may be of practical interest, but such a discovery could not change the subject-matter of cybernetics.

The development of sciences shows the apparent tendency of transformation of first type sciences to the sciences of the second and third types. Thus, the idea of development is not excluded from the classification of sciences, acquiring another meaning. Thus, hypotheses of the development of nature from the perspective of *Naturphilosophie* are replaced by the Science of science's facts.

The borderlines between the sciences of the first, second, and third types are not always sharp. There can be intermediate variants, when definite properties and relational complexes are studied best in some specific objects. One can argue whether modern physics and chemistry refer to the second type or some intermediate variant.

As the notion of relation is fundamental for understanding the structure concept, the position expressed by Bogdan Dyankov, according to which such sciences as mathematics, cybernetics, formal logic and dialectics, investigate not any particular forms of motion, but certain structures of material reality [10], is close to the given point of view. We would like to mention here that reference to the material character of reality is not always proper. As to formal logic, the reality it studies has an ideal character, and mathematics, as well as dialectics, studies any kind of reality, i.e. the investigated relations are absolutely indifferent to the character of the related objects. One can probably say the same about cybernetics, though here this problem is somewhat more complicated. The idea of differentiation of sciences according to the intention of their studies of things, properties and relations found a positive reply in the works of A. Polikarov [12]. In his later works [13–15] he has used it to construct his own system of classification of the sciences. This idea comes as the initial *fundamentum divisionis*. Another *fundamentum divisionis* divides sciences into fundamental and applied. A. Polikarov calls the sciences of things, irrespective of their properties and relations, *concrete* sciences, as distinct from the sciences of properties and relations, which are included in the class of *abstract* sciences. Abstract sciences are sometimes called *formal* ([14], p. 144, [15], p. 114). Logic, mathematics, cybernetics, informatics, are included in the abstract fundamental sciences. The abstract applied sciences are applied mathematics and, technical cybernetics. The concrete fundamental sciences include astronomy, physics, chemistry, geology, biology (natural sciences) and psychology, linguistics, economy, law, history (social sciences).

It is important that the division of the sciences into natural and social concerns only concrete sciences. However it is still widely used for all sciences. The division of sciences into social, natural and technical is considered complete. And this is a great mistake as is seen from the classification given by A. Polikarov. Technique does not embrace all the applied concrete sciences. Medicine, pedagogics, agricultural sciences also should refer to them.

A. Polikarov calls the subject-matter of the concrete sciences *natural*, and that of the abstract sciences *artificial*. And here we should like to object. The integrity, which unites different properties studied by modern physics or chemistry, has no less a natural character than the integrity of atoms or molecules. The same may be said *mutatis mutandis* about the relations studied in the framework of mathematics and cybernetics.

There is some doubt as to the term 'concrete science' as opposed to 'abstract science'. Here one may speak only about different types of abstraction. When studying molecules (in the 19th century) physics abstracted from all those properties and relations which were related to atoms, referring them to the sphere of chemistry. And finally, taking the above consideration into account modern physics and chemistry cannot be included in the class of 'concrete' sciences. From our viewpoint they may be thought of at least as sciences of an intermediate type, tending to turn into sciences of definite sets of properties.

The use of the notions 'abstract' and 'concrete' instead of 'things', 'properties', and 'relations', narrows the range of variants which can be taken into consideration when developing some classification. Thus, for instance, the sciences of intermediate type disappear and sometimes the essential difference between sciences of properties and sciences of relations disappears as well.

The importance of any classification grows depending on the extent to which it can be used to forecast further development. The tendencies of the transition from sciences of things to sciences of properties and relations were mentioned above. Another tendency is differentiation. Once integral, a science may fall into a number of separate sciences. In connection with this an interesting question arises: Will the type of the former science preserve itself in the course and result of its disintegration? Without considering this question in general, we shall give a typical example. Developing sciences which study classes of things disintegrate according to one or another pattern depending on the degree of wholeness of parts of the investigated object or of its properties and relations which are of a particular interest. Thus, geography naturally disintegrates into geography of Land and geography of Ocean. Further a whole class of sciences arises. The richer and the more variable this class is, the greater is the extent of the application of the general classification to it. Thus sciences of Ocean can be divided into three classes [16]. The first ones, referring to the geography of world oceans, study world oceans as a whole in all the versatility of its properties and relations. The others, included into oceanology, study certain complexes of properties of the sea medium, processes and phenomena taking place in the world ocean. These are the physics of the sea, the chemistry of the sea, the geology of the sea, geophysics of the sea, etc. Finally, the third group, which may be called 'management oceanology' includes the complex of the sciences, connected with the study of economic, economic-ecological, political and other relations in the sphere of the world ocean utilization management. One can refer here to the disciplines which appeared during the last 15–20 years such as marine economy, marine econology [17], marine sociology, marine law, etc. We see that the classes of sciences about the world ocean correspond to the classes of sciences in general.

One should admit that distinguishing between the above-mentioned three types of sciences without considering sciences of intermediate type, does not always give us the possibility to characterize this or that science adequately enough. Sometimes the given structure reminds us of the robber Procrustes' bed. To make our scheme more flexible it is necessary to assume different kinds of elaborations in it. These are treated in our work [18]. There the categories 'thing', 'properties', 'relations' are used for the explication of different procedures of scientific investigations. Each of these procedures is defined by the order of transition from one of the categories of our triad to others.

The simplest variant of investigation, characteristic of the empirical level of knowledge, is the movement of thought from fixing some things to defining the properties and relations of these things.

The next, higher level of investigation presumes the establishment of relations not indirectly between objects, but between their properties. The investigation of this type brings analysis, for example, to the Boyle-Mariotte law, in which the relations between the properties of ideal gas are determined.

The third type of the procedures suggests the transition from things to relations between them and characterizing those relations.

The common feature of all these types of procedures is the initial point – from things to properties and relations. The following three types are combined by the fact that thought goes *vice versa* – from properties to things and relations.

The fourth type of the procedures implies the search for a substratum for a given set of properties, and the determination of a relation between these properties. Such is the search for substrata of thermal, electrical, magnetic properties in physics. The fifth type suggest that the relations should be established not indirectly between properties, but between those objects on which these properties are realized.

The sixth type of investigations presupposes, as an initial point, the fixation of a property. Further one looks for a relation satisfying this property and then, the things on which the obtained relation is realised. This procedure expresses a typical pattern of a system's construction [19].

The three last types of procedures begin with a fixed relation. The same holds for the seventh one. Here the given relations are interpreted on the one hand on things, and on the other, on their properties. Usually in such cases, things are identified with their properties. We came across such a situation when applying equations to describe some processes.

In the eighth type of procedures, the relation is interpreted for some objects and then the properties of these objects are investigated. This procedure is characteristic of the process of mathematization of scientific knowledge, when some relations are interpreted on mathematical objects (models) and then the properties of these objects are studied. The ninth, the last type of procedures, suggests that the initial relation is fixed, then properties satisfying this relation are looked for, and further, things which are substrata of these properties. Such a procedure defines another pattern of a system's construction, dual in relation to the sixth procedure. A typical example of this kind of investigation is the process of creating the periodic system of the elements by D. I. Mendeleev, which was explored in detail by B. M. Kedrov [20].

How is the given typology of procedures of scientific investigations correlated with the sciences' classification? It is quite obvious there is no identity. First of all, in the frame of the same science different procedures can be used. However, sciences may be classified according to those procedures which are the most important for a given science. We have noted above some examples of sciences for which the investigation procedures under consideration are typical.

Another difficulty is more essential. One and the same procedure is used in different sciences. The difference of sciences in this case can be expressed using categories 'definite' and 'indefinite'. With the help of these categories those things, properties, and relations which define the sequence of steps in the investigation procedures are characterised. Every procedure has 8 formally possible combinations, and the number thus equals $9 \times 8 = 72$. However, not all of them reflect a real cognitive situation. For example, if in using the first procedure we considered indefinite relations and properties as initial objects, which are the subject of investigation, then we should not receive any knowledge as a result. Number the realised combinations, each of which in practice denotes existing or really possible types of science. Though the number of such types is much less than 72, the enumeration of them gives us much greater variety than the one mentioned above.

Mark every science by two numbers. The first one, Roman, denotes the number of the procedure, the second one, an Arabic numeral, denotes the type of science in the frame of this procedure.

I.1. The sciences of definite things with indefinite properties and relations. It is just the type which was considered above in the case of sciences of things or of forms of motion corresponding to them.

I.2. The sciences which investigate different properties of given objects and definite types of relations between them. As an example, one can consider some variants of mathematized theories which make use of the given mathematical apparatus and, naturally, can investigate only those relations in the field of objects of interest which permit their elaboration with the help of the given mathematical apparatus. Such is mathematical economy.

I.3. Dual in relation to the previous one. These are sciences which investigate different relations of the given range of objects, but only the fixed type of their properties. In particular, sciences which are interested only in those properties of a given set of objects which permit measurement, are in this situation. Such is, for example, the theory of information or a quite new science, quantitative study (measuring) of art works [21].

I.4. The sciences of fixed things, properties and relations. Such sciences usually arise as a result of the interaction of two sciences, one of which provides an object, another a method through which investigated properties and relations are fixed. For example, biophysics studies only biological objects, but it considers only such properties and relations of the objects that allow their investigation by physical means. Mathematical linguistics is another example of a science of this type. Here one can also refer mathematical logic, bionics, and a good many other sciences.

II.1. Sciences possessing a fixed object area and studying any properties of objects of that area and any relations between them. For such sciences, a higher level of abstraction is characteristic in comparison with sciences of type I.1. The relations are placed not between the original objects but between their properties. To the sciences of this type, one can refer general linguistics. It studies a certain area of objects: words and sentences of natural languages. Any properties of these units are of interest for general linguistics, but it does not investigate the relations between concrete language units, which are studied with the competence of the lexicology of a certain concrete language. General linguistics studies only different relations between properties of language units.

II.2. Science of this type differs from others by the fact that it considers relations of a strictly fixed type. To science of this type, one may refer structural linguistics in the sense of S. K. Shaumyan, in which the relations between certain language characteristics are considered only from the aspect of a certain model [22].

II.3. A circle of objects and their properties are fixed without defining any relations between these properties. For example, anatomy and psychology of a man and of animals studies some definite objects only, and not all their properties, only some class of them. But all the relations between the properties of this class are of interest. Psychology, geology and some other sciences are included into this class of sciences.

II.4. In distinction from the previous type of science, here the relations between investigated properties are fixed. Such a fixation takes place in the process of mathematization which thus transfers sciences of type II.3. into sciences of type II.4.

III.1. In a science of this type, things are given and any relations between these things and any properties of these relations are studied. But the study of the properties of the objects themselves does not enter the competence of a science of this type. An example of a science of the type has not been found, nevertheless there is no ground to reject the possibility of such a science.

III.2. Here things, their relations, and properties of the relations are fixed. Sciences which are connected with the application of certain formal constructions in a definite content area mostly refer to this type. Thus for example, the application of group theory in theoretical physics fixes an object area, a certain type of investigated relations, namely group-theoretical ones, and certain properties of those relations.

III.3. Contrary to the previous case, any relations are investigated; but only certain properties are fixed, and in view of them relations are studied. Such is the theory of models, developed in the works by A. Tarski and A. I. Maltsev [23]. In the notion of a model, an initial set, i.e. an object area, is supposed to be fixed, and then any relations in this object area are considered, but within the frameworks of the properties that permit the application of the model theory method.

IV.1. There are defined properties. The relations between the properties and their substrata are looked for. Here one can refer applied mathematical disciplines which study a certain type of properties, for example applied vector analysis. The relations between selected properties are studied here and the interpretations are the objects looked for.

IV.2. Properties and their relations are defined here. But substrata of these properties remain undefined. This is characteristic of an applied philosophy of the type of 'dialectics of nature'. It fixes some properties essential for it only, the relation between them being defined within its framework also. For example – 'ideal' – 'material'. In the process of its application, substrata of the properties are looked for.

IV.3. An initial issue here is some definite properties. Their substrata are also defined. Various relations between those properties are looked for. A proper example has not been found.

IV.4. Contrary to the previous case, the relations between properties are also fixed. A proper example has not been found.

V.1. This science is interested in any relations between any objects as these relations possess some previously fixed properties. Such is gerontology if it is understood not as a science of really existing long-living men but as a science of ageing in general. Then the circle of objects of its interest would be quite indefinite.

V.2. In distinction from the previous case, here the class of relations of interest should be fixed. Mathematized gerontology would be placed here.

V.3. One rigorously fixes both the properties studied and the objects upon which these properties are realized and relations between the objects. An example of the kind is historic chronology as a branch of scientific knowledge.

VI.1. The property is fixed. Further, various relations, possessing this property and various objects upon which these relations are realised, are investigated. If one takes 'making a system' as a fixed property, then one can take General systems theory (GST) as an example of such a branch of scientific knowledge. This theory considers various methods of forming systems on different substrata, which thus remain undefined. The proposed variants of GST (L. von Bertalanffy, M. Mesarovitch, Yu. Urmantsev etc.) differ in the ways of systems presentation of objects, i.e. in relations satisfying the fixed property. Each of these variants of GST fixes this relation. If one takes GST as a kind of metatheory, studying various variants of constructing concrete GST, the mentioned relations and substrata would become indefinite. Thus GST as a metatheory should be referred to science type VI.1.

VI.2. A property and a relation satisfying it are fixed. A substratum remains indefinite. One refers here the above-mentioned concrete variants of GST.

VI.3. Properties, relations satisfying it, and substrata of those relations are fixed. Such are special systems theories, which concern this or that concrete object area. For example a theory of biological systems, a theory of social systems, semiotics, etc.

VII.1. A relation is fixed, which is further interpreted on the undefined area of things and properties. One refers here cybernetics, mathematics as a science of quantitative relations, and other sciences which have been mentioned above as the sciences of relations. But this is only part of them. The rest would better refer to the following types.

VII.2. A fixed relation is established between some definite properties and indefinite things. We think of geometry to be of this character, as it studies certain relations between some definite spatial properties in indefinite things.

VII.3. Here one fixes relations and an object area. Properties between which these relations are fixed, may be different. As an example one may give the theory of equations of mathematical physics. The equations represent some fixed relations and permit various interpretations of their variables. However, all these interpretations must be connected with physical objects, i.e. some definite object area.

VIII.1. A fixed relation is realized on an indefinite object area and then their properties are investigated. Thus science investigating the properties of mathematized branches of knowledge could be built.

VIII.2. Contrary to the previous case, one studies not any, but only fixed properties. Such scientific disciplines can be treated as branches of the previous type.

VIII.3. Here, one fixes not properties but only an object area. As an example one can propose a 'theory of mathematized humanities' or 'cybernetics in humanities'.

VIII.4. Everything is fixed. Certain relations in a definite object area and some definite properties of these objects are studied. Examples are not known.

The last, the ninth procedure (as was noted above) is dual to the sixth in respect of transformation of a 'property' – 'relation'. This means that the description of one of them comes from the description of the other by substituting the term 'relation' for 'property' and *vice versa*. Accordingly we have two dual definitions of the concept of a system.

The sixth procedure gives the definition of a system as an object on which a relation with a fixed property is realized. The ninth one gives the definition of a system as an object on which a property (properties) with a fixed relation is realized. The detailed interpretation of these definitions as generalizations of those given by different authors is presented in the preceding works [24–26].

It is essentially that the definitions of systems dual to one another are complementary to each other. Here one can speak of a *sui generis* generalization of a complementarity principle by N. Bohr [27]. It is logical to suppose that this complementary character would take place for scientific tendencies having dual characteristics in respect of the mentioned transformations. If it is so, then as an example of sciences, whose characteristics are obtained from scheme IX, we obtain the same ones which are obtained from scheme VI.

IX.1. A relation is fixed. Further one studies properties of the relation and various objects on which these properties are realized. Take as a fixed relation 'integrity', 'interaction'. There exist different sets of properties, which, being in this relation, reveal the systematic character of objects investigated.

One usually takes such properties as integrity, complexity, hierarchy. But one may take others. The scientific discipline studying these sets of properties and their realizations on various objects is general systems theory as metatheory. This is another presentation of this theory which is dual to the above considered one.

IX.2. If a system-making property is fixed, we study some concrete GST, for example that of L. von Bertalanffy or Yu. Urmantsev.

IX.3. If the substrata are also fixed, we obtain special systems theories, for example a theory of biological systems, a theory of social systems, semiotics etc.

Thus we can distinguish 29 classes of sciences, some of them probably empty for the time being. This number has no upper growth limit. To our mind other detailed elaboration is useful which would make the variety of the types of sciences still richer. Such details are connected first of all with the notion of indefiniteness.

One may note that when indefinite objects (relations, properties) were cited above, this indefiniteness was not understood in one and the same sense. Sometimes it was just some object, no one knows which, but probably not an arbitrary one, not *any* object. Such are the objects which can be substrata of a control which are studied in cybernetics.

Objects which can be presented as a system, are quite another matter. These are namely *any*, what ever we took, *arbitrary* objects. Distinguishing just between indefiniteness and arbitrariness we get much greater versatility of the types of scientific knowledge than that which was given above. This problem is considered in [28]. There we gave another formalization of procedures of scientific research and showed that each of the procedures can yield 729 types of scientific knowledge. The use of a formal apparatus, the language of ternary description, gives us the possibility to develop this classification, gaining more detailed patterning of characteristic features of scientific knowledge.

In conclusion, it is worth noting that our given classification of scientific research is constructed only within one of the schemes of systems approach, expressed in the sixth procedure. One fixes a system-making property – the concept of classification as a system – "an adequate reflection of characteristic peculiarities of the types of scientific research and scientific knowledge." Then the formal structure satisfying this concept is built, and finally, a substratum is looked for – the concrete sciences, scientific tendencies, theories on which this structure is realized. Thus the term "system approach" which is used in the title of this article, reflects a definite strategy of research which is based on the corresponding system theory [26]. It is noteworthy that this strategy is combined easily enough with more traditional approaches [29].

Odessa State University, Ukraine

REFERENCES

1. Engels Fr. *Herr Eugen Dührings umwälzung der Wissenschaft.* Marx/Engels, *Gesamtausgabe.* M., 1935.
2. Engels Fr. *Dialektik der Natur.* Marx/Engels, *Gesamtausgabe.* – M., 1935.
3. Кедров Б. М. Энгельс и естествознание. Раздел четвертый. – М.: ОГИЗ, 1947.
4. Кедров Б. М. Классификация наук. Часть I. М.: Изд-во ВПШ и АПН при ЦК КПСС, 1961.
5. Кедров Б. М. Классификация наук. Часть II. – М.: Наука, 1965.
6. Кедров Б. М. Классификация наук. Прогноз К. Маркса о науке будущего. – М.: Мысль, 1985.
7. Ляпунов А. А., Соболев С. Л., Китов А. И. Основные черты кибернетики // Вопросы философии. – 1955. – № 4.
8. Кольман А. Кибернетика. – М.: Знание, 1956.
9. Уемов А. И. Реальный смысл проблем кибернетики и их извращение в буржуазной науке // Учен. записки Ивановск. гос. пед. ин-та. Т. 8. Иваново, 1956. – С. 166–199.
10. Дянков Б. Относно предмета на науката и принципа на класификация на науките // Известия на института по философия. Т. УП, кн. 2/1962. – С. 56–82.
11. Уемов А. И. Некоторые тенденции в развитии естественных наук и проблема их классификации // Вопросы философии. – 1961. – № 8. – С. 66–75.
12. Поликаров А. По някои общи методологически въпроси на природните науки. Българска академия на науките. Известия на института по философия. Том VIII, кн. 1.
13. Polikarov A. *Klassifizierungstypen der Wissenschaften* in *Wissenschaftliche Zeitschrift der Karl-Marx-Universität, Gesellschafts und Sprachwissenschaftliche Reihe,* Leipsig. *17* 1968. Heft 5, – S. 565–568.
14. Поликаров А. Методология на научнато познание. – София: Наука и изкуство, 1973, с. 138–153.
15. Поликаров А. Проблеми на научнато познание от методологична гледна точка. –София: Наука и изкуство, 1977, с. 107–123.
16. Степанов В. Н., Уемов А. И. К разработке классификации наук об океане // Известия всесоюзного географического общества. – 1987. – № 4. – С. 337–342.
17. Мелешкин М. Т. Эконологические проблемы мирового океана. – М.: Экономика, 1981.
18. Уемов А. И. Валенчик Р. Формальная типологид научного знанид и проблема

его единства // Философид и естествознание. К семидесятилетию академика Kegpoba. – М.: Наука, 1974. – С. 133–151.

19. Уемов А. И. Логический анализ системного подхода к объектам и его место cpegu gpyrux uетоgов исследования // Системные исследования. Ежегодник 1969. – М., 1969. – С. 80–96.
20. Кедров В. М. Микроанатомия великого открытия. – М., 1970.
21. Семиотика и искусствометрия / Под ред. Ю. М. Лотмана и В. М. Петрова. – М.: Мир, 1972.
22. Шаумян С. К. Структурная лингвистика. – М., 1968.
23. Мальцев А. И. Алгебраические системы. – М.: Наука, 1970.
24. Проблемы формального анализа систем. – М. Высшая шмола, 1968.
25. Логика и методология системных исследований. Киев-Одесса: Вища школа, 1977.
26. Уемов А. И. Системный подход и общая теорuд систем. – М.: Мысль, 1978.
27. Комарчев В. А., Кошарский Б. Д., Поликарпов Г. А., Уемов А. И. Дополнительность. Концепция, отношение, принцип? Принцип дрполнительности и материалистическая диалектика. – М.: Наука, 1976. – С. 92–101.
28. Уемов А. И. Формальные аспекты систематизации научного знания и процедур его развития. Системный анализ и научное знание. – М.: Наука, 1978. – С. 95–141.
29. Уемов А. И. Барабаш В. С., Коврига Е. С. Классификация и кодирование научных исследований Мирового океана. Эконология Мирового океана. – Киев: Институт экономики АН УССР, 1981. – С. 43–50.

AZARYA POLIKAROV: SELECTED BOOKS AND ARTICLES

I. BOOKS

A. *In Bulgarian*

(*) indicates summaries in foreign languages – Russian, English, or German.

Matter and Knowledge (Publ. House of the Bulgarian Academy of Sciences. Sofia, 1961), 188 pp. (*)

Relativity and Quanta. Ibid. 1963, 372 pp. (*)

Methodology of Scientific Knowledge, Volume 1 (Publ. House "Nauka i Izkustvo". Sofia, 1972), 278 pp. (*)

Methodology of Scientific Knowledge. Ibid., Volume 2 (Sofia, 1973), 312 pp. (*)

Problems of Scientific Knowledge from a Methodological Point of View (Publ. House "Nauka i Izkustvo". Sofia, 1977), 456 pp. (*)

Physics of the Twentieth Century (Partizdat. Sofia, 1977), 238 pp.

Essays on the Methodology of Science (Publ. House "Nauka i Izkustvo". Sofia, 1981), 227 pp.

Einstein: Various Perspectives. Ibid., 1985, 182 pp.

Orientation in the Methodology of Science. Ibid., 1987, 240 pp. (*)

Scientific Revolutions and Progress (Publ. House of Sofia University. Sofia, 1993), 288 pp.(*)

Revolutions in Physics ("Prof. Marin Drinov". Academic Publ. House. Sofia, 1996), 182 pp.(*)

B. *In English*

Science and Philosophy (Publ. House of the Bulgarian Academy of Sciences. Sofia, 1973), 134 pp.

Methodological Problems of Science. Ibid., 1983, 328 pp.

C. *In German*

Moderne Physik – Weltbild und Denkstil (Manz Verlag. München, 1966), 100 pp.

D. *In Russian*

Relativity and Quanta (Progress. Moscow, 1966), 500 pp.

E. *In Slovak*

Problemy vedeckeho Poznania (Pravda. Bratislava, 1980), 489 pp.

D. Ginev and R. S. Cohen (eds.), Issues and Images in the Philosophy of Science, 391–396.

II. ARTICLES

A. *In Bulgarian*

'On the Classification of the Hypotheses of Quasars', *Izvestia* (Proceedings) of the Section of Astronomy **2** (1967), pp. 157–66 (with M. Kalinkov).
'On the Methodology of Solving Scientific Problems', *Problems of the Science of Science* (Publ. House of the Bulgarian Academy of Sciences. Sofia, 1970), pp. 60–118.
'Creative Thinking and Heuristics', *Vselena* (Universe). Scientific Almanach (Publ. House "Nauka i Izkustvo". Sofia, 1971), pp. 129–35.
'Leibniz – A Great Rationalist', in G. W. Leibniz, *Nouveaux essais sur l'entendement humain. Ibid.*, 1974, pp. 726–47.
'On the Scientific Revolutions during the 19th and the 20th Century', *Filossofska Misl* **32** (1976), 12, pp. 83–6.
'Conceptions about the Factors of Scientific Creativity', *Psikhologia* (1978), 1, pp. 6–14.
'Main Stages in the Framing of Quantum Mechanics', in A. Polikarov (ed.) *Methodological Problems of Quantum Physics* (Publ. House "Nauka i Izkustvo". Sofia, 1979), pp. 7–27. (*)
'The Psychocentric Trend in Heuristics', in *Problems of Scientific Talent. Ibid.*, 1979, pp. 248–64.
'Concerning the Development of Science', in A. Angelov *et al.* (eds.) *Problems and Prospects of Sociology of Science. Ibid.*, 1981, pp. 96–110.
'Hypothetico-deductive Theories in Physics', *Filossofska Misl* **37** (1981), 3, pp. 52–64.
'Topical and Prospective Philosophical Problems associated with the Scientific Revolutions in Our Time', *ibid* **39** (1983), 2, pp. 29–39.
'Einstein and the Unity of the Physical Worldview', *ibid.* **8** (1983), pp. 65–78.
'Harmony of the World beyond "Common Sense"', *Problemi na kulturata* **2** (1983), 1, pp. 13–24.
'The Undeciphered Creativity', in St. Stoev, *The Consciousness and Unconscious in Scientific Creativity* (Publ. House Bulg. Acad. Sci. Sofia, 1983), pp. 5–11.
'Niels Bohr's Crazy Ideas', *Filossofska Misl* **41** (1985), 9, pp. 25–35.
'Science, Methodology and History of Science and of Methodology', *Topical Problems of Science* (Bulg. Acad. Sci. Sofia, 1986), 1, 36 pp. (with D. Ginev).
'Main Trends in the Growth of Science', in A. Polikarov *et al.* (eds.), *Global Problems* (Publ. House "Nauka i Izkustvo". Sofia, 1986), pp. 48–61.
'Impact of Philosophical Ideas on the Development of Science', in S. Sarkissyan (ed.), *Formation of a Scientific Worldview through Education in Science* (Publ. House "Narodna Prosveta". Sofia, 1987), pp. 40–62.
'Niels Bohr and Atomic Physics', in A. Polikarov (ed.) *Niels Bohr and Atomic Physics* (Publ. House "Nauka i Izkustvo". Sofia, 1987), pp. 5–27.
'Philosophical Categories', in *Marxist-Leninist Philosophy*, Ch. 6, *ibid.*, 1987, pp. 150–76.
'On the Issue of the Critical Functions of Methodology of Science', in *Philosophical Knowledge. Essence, Development, Functions. Ibid.*, 1988, pp. 176–84 (with D. Ginev).
'Concerning the Specificity of the Growth of Philosophy', *Filossofska Misl* **45** (1989), 7, pp. 48–56.
'Development of Science, Scientific Revolutions, and Incommensurability of Superseding Theories'. *Ibid.* **46** (1990), 4, pp. 3–15.
'The Joke', *Kultura* (1990), 6, pp. 35–9.
'Philosophy of Science and Some Open Issues of Science', *Filosofska Misl* **47** (1991), 4, pp. 26–31.
'Scientific Revolutions, Revolutions in Physics', *J. Bulg. Acad. Sci.* **105** (1992), 4, pp. 3–14.
'Science: from the Possible to the Real', *Priroda* (1993), 3, pp. 76–9.
'Which are the Factors of Scientific Development?', *J. Bulg. Acad. Sci.* **107** (1994), 5/6, pp. 5–17.
'Metamorphoses of Atomism', *ibid.* **109** (1996).

'Controversies in Philosophy of Science', *ibid.*
'Relativity: A Near Retrospection', in *Albert Einstein. Mein Weltbild* ("Lyubomudrie". Sofia, 1996).

B. *In English*

'Nature, Knowledge, Dialectics', in *Science, Technology, Man* (Publ. House of the Bulgarian Academy of Sciences. Sofia, 1973), pp. 45–53.
'The Divergent-Convergent Method', in R. S. Cohen, M. Wartofsky (eds.) *Methodological and Historical Essays in the Natural and Social Sciences. Boston Studies in the Philosophy of Science*, Vol. 14 (D. Reidel, Dordrecht, 1974), pp. 213–33.
'Determinism in Physics', *Soviet Studies in Philosophy* **13** (1974), pp. 67–85 [Transl. from Russian. Cf. Russian articles 7 and 12].
'Interaction of Physics and Other Sciences at a Methodological Level', *C. R. Bulg. Acad. Sci.* **28** (1975), pp. 1151–4.
'Multidisciplinary Approach to Heuristics', *ibid.* **29** (1976), pp. 1081–4.
'The Project of a Synthetic Conception of the Development of Science', *Teorie razvoje vedi* **2** (1978), 4, pp. 61–71.
'Ten Arguments in Favour of the Relevance of Philosophy for the Sciences', *Intern. Quarterly of Philosophy* **18** (1978), 3, pp. 20–5.
'Overturn and Continuity of the Hypotheses in the Framing of the Theory of Relativity', in H. Nelkowski *et al.* (eds.) *Einstein Symposium Berlin. Lecture Notes in Physics*, Vol. 100 (Springer Verlag. Berlin, 1979), pp. 397–411.
'Einstein's Conception of Physical Theory', *Epistemologia* **2** (1979), pp. 99–122.
'Psychology of Scientific Creativity', *GATE* (Sofia) **2** (1980), 1, pp. 91–107.
'From the Photon to the Universe. Einstein's Discoveries', *Man, Evolution, Cosmos* **2** (1983), 1, pp. 3–13.
'Leading Role of Philosophical Ideas in Natural Sciences', in *Recueil de texts, contribution a la philosophie et la science . . . Acad. serbe des sciences et des arts*, Vol. 91 (Beograd, 1983), pp. 179–84.
'On Possibilities and Restrictions in Science', in 7th Intern. Congress of Logic, Methodology and Philosophy of Science (Salzburg, 1983), Vol. 3, pp. 204–5.
'The Situation in Atomic Physics and Bohr's Methodology', in *Contemporary Bulgarian Philosophy* (Sofia-Press. Sofia, 1983), pp. 128–44.
'Pro Methodology of Science', in A. Mercier and M. Slivar (eds.) *Philosophes critiques d'eux memes*, Vol. 11 (Verlag Peter Lang. Bern, 1984), pp. 211–22.
'Methodological Problems and Approaches in Artificial Intelligence', in W. Bibel, B. Petkoff (eds.) *Artificial Intelligence: Methodology, Systems, Applications* (Elsevier Science Publishers B. V., North-Holland, Amsterdam, 1985), pp. 11–7.
'Methodological Problems', *C. R. Bulg. Acad. Sci.* **39** (1986), 1, pp. 7–9.
'Methodological Statements', *ibid.* 2, pp. 5–8.
'On the Solution of Methodological Problems', *ibid* 3, pp. 5–7.
'On the Formulation of Complementarity', *ibid.* 8, pp. 5–8.
'On the "Diffusion" of Methodological Elements into Science', *Epistemologia* **9** (1986), pp. 265–80 (with D. Ginev).
'Dominants in the Development of Physical Knowledge', *Czechoslovak J. Phys.* **B36** (1986), pp. 52–3 (with D. Ginev).
'Modular Conception of Physical Theory', in Abstracts 8. Intern. Congress of Logic, Methodology and Philosophy of Science (Moscow, 1987), Vol. 2, pp. 111–4.
'The Scientification of Methodology of Science', *Z. allg. Wissenschaftstheorie* **19** (1988), pp. 18–27 (with D. Ginev).
'On the Nature of Einstein's Realism', *Epistemologia* **12** (1989), pp. 277–304.
'Remarks on Logical Empiricism and Some of A. J. Ayer's Achievements: Some Fifty Years

Later', in L. E. Hahn (ed.) *The Philosophy of A. J. Ayer*. Open Court Publ. Co. La Salle, IL 1992, pp. 407–24 (with D. Ginev).

'Is There an Incommensurability between Superseding Theories?', *J. Gen. Phil. Sci.* **24** (1993), pp. 127–46.

'On Various Kinds of Scientific Revolutions in Physics', *Epistemologia* **16** (1993), pp. 213–34.

'Some Questions Concerning Limitations of the Range of Validity of Kuhn's Model of the History of Science', in K. Gavroglu *et al.* (eds.) *Physics, Philosophy and Scientific Community. Boston Studies in the Philosophy of Science*, Vol. 163 (Kluwer Academic, Dordrecht, 1995), pp. 235–40.

'Concerning the Integration of Sciences: Kinds and Stages', *J. Gen. Phil. Sci.* **26** (1995), pp. 297–312.

C. *In German*

'Über die Kategorie Materie', *Deutsche Zeitschrift für Philosophie* **4** (1956), 5/6, pp. 539–49.

'Zur philosophischen Fragestellung des kosmologischen Problems' (*Wiss. Z. Karl-Marx-Univ. Leipzig*) **7**, 1957/58, pp. 13–21.

'Die Wechselwirkung zwischen Philosophie und Naturwissenschaften' (*Wiss. Z. Techn. Hochschule Dresden*) **9** (1959/60), 1, pp. 239–53.

'Zum Problem der Systematisierung der philosophischen Kategorien' (*Wiss. Z. Humboldt-Univ. Berlin, Ges.-Sprachw. R.*) **9** (1959/60), pp. 4, 361–70.

'Ist die Widerspiegelung eine allgemeine Eigenschaft der Materie?' *Naturwissenschaft und Philosophie* (Akademie-Verlag. Berlin, 1960), pp. 283–303.

'Philosophische Diskussionen über die Relativitätstheorie', *Forschen und Wirken* **3** (Verlag der Wissenschaften, Berlin, 1960), pp. 33–59.

'Zum heutigen Stand des kosmologischen Problems', *Actes du Symposium intern. R. J. Boscovic 1961* (Beograd, 1962), pp. 195–204.

'Über die Deutung der Quantenmechanik' (*Wiss. Z. Humboldt-Univ. Berlin, Math.-Nat. R.*) **11** (1962), pp. 1–21.

'Zum physikalischen Kausalgesetz', *ibid.*, pp. 711–30.

'Über die Klasse der möglichen Lösungen des Kausalitätsproblems in der Physik' (*Wiss. Z. Humboldt-Univ. Berlin, Math.-Nat. R.*) **12** (1963), pp. 439–44.

'Zum Problem der Thermodynamik des Universums', *Deutsche Zeitschrift für Philosophie* **11** (1963), pp. 2, 197–211.

'Zur Kritik der alternativistischen Auffassung', *C. R. Acad. Bulg. Sci.* **16** (1963), pp. 457–60.

'Zum Problem der Deutung des Einsteinschen Äquivalenzsatzes von Masse und Energie' (*Wiss. Z. Humboldt-Univ. Berlin*), *Math.-Nat. R.* **13** (1964), pp. 123–6.

'Elemente der Heuristik', *Rostocker philosophische Manuskripte* **3** (1966), pp. 119–23.

'Klassifizierung der Wissenschaften' (*Wiss. Z. Karl-Marx-Univ. Leipzig*) **17** (1968), pp. 565–68.

'Heuristisches Herangehen an die Lösung wissenschaftlicher Probleme', in *Wissenschaft, Ideologie, Philosophie* (Akademie-Verlag, Berlin, 1973), pp. 159–79.

'Die Dialektik der stationären Prozesse', in W. R. Beyer (Hg.) *Hegel-Jahrbuch 1974* (Köln, 1975), pp. 343–6.

'Was sind Monaden?', in *Aktes des II. internationalen Leibniz-Kongresses* **3** (Franz Steiner Verlag, Wiesbaden, 1975), pp. 293–6.

'Zum Konversionsprinzip', *C.R. Acad. Bulg. Sci.* **29** (1976), pp. 291–4 (mit L. Kreiser).

'Zum Problem der dialektisch-materialistischen Modelle auf dem Gebiet der Physik', *C. R. Acad. Bulg. Sci.* **30** (1977), pp. 789–92.

'Strukturmodelle der Wissenschaftsentwicklung', in F. Rapp (Hg.) *Naturverständnis und Naturbeherrschung* (Wilhelm Fink Verlag, München, 1981), pp. 111–28.

'Über den adaptiven Charakter des Traumes', *C. R. Acad. Bulg. Sci.* **35** (1982), pp. 407–10 (mit Chr. Toegel).

'Weltanschauliche Funktionen heutiger Naturwissenschaft', in H. J. Sandkueler, H. H. Holz (Hg.) *Dialektik* **12** (Pahl-Rugenstein Verl., Köln, 1986), pp. 241–54.

'Die Methodologisierung der Wissenschaftsgeschichte und die Historisierung der Wissenschaftsmethodologie', in Chr. Toegel (Hg.) *Struktur und Dynamik wissenschaftlicher Theorien* (Verlag Peter Lang. Ff./M., 1986), pp. 9–25 (mit D. Ginev).

'Notizen über Wittgensteins Stellung zur Philosophie', in H. Nagl-Docekal (Hg.) *Ludwig Wittgenstein und die Philosophie des 20. Jahrhunderts.* Verein "Freunde des Hauses Wittgenstein" (Wien, 1989), pp. 71–6.

'Über den Charakter von Einsteins philosophischem Realismus', *Philosophia Naturalis* **26** (1989), 1, pp. 135–58.

'Einsteins philosophischer Realismus', in W. Muschik, W. R. Shea (Hg.) *Philosophie, Physik, Wissenschaftsgeschichte.* TUB-Dokumentation Kongresse und Tagungen, *H. 45* (Berlin, 1989), pp. 155–61.

'Elf Thesen über Marx elfte Feuerbach These', *Filossofka Misl* **47** (1991), pp. 5, 138–9. [Parallel text in Bulgarian]

D. *In Russian*

'The Principle of Causality in Contemporary Physics', *Annual of Sofia Univ.* **47** (1951/52), 1, pp. 226–76.

'On the Supraluminous Radiation of Cosmic Particles in the Earth's Atmosphere', *C. R. Acad. Bulg. Sci.* **7** (1954), 2, pp. 29–32.

'A General Characteristics of Contemporary Idealism and the Directions of Its Criticism', *Filossofskie Nauki* (1961), 3, pp. 84–97.

'Logical Structures of the Solutions of Scientific Problems', *ibid.* (1965), 1, pp. 26–31.

'On the Methodology of Solving Scientific Problems', in *Dialectics and Contemporary Natural Science* (Moscow, 1970), pp. 70–6, 431–2, 439. (*)

'A Draft for Classifying Interdisciplinary Branches', *C. R. Acad. Bulg. Sci.* **25** (1972), pp. 289–92.

'On the Conception of Determinism in Physics', *ibid*, pp. 1311–4.

'On the Issue of the Relationship between Empirical and Theoretical Knowledge', in *Problems of Philosophy and Methodology of Contemporary Science* (Publ. House "Nauka". Moscow, 1973), pp. 329–40.

'The Causality Principle and the Theory of Reflection', in *Leninist Theory of Reflection and Contemporary Science*, Vol. 2 (Publ. House "Nauka i Izkustvo". Sofia, 1973), pp. 77–91.

'On the Problem of Classification of the Sciences' (Erevan State Univ., 1973), **3**, pp. 58–67.

'On the Problem of the Essence of Time', *Izvestia* (Proceedings) of the Institute of Philosophy, **21** (1973), pp. 125–45 (with Ya. F. Askin).

'Determinism and Indeterminism in Physics', in *Contemporary Determinism and Science*, Vol. 2 ('Nauka', Sibirian Division, Novossibirsk, 1975), pp. 76–106.

'Confrontation of Scientific Trends and the Development of Science', *C. R. Acad. Bulg. Sci.* **28** (1975), pp. 1295–8.

'Correlation between the Most Important Scientific Achievements and Scientific Information', *ibid.* **29** (1976), pp. 1397–400.

'Concerning Kuhn's Conception of the Growth of Science', *Filossofskie Nauki* (1976), 4, pp. 106–13.

'On the Hypothetico-Deductive Model of Scientific Theories', *C. R. Acad. Bulg. Sci.* **31** (1978), pp. 1097–100.

'Concerning the Classification of Physical Theories', *ibid.* **32** (1979), pp. 410–22.

'On the Philosophical Problems of Cosmology', in *Place and Role of Philosophy and Sciences in the Contemporary World* (Sofia, 1980), pp. 127–34.

'On the Framing of Derivative Alternative Theories in Physics', *C. R. Acad. Bulg. Sci.* **34** (1981), pp. 907–10.

'Methodologies, Corresponding to Definite Scientific Situations', in *Anthology of Contemporary Bulgarian Philosophic Thought* (Publ. House Bulg. Acad. Sci. Sofia, 1983), pp. 97–103.
'Historical Dynamics of the Scientific Development and the Didactics of the Syllabus', '*Sovremenaja Visshaja Shkola*' 1987, **3**(59), pp. 103–11 (with D. Ginev).
'On the Model of an Inflationary Universe', *C. R. Acad. Bulg. Sci.* **41** (1988), 3, pp. 5–6.
'Science and Philosophy', *Voprossi Filossofii* (Erevan, 1988), **6/7**, pp. 39–45.
'A Modular Conception of Scientific Theory', in E. A. Mamchur *et al.* (eds.) *Natural Science: System and Dynamics* ('Nauka', Moscow, 1990), pp. 191–202.

E. *In French*

'Sur le probleme de l'essence du temps', in J. Zeman (ed.) *Time in Science and Philosophy* (Academia. Prague, 1971), pp. 261–4.

NAME INDEX

Boston Studies in the Philosophy of Science

Editor: Robert S. Cohen, *Boston University*

1. M.W. Wartofsky (ed.): *Proceedings of the Boston Colloquium for the Philosophy of Science, 1961/1962.* [Synthese Library 6] 1963 ISBN 90-277-0021-4
2. R.S. Cohen and M.W. Wartofsky (eds.): *Proceedings of the Boston Colloquium for the Philosophy of Science, 1962/1964.* In Honor of P. Frank. [Synthese Library 10] 1965 ISBN 90-277-9004-0
3. R.S. Cohen and M.W. Wartofsky (eds.): *Proceedings of the Boston Colloquium for the Philosophy of Science, 1964/1966.* In Memory of Norwood Russell Hanson. [Synthese Library 14] 1967 ISBN 90-277-0013-3
4. R.S. Cohen and M.W. Wartofsky (eds.): *Proceedings of the Boston Colloquium for the Philosophy of Science, 1966/1968.* [Synthese Library 18] 1969 ISBN 90-277-0014-1
5. R.S. Cohen and M.W. Wartofsky (eds.): *Proceedings of the Boston Colloquium for the Philosophy of Science, 1966/1968.* [Synthese Library 19] 1969 ISBN 90-277-0015-X
6. R.S. Cohen and R.J. Seeger (eds.): *Ernst Mach, Physicist and Philosopher.* [Synthese Library 27] 1970 ISBN 90-277-0016-8
7. M. Čapek: *Bergson and Modern Physics.* A Reinterpretation and Re-evaluation. [Synthese Library 37] 1971 ISBN 90-277-0186-5
8. R.C. Buck and R.S. Cohen (eds.): *PSA 1970.* Proceedings of the 2nd Biennial Meeting of the Philosophy and Science Association (Boston, Fall 1970). In Memory of Rudolf Carnap. [Synthese Library 39] 1971 ISBN 90-277-0187-3; Pb 90-277-0309-4
9. A.A. Zinov'ev: *Foundations of the Logical Theory of Scientific Knowledge (Complex Logic).* Translated from Russian. Revised and enlarged English Edition, with an Appendix by G.A. Smirnov, E.A. Sidorenko, A.M. Fedina and L.A. Bobrova. [Synthese Library 46] 1973 ISBN 90-277-0193-8; Pb 90-277-0324-8
10. L. Tondl: *Scientific Procedures.* A Contribution Concerning the Methodological Problems of Scientific Concepts and Scientific Explanation.Translated from Czech. [Synthese Library 47] 1973 ISBN 90-277-0147-4; Pb 90-277-0323-X
11. R.J. Seeger and R.S. Cohen (eds.): *Philosophical Foundations of Science.* Proceedings of Section L, 1969, American Association for the Advancement of Science. [Synthese Library 58] 1974 ISBN 90-277-0390-6; Pb 90-277-0376-0
12. A. Grünbaum: *Philosophical Problems of Space and Times.* 2nd enlarged ed. [Synthese Library 55] 1973 ISBN 90-277-0357-4; Pb 90-277-0358-2
13. R.S. Cohen and M.W. Wartofsky (eds.): *Logical and Epistemological Studies in Contemporary Physics.* Proceedings of the Boston Colloquium for the Philosophy of Science, 1969/72, Part I. [Synthese Library 59] 1974 ISBN 90-277-0391-4; Pb 90-277-0377-9
14. R.S. Cohen and M.W. Wartofsky (eds.): *Methodological and Historical Essays in the Natural and Social Sciences.* Proceedings of the Boston Colloquium for the Philosophy of Science, 1969/72, Part II. [Synthese Library 60] 1974 ISBN 90-277-0392-2; Pb 90-277-0378-7
15. R.S. Cohen, J.J. Stachel and M.W. Wartofsky (eds.): *For Dirk Struik.* Scientific, Historical and Political Essays in Honor of Dirk J. Struik. [Synthese Library 61] 1974 ISBN 90-277-0393-0; Pb 90-277-0379-5
16. N. Geschwind: *Selected Papers on Language and the Brains.* [Synthese Library 68] 1974 ISBN 90-277-0262-4; Pb 90-277-0263-2
17. B.G. Kuznetsov: *Reason and Being.* Translated from Russian. Edited by C.R. Fawcett and R.S. Cohen. 1987 ISBN 90-277-2181-5

18. P. Mittelstaedt: *Philosophical Problems of Modern Physics*. Translated from the revised 4th German edition by W. Riemer and edited by R.S. Cohen. [Synthese Library 95] 1976
ISBN 90-277-0285-3; Pb 90-277-0506-2
19. H. Mehlberg: *Time, Causality, and the Quantum Theory*. Studies in the Philosophy of Science. Vol. I: *Essay on the Causal Theory of Time*. Vol. II: *Time in a Quantized Universe*. Translated from French. Edited by R.S. Cohen. 1980
Vol. I: ISBN 90-277-0721-9; Pb 90-277-1074-0
Vol. II: ISBN 90-277-1075-9; Pb 90-277-1076-7
20. K.F. Schaffner and R.S. Cohen (eds.): *PSA 1972*. Proceedings of the 3rd Biennial Meeting of the Philosophy of Science Association (Lansing, Michigan, Fall 1972). [Synthese Library 64] 1974 ISBN 90-277-0408-2; Pb 90-277-0409-0
21. R.S. Cohen and J.J. Stachel (eds.): *Selected Papers of Léon Rosenfeld*. [Synthese Library 100] 1979 ISBN 90-277-0651-4; Pb 90-277-0652-2
22. M. Čapek (ed.): *The Concepts of Space and Time*. Their Structure and Their Development. [Synthese Library 74] 1976 ISBN 90-277-0355-8; Pb 90-277-0375-2
23. M. Grene: *The Understanding of Nature*. Essays in the Philosophy of Biology. [Synthese Library 66] 1974 ISBN 90-277-0462-7; Pb 90-277-0463-5
24. D. Ihde: *Technics and Praxis*. A Philosophy of Technology. [Synthese Library 130] 1979
ISBN 90-277-0953-X; Pb 90-277-0954-8
25. J. Hintikka and U. Remes: *The Method of Analysis*. Its Geometrical Origin and Its General Significance. [Synthese Library 75] 1974 ISBN 90-277-0532-1; Pb 90-277-0543-7
26. J.E. Murdoch and E.D. Sylla (eds.): *The Cultural Context of Medieval Learning*. Proceedings of the First International Colloquium on Philosophy, Science, and Theology in the Middle Ages, 1973. [Synthese Library 76] 1975 ISBN 90-277-0560-7; Pb 90-277-0587-9
27. M. Grene and E. Mendelsohn (eds.): *Topics in the Philosophy of Biology*. [Synthese Library 84] 1976 ISBN 90-277-0595-X; Pb 90-277-0596-8
28. J. Agassi: *Science in Flux*. [Synthese Library 80] 1975
ISBN 90-277-0584-4; Pb 90-277-0612-3
29. J.J. Wiatr (ed.): *Polish Essays in the Methodology of the Social Sciences*. [Synthese Library 131] 1979 ISBN 90-277-0723-5; Pb 90-277-0956-4
30. P. Janich: *Protophysics of Time*. Constructive Foundation and History of Time Measurement. Translated from German. 1985 ISBN 90-277-0724-3
31. R.S. Cohen and M.W. Wartofsky (eds.): *Language, Logic, and Method*. 1983
ISBN 90-277-0725-1
32. R.S. Cohen, C.A. Hooker, A.C. Michalos and J.W. van Evra (eds.): *PSA 1974*. Proceedings of the 4th Biennial Meeting of the Philosophy of Science Association. [Synthese Library 101] 1976 ISBN 90-277-0647-6; Pb 90-277-0648-4
33. G. Holton and W.A. Blanpied (eds.): *Science and Its Public*. The Changing Relationship. [Synthese Library 96] 1976 ISBN 90-277-0657-3; Pb 90-277-0658-1
34. M.D. Grmek, R.S. Cohen and G. Cimino (eds.): *On Scientific Discovery*. The 1977 Erice Lectures. 1981 ISBN 90-277-1122-4; Pb 90-277-1123-2
35. S. Amsterdamski: *Between Experience and Metaphysics*. Philosophical Problems of the Evolution of Science. Translated from Polish. [Synthese Library 77] 1975
ISBN 90-277-0568-2; Pb 90-277-0580-1
36. M. Marković and G. Petrović (eds.): *Praxis*. Yugoslav Essays in the Philosophy and Methodology of the Social Sciences. [Synthese Library 134] 1979
ISBN 90-277-0727-8; Pb 90-277-0968-8

37. H. von Helmholtz: *Epistemological Writings.* The Paul Hertz / Moritz Schlick Centenary Edition of 1921. Translated from German by M.F. Lowe. Edited with an Introduction and Bibliography by R.S. Cohen and Y. Elkana. [Synthese Library 79] 1977
ISBN 90-277-0290-X; Pb 90-277-0582-8
38. R.M. Martin: *Pragmatics, Truth and Language.* 1979
ISBN 90-277-0992-0; Pb 90-277-0993-9
39. R.S. Cohen, P.K. Feyerabend and M.W. Wartofsky (eds.): *Essays in Memory of Imre Lakatos.* [Synthese Library 99] 1976 ISBN 90-277-0654-9; Pb 90-277-0655-7
40. Not published.
41. Not published.
42. H.R. Maturana and F.J. Varela: *Autopoiesis and Cognition.* The Realization of the Living. With a Preface to 'Autopoiesis' by S. Beer. 1980
ISBN 90-277-1015-5; Pb 90-277-1016-3
43. A. Kasher (ed.): *Language in Focus: Foundations, Methods and Systems.* Essays in Memory of Yehoshua Bar-Hillel. [Synthese Library 89] 1976
ISBN 90-277-0644-1; Pb 90-277-0645-X
44. T.D. Thao: *Investigations into the Origin of Language and Consciousness.* 1984
ISBN 90-277-0827-4
45. F.G.-I. Nagasaka (ed.): *Japanese Studies in the Philosophy of Science.* 1997
ISBN 0-7923-4781-1
46. P.L. Kapitza: *Experiment, Theory, Practice.* Articles and Addresses. Edited by R.S. Cohen. 1980 ISBN 90-277-1061-9; Pb 90-277-1062-7
47. M.L. Dalla Chiara (ed.): *Italian Studies in the Philosophy of Science.* 1981
ISBN 90-277-0735-9; Pb 90-277-1073-2
48. M.W. Wartofsky: *Models.* Representation and the Scientific Understanding. [Synthese Library 129] 1979 ISBN 90-277-0736-7; Pb 90-277-0947-5
49. T.D. Thao: *Phenomenology and Dialectical Materialism.* Edited by R.S. Cohen. 1986
ISBN 90-277-0737-5
50. Y. Fried and J. Agassi: *Paranoia.* A Study in Diagnosis. [Synthese Library 102] 1976
ISBN 90-277-0704-9; Pb 90-277-0705-7
51. K.H. Wolff: *Surrender and Cath.* Experience and Inquiry Today. [Synthese Library 105] 1976 ISBN 90-277-0758-8; Pb 90-277-0765-0
52. K. Kosík: *Dialectics of the Concrete.* A Study on Problems of Man and World. 1976
ISBN 90-277-0761-8; Pb 90-277-0764-2
53. N. Goodman: *The Structure of Appearance.* [Synthese Library 107] 1977
ISBN 90-277-0773-1; Pb 90-277-0774-X
54. H.A. Simon: *Models of Discovery* and Other Topics in the Methods of Science. [Synthese Library 114] 1977 ISBN 90-277-0812-6; Pb 90-277-0858-4
55. M. Lazerowitz: *The Language of Philosophy.* Freud and Wittgenstein. [Synthese Library 117] 1977 ISBN 90-277-0826-6; Pb 90-277-0862-2
56. T. Nickles (ed.): *Scientific Discovery, Logic, and Rationality.* 1980
ISBN 90-277-1069-4; Pb 90-277-1070-8
57. J. Margolis: *Persons and Mind.* The Prospects of Nonreductive Materialism. [Synthese Library 121] 1978 ISBN 90-277-0854-1; Pb 90-277-0863-0
58. G. Radnitzky and G. Andersson (eds.): *Progress and Rationality in Science.* [Synthese Library 125] 1978 ISBN 90-277-0921-1; Pb 90-277-0922-X

59. G. Radnitzky and G. Andersson (eds.): *The Structure and Development of Science*. [Synthese Library 136] 1979 ISBN 90-277-0994-7; Pb 90-277-0995-5
60. T. Nickles (ed.): *Scientific Discovery*. Case Studies. 1980 ISBN 90-277-1092-9; Pb 90-277-1093-7
61. M.A. Finocchiaro: *Galileo and the Art of Reasoning*. Rhetorical Foundation of Logic and Scientific Method. 1980 ISBN 90-277-1094-5; Pb 90-277-1095-3
62. W.A. Wallace: *Prelude to Galileo*. Essays on Medieval and 16th-Century Sources of Galileo's Thought. 1981 ISBN 90-277-1215-8; Pb 90-277-1216-6
63. F. Rapp: *Analytical Philosophy of Technology*. Translated from German. 1981 ISBN 90-277-1221-2; Pb 90-277-1222-0
64. R.S. Cohen and M.W. Wartofsky (eds.): *Hegel and the Sciences*. 1984 ISBN 90-277-0726-X
65. J. Agassi: *Science and Society*. Studies in the Sociology of Science. 1981 ISBN 90-277-1244-1; Pb 90-277-1245-X
66. L. Tondl: *Problems of Semantics*. A Contribution to the Analysis of the Language of Science. Translated from Czech. 1981 ISBN 90-277-0148-2; Pb 90-277-0316-7
67. J. Agassi and R.S. Cohen (eds.): *Scientific Philosophy Today*. Essays in Honor of Mario Bunge. 1982 ISBN 90-277-1262-X; Pb 90-277-1263-8
68. W. Krajewski (ed.): *Polish Essays in the Philosophy of the Natural Sciences*. Translated from Polish and edited by R.S. Cohen and C.R. Fawcett. 1982 ISBN 90-277-1286-7; Pb 90-277-1287-5
69. J.H. Fetzer: *Scientific Knowledge*. Causation, Explanation and Corroboration. 1981 ISBN 90-277-1335-9; Pb 90-277-1336-7
70. S. Grossberg: *Studies of Mind and Brain*. Neural Principles of Learning, Perception, Development, Cognition, and Motor Control. 1982 ISBN 90-277-1359-6; Pb 90-277-1360-X
71. R.S. Cohen and M.W. Wartofsky (eds.): *Epistemology, Methodology, and the Social Sciences*. 1983. ISBN 90-277-1454-1
72. K. Berka: *Measurement*. Its Concepts, Theories and Problems. Translated from Czech. 1983 ISBN 90-277-1416-9
73. G.L. Pandit: *The Structure and Growth of Scientific Knowledge*. A Study in the Methodology of Epistemic Appraisal. 1983 ISBN 90-277-1434-7
74. A.A. Zinov'ev: *Logical Physics*. Translated from Russian. Edited by R.S. Cohen. 1983 [*see also* Volume 9] ISBN 90-277-0734-0
75. G-G. Granger: *Formal Thought and the Sciences of Man*. Translated from French. With and Introduction by A. Rosenberg. 1983 ISBN 90-277-1524-6
76. R.S. Cohen and L. Laudan (eds.): *Physics, Philosophy and Psychoanalysis*. Essays in Honor of Adolf Grünbaum. 1983 ISBN 90-277-1533-5
77. G. Böhme, W. van den Daele, R. Hohlfeld, W. Krohn and W. Schäfer: *Finalization in Science*. The Social Orientation of Scientific Progress. Translated from German. Edited by W. Schäfer. 1983 ISBN 90-277-1549-1
78. D. Shapere: *Reason and the Search for Knowledge*. Investigations in the Philosophy of Science. 1984 ISBN 90-277-1551-3; Pb 90-277-1641-2
79. G. Andersson (ed.): *Rationality in Science and Politics*. Translated from German. 1984 ISBN 90-277-1575-0; Pb 90-277-1953-5
80. P.T. Durbin and F. Rapp (eds.): *Philosophy and Technology*. [*Also* Philosophy and Technology Series, Vol. 1] 1983 ISBN 90-277-1576-9

Boston Studies in the Philosophy of Science

81. M. Marković: *Dialectical Theory of Meaning*. Translated from Serbo-Croat. 1984 ISBN 90-277-1596-3
82. R.S. Cohen and M.W. Wartofsky (eds.): *Physical Sciences and History of Physics*. 1984. ISBN 90-277-1615-3
83. É. Meyerson: *The Relativistic Deduction*. Epistemological Implications of the Theory of Relativity. Translated from French. With a Review by Albert Einstein and an Introduction by Milić Čapek. 1985 ISBN 90-277-1699-4
84. R.S. Cohen and M.W. Wartofsky (eds.): *Methodology, Metaphysics and the History of Science*. In Memory of Benjamin Nelson. 1984 ISBN 90-277-1711-7
85. G. Tamás: *The Logic of Categories*. Translated from Hungarian. Edited by R.S. Cohen. 1986 ISBN 90-277-1742-7
86. S.L. de C. Fernandes: *Foundations of Objective Knowledge*. The Relations of Popper's Theory of Knowledge to That of Kant. 1985 ISBN 90-277-1809-1
87. R.S. Cohen and T. Schnelle (eds.): *Cognition and Fact*. Materials on Ludwik Fleck. 1986 ISBN 90-277-1902-0
88. G. Freudenthal: *Atom and Individual in the Age of Newton*. On the Genesis of the Mechanistic World View. Translated from German. 1986 ISBN 90-277-1905-5
89. A. Donagan, A.N. Perovich Jr and M.V. Wedin (eds.): *Human Nature and Natural Knowledge*. Essays presented to Marjorie Grene on the Occasion of Her 75th Birthday. 1986 ISBN 90-277-1974-8
90. C. Mitcham and A. Hunning (eds.): *Philosophy and Technology II*. Information Technology and Computers in Theory and Practice. [*Also* Philosophy and Technology Series, Vol. 2] 1986 ISBN 90-277-1975-6
91. M. Grene and D. Nails (eds.): *Spinoza and the Sciences*. 1986 ISBN 90-277-1976-4
92. S.P. Turner: *The Search for a Methodology of Social Science*. Durkheim, Weber, and the 19th-Century Problem of Cause, Probability, and Action. 1986. ISBN 90-277-2067-3
93. I.C. Jarvie: *Thinking about Society*. Theory and Practice. 1986 ISBN 90-277-2068-1
94. E. Ullmann-Margalit (ed.): *The Kaleidoscope of Science*. The Israel Colloquium: Studies in History, Philosophy, and Sociology of Science, Vol. 1. 1986 ISBN 90-277-2158-0; Pb 90-277-2159-9
95. E. Ullmann-Margalit (ed.): *The Prism of Science*. The Israel Colloquium: Studies in History, Philosophy, and Sociology of Science, Vol. 2. 1986 ISBN 90-277-2160-2; Pb 90-277-2161-0
96. G. Márkus: *Language and Production*. A Critique of the Paradigms. Translated from French. 1986 ISBN 90-277-2169-6
97. F. Amrine, F.J. Zucker and H. Wheeler (eds.): *Goethe and the Sciences: A Reappraisal*. 1987 ISBN 90-277-2265-X; Pb 90-277-2400-8
98. J.C. Pitt and M. Pera (eds.): *Rational Changes in Science*. Essays on Scientific Reasoning. Translated from Italian. 1987 ISBN 90-277-2417-2
99. O. Costa de Beauregard: *Time, the Physical Magnitude*. 1987 ISBN 90-277-2444-X
100. A. Shimony and D. Nails (eds.): *Naturalistic Epistemology*. A Symposium of Two Decades. 1987 ISBN 90-277-2337-0
101. N. Rotenstreich: *Time and Meaning in History*. 1987 ISBN 90-277-2467-9
102. D.B. Zilberman: *The Birth of Meaning in Hindu Thought*. Edited by R.S. Cohen. 1988 ISBN 90-277-2497-0
103. T.F. Glick (ed.): *The Comparative Reception of Relativity*. 1987 ISBN 90-277-2498-9

104. Z. Harris, M. Gottfried, T. Ryckman, P. Mattick Jr, A. Daladier, T.N. Harris and S. Harris: *The Form of Information in Science*. Analysis of an Immunology Sublanguage. With a Preface by Hilary Putnam. 1989 ISBN 90-277-2516-0
105. F. Burwick (ed.): *Approaches to Organic Form*. Permutations in Science and Culture. 1987 ISBN 90-277-2541-1
106. M. Almási: *The Philosophy of Appearances*. Translated from Hungarian. 1989 ISBN 90-277-2150-5
107. S. Hook, W.L. O'Neill and R. O'Toole (eds.): *Philosophy, History and Social Action*. Essays in Honor of Lewis Feuer. With an Autobiographical Essay by L. Feuer. 1988 ISBN 90-277-2644-2
108. I. Hronszky, M. Fehér and B. Dajka: *Scientific Knowledge Socialized*. Selected Proceedings of the 5th Joint International Conference on the History and Philosophy of Science organized by the IUHPS (Veszprém, Hungary, 1984). 1988 ISBN 90-277-2284-6
109. P. Tillers and E.D. Green (eds.): *Probability and Inference in the Law of Evidence*. The Uses and Limits of Bayesianism. 1988 ISBN 90-277-2689-2
110. E. Ullmann-Margalit (ed.): *Science in Reflection*. The Israel Colloquium: Studies in History, Philosophy, and Sociology of Science, Vol. 3. 1988 ISBN 90-277-2712-0; Pb 90-277-2713-9
111. K. Gavroglu, Y. Goudaroulis and P. Nicolacopoulos (eds.): *Imre Lakatos and Theories of Scientific Change*. 1989 ISBN 90-277-2766-X
112. B. Glassner and J.D. Moreno (eds.): *The Qualitative-Quantitative Distinction in the Social Sciences*. 1989 ISBN 90-277-2829-1
113. K. Arens: *Structures of Knowing*. Psychologies of the 19th Century. 1989 ISBN 0-7923-0009-2
114. A. Janik: *Style, Politics and the Future of Philosophy*. 1989 ISBN 0-7923-0056-4
115. F. Amrine (ed.): *Literature and Science as Modes of Expression*. With an Introduction by S. Weininger. 1989 ISBN 0-7923-0133-1
116. J.R. Brown and J. Mittelstrass (eds.): *An Intimate Relation*. Studies in the History and Philosophy of Science. Presented to Robert E. Butts on His 60th Birthday. 1989 ISBN 0-7923-0169-2
117. F. D'Agostino and I.C. Jarvie (eds.): *Freedom and Rationality*. Essays in Honor of John Watkins. 1989 ISBN 0-7923-0264-8
118. D. Zolo: *Reflexive Epistemology*. The Philosophical Legacy of Otto Neurath. 1989 ISBN 0-7923-0320-2
119. M. Kearn, B.S. Philips and R.S. Cohen (eds.): *Georg Simmel and Contemporary Sociology*. 1989 ISBN 0-7923-0407-1
120. T.H. Levere and W.R. Shea (eds.): *Nature, Experiment and the Science*. Essays on Galileo and the Nature of Science. In Honour of Stillman Drake. 1989 ISBN 0-7923-0420-9
121. P. Nicolacopoulos (ed.): *Greek Studies in the Philosophy and History of Science*. 1990 ISBN 0-7923-0717-8
122. R. Cooke and D. Costantini (eds.): *Statistics in Science*. The Foundations of Statistical Methods in Biology, Physics and Economics. 1990 ISBN 0-7923-0797-6
123. P. Duhem: *The Origins of Statics*. Translated from French by G.F. Leneaux, V.N. Vagliente and G.H. Wagner. With an Introduction by S.L. Jaki. 1991 ISBN 0-7923-0898-0
124. H. Kamerlingh Onnes: *Through Measurement to Knowledge*. The Selected Papers, 1853-1926. Edited and with an Introduction by K. Gavroglu and Y. Goudaroulis. 1991 ISBN 0-7923-0825-5

Boston Studies in the Philosophy of Science

125. M. Čapek: *The New Aspects of Time: Its Continuity and Novelties*. Selected Papers in the Philosophy of Science. 1991 ISBN 0-7923-0911-1
126. S. Unguru (ed.): *Physics, Cosmology and Astronomy, 1300-1700*. Tension and Accommodation. 1991 ISBN 0-7923-1022-5
127. Z. Bechler: *Newton's Physics on the Conceptual Structure of the Scientific Revolution*. 1991 ISBN 0-7923-1054-3
128. É. Meyerson: *Explanation in the Sciences*. Translated from French by M-A. Siple and D.A. Siple. 1991 ISBN 0-7923-1129-9
129. A.I. Tauber (ed.): *Organism and the Origins of Self*. 1991 ISBN 0-7923-1185-X
130. F.J. Varela and J-P. Dupuy (eds.): *Understanding Origins*. Contemporary Views on the Origin of Life, Mind and Society. 1992 ISBN 0-7923-1251-1
131. G.L. Pandit: *Methodological Variance*. Essays in Epistemological Ontology and the Methodology of Science. 1991 ISBN 0-7923-1263-5
132. G. Munévar (ed.): *Beyond Reason*. Essays on the Philosophy of Paul Feyerabend. 1991 ISBN 0-7923-1272-4
133. T.E. Uebel (ed.): *Rediscovering the Forgotten Vienna Circle*. Austrian Studies on Otto Neurath and the Vienna Circle. Partly translated from German. 1991 ISBN 0-7923-1276-7
134. W.R. Woodward and R.S. Cohen (eds.): *World Views and Scientific Discipline Formàtion*. Science Studies in the [former] German Democratic Republic. Partly translated from German by W.R. Woodward. 1991 ISBN 0-7923-1286-4
135. P. Zambelli: *The Speculum Astronomiae and Its Enigma*. Astrology, Theology and Science in Albertus Magnus and His Contemporaries. 1992 ISBN 0-7923-1380-1
136. P. Petitjean, C. Jami and A.M. Moulin (eds.): *Science and Empires*. Historical Studies about Scientific Development and European Expansion. ISBN 0-7923-1518-9
137. W.A. Wallace: *Galileo's Logic of Discovery and Proof*. The Background, Content, and Use of His Appropriated Treatises on Aristotle's *Posterior Analytics*. 1992 ISBN 0-7923-1577-4
138. W.A. Wallace: *Galileo's Logical Treatises*. A Translation, with Notes and Commentary, of His Appropriated Latin Questions on Aristotle's *Posterior Analytics*. 1992 ISBN 0-7923-1578-2
Set (137 + 138) ISBN 0-7923-1579-0
139. M.J. Nye, J.L. Richards and R.H. Stuewer (eds.): *The Invention of Physical Science*. Intersections of Mathematics, Theology and Natural Philosophy since the Seventeenth Century. Essays in Honor of Erwin N. Hiebert. 1992 ISBN 0-7923-1753-X
140. G. Corsi, M.L. dalla Chiara and G.C. Ghirardi (eds.): *Bridging the Gap: Philosophy, Mathematics and Physics*. Lectures on the Foundations of Science. 1992 ISBN 0-7923-1761-0
141. C.-H. Lin and D. Fu (eds.): *Philosophy and Conceptual History of Science in Taiwan*. 1992 ISBN 0-7923-1766-1
142. S. Sarkar (ed.): *The Founders of Evolutionary Genetics*. A Centenary Reappraisal. 1992 ISBN 0-7923-1777-7
143. J. Blackmore (ed.): *Ernst Mach – A Deeper Look*. Documents and New Perspectives. 1992 ISBN 0-7923-1853-6
144. P. Kroes and M. Bakker (eds.): *Technological Development and Science in the Industrial Age*. New Perspectives on the Science–Technology Relationship. 1992 ISBN 0-7923-1898-6
145. S. Amsterdamski: *Between History and Method*. Disputes about the Rationality of Science. 1992 ISBN 0-7923-1941-9

Boston Studies in the Philosophy of Science

146. E. Ullmann-Margalit (ed.): *The Scientific Enterprise.* The Bar-Hillel Colloquium: Studies in History, Philosophy, and Sociology of Science, Volume 4. 1992 ISBN 0-7923-1992-3
147. L. Embree (ed.): *Metaarchaeology.* Reflections by Archaeologists and Philosophers. 1992 ISBN 0-7923-2023-9
148. S. French and H. Kamminga (eds.): *Correspondence, Invariance and Heuristics.* Essays in Honour of Heinz Post. 1993 ISBN 0-7923-2085-9
149. M. Bunzl: *The Context of Explanation.* 1993 ISBN 0-7923-2153-7
150. I.B. Cohen (ed.): *The Natural Sciences and the Social Sciences.* Some Critical and Historical Perspectives. 1994 ISBN 0-7923-2223-1
151. K. Gavroglu, Y. Christianidis and E. Nicolaidis (eds.): *Trends in the Historiography of Science.* 1994 ISBN 0-7923-2255-X
152. S. Poggi and M. Bossi (eds.): *Romanticism in Science.* Science in Europe, 1790–1840. 1994 ISBN 0-7923-2336-X
153. J. Faye and H.J. Folse (eds.): *Niels Bohr and Contemporary Philosophy.* 1994 ISBN 0-7923-2378-5
154. C.C. Gould and R.S. Cohen (eds.): *Artifacts, Representations, and Social Practice.* Essays for Marx W. Wartofsky. 1994 ISBN 0-7923-2481-1
155. R.E. Butts: *Historical Pragmatics.* Philosophical Essays. 1993 ISBN 0-7923-2498-6
156. R. Rashed: *The Development of Arabic Mathematics: Between Arithmetic and Algebra.* Translated from French by A.F.W. Armstrong. 1994 ISBN 0-7923-2565-6
157. I. Szumilewicz-Lachman (ed.): *Zygmunt Zawirski: His Life and Work.* With Selected Writings on Time, Logic and the Methodology of Science. Translations by Feliks Lachman. Ed. by R.S. Cohen, with the assistance of B. Bergo. 1994 ISBN 0-7923-2566-4
158. S.N. Haq: *Names, Natures and Things.* The Alchemist Jābir ibn Ḥayyān and His *Kitāb al-Aḥjār* (Book of Stones). 1994 ISBN 0-7923-2587-7
159. P. Plaass: *Kant's Theory of Natural Science.* Translation, Analytic Introduction and Commentary by Alfred E. and Maria G. Miller. 1994 ISBN 0-7923-2750-0
160. J. Misiek (ed.): *The Problem of Rationality in Science and its Philosophy.* On Popper vs. Polanyi. The Polish Conferences 1988–89. 1995 ISBN 0-7923-2925-2
161. I.C. Jarvie and N. Laor (eds.): *Critical Rationalism, Metaphysics and Science.* Essays for Joseph Agassi, Volume I. 1995 ISBN 0-7923-2960-0
162. I.C. Jarvie and N. Laor (eds.): *Critical Rationalism, the Social Sciences and the Humanities.* Essays for Joseph Agassi, Volume II. 1995 ISBN 0-7923-2961-9
Set (161–162) ISBN 0-7923-2962-7
163. K. Gavroglu, J. Stachel and M.W. Wartofsky (eds.): *Physics, Philosophy, and the Scientific Community.* Essays in the Philosophy and History of the Natural Sciences and Mathematics. In Honor of Robert S. Cohen. 1995 ISBN 0-7923-2988-0
164. K. Gavroglu, J. Stachel and M.W. Wartofsky (eds.): *Science, Politics and Social Practice.* Essays on Marxism and Science, Philosophy of Culture and the Social Sciences. In Honor of Robert S. Cohen. 1995 ISBN 0-7923-2989-9
165. K. Gavroglu, J. Stachel and M.W. Wartofsky (eds.): *Science, Mind and Art.* Essays on Science and the Humanistic Understanding in Art, Epistemology, Religion and Ethics. Essays in Honor of Robert S. Cohen. 1995 ISBN 0-7923-2990-2
Set (163–165) ISBN 0-7923-2991-0
166. K.H. Wolff: *Transformation in the Writing.* A Case of Surrender-and-Catch. 1995 ISBN 0-7923-3178-8

Boston Studies in the Philosophy of Science

190. K. Lehrer and J.C. Marek (eds.): *Austrian Philosophy Past and Present.* Essays in Honor of Rudolf Haller. 1996 ISBN 0-7923-4347-6
191. J.L. Lagrange: *Analytical Mechanics.* Translated and edited by Auguste Boissonade and Victor N. Vagliente. Translated from the *Mécanique Analytique, novelle édition* of 1811. 1997 ISBN 0-7923-4349-2
192. D. Ginev and R.S. Cohen (eds.): *Issues and Images in the Philosophy of Science.* Scientific and Philosophical Essays in Honour of Azarya Polikarov. 1997 ISBN 0-7923-4444-8
193. R.S. Cohen, M. Horne and J. Stachel (eds.): *Experimental Metaphysics.* Quantum Mechanical Studies for Abner Shimony, Volume One. 1997 ISBN 0-7923-4452-9
194. R.S. Cohen, M. Horne and J. Stachel (eds.): *Potentiality, Entanglement and Passion-at-a-Distance.* Quantum Mechanical Studies for Abner Shimony, Volume Two. 1997 ISBN 0-7923-4453-7; Set 0-7923-4454-5
195. R.S. Cohen and A.I. Tauber (eds.): *Philosophies of Nature: The Human Dimension.* 1997 ISBN 0-7923-4579-7
196. M. Otte and M. Panza (eds.): *Analysis and Synthesis in Mathematics.* History and Philosophy. 1997 ISBN 0-7923-4570-3
197. forthcoming
198. D. Baird, R.I.G. Hughes and A. Nordmann (eds.): *Heinrich Hertz: Classical Physicist, Modern Philosopher.* 1997 ISBN 0-7923-4653-X

Also of interest:

R.S. Cohen and M.W. Wartofsky (eds.): *A Portrait of Twenty-Five Years Boston Colloquia for the Philosophy of Science, 1960-1985.* 1985 ISBN Pb 90-277-1971-3

Previous volumes are still available.

KLUWER ACADEMIC PUBLISHERS – DORDRECHT / BOSTON / LONDON

MIX
Papier aus verantwortungsvollen Quellen
Paper from responsible sources
FSC® C105338

If you have any concerns about our products,
you can contact us on
ProductSafety@springernature.com

In case Publisher is established outside the EU,
the EU authorized representative is:
Springer Nature Customer Service Center GmbH
Europaplatz 3, 69115 Heidelberg, Germany

Printed by Libri Plureos GmbH
in Hamburg, Germany